AF598125

Methods in Molecular Biology™

Series Editor
John M. Walker
School of Life Sciences
University of Hertfordshire
Hatfield, Hertfordshire, AL10 9AB, UK

For further volumes:
http://www.springer.com/series/7651

Nanomaterial Interfaces in Biology

Methods and Protocols

Edited by

Paolo Bergese

Dipartimento di Ingegneria Meccanicae Industriale, Università degli Studi di Brescia, Brescia, Italy

Kimberly Hamad-Schifferli

Department of Mechanical Engineering, Massachusetts Institute of Technology
Cambridge, MA, USA

Editors
Paolo Bergese
Dipartimento di Ingegneria Meccanicae Industriale
Università degli Studi di Brescia
Brescia, Italy

Kimberly Hamad-Schifferli
Department of Mechanical Engineering
Massachusetts Institute of Technology
Cambridge, MA, USA

ISSN 1064-3745 ISSN 1940-6029 (electronic)
ISBN 978-1-62703-461-6 ISBN 978-1-62703-462-3 (eBook)
DOI 10.1007/978-1-62703-462-3
Springer New York Heidelberg Dordrecht London

Library of Congress Control Number: 2013943590

Printed on acid-free paper

Humana Press is a brand of Springer
Springer is part of Springer Science+Business Media (www.springer.com)

Preface

The intersection of nanotechnology with biology has given rise to numerous ideas for new ways to use nanotechnology for biological applications. Nanomaterials possess unique size- and material-dependent properties, which make them attractive for improving regular biomedical fields, such as drug delivery, imaging, therapy, and diagnostics as well as next-level developments, comprising activation/deactivation, mimicking, and implementation of biomolecular systems and functions. Consequently nanotechnology has held great potential for novel and unique capabilities, and stirred the imagination of many scientists.

While nanotechnology and nanoscience has held great promise for revolutionizing biology, it has been hindered by the fact that when nanomaterials are interfaced to biomolecules or put into biological environments, many undesirable side effects result, such as aggregation and nonspecific adsorption, which can compromise biological function or give rise to negative biological responses. This can be largely attributed to a range of interface and intermolecular interactions between the nanomaterials with the biomolecules as well as the solvent that mediates their interactions. These interactions are difficult to control, predict, and prevent. Interface effects of inorganic surfaces have been a major issue historically, manifesting as surface fouling of medical device implants and stents. Unfortunately, these effects worsen or give rise to new and unexpected complications for nanoscale materials because surface volume ratios are exponentially higher, and nanoscale surfaces have different physical and chemical properties.

Therefore, despite the fact that we now have a high degree of control over the synthetic properties of the nanomaterials, similar control over their interfaces to biology has yet to be achieved. This is crucial as their biological interface ultimately determines their biological identity and fate. In the last 10 years, the biological–nanomaterial interface has created unprecedented challenges not only for finding useful ways to exploit nanomaterials in biology but also in their unintentional consequences, such as environmental and toxicological effects. For example surface fouling, nonspecific adsorption, or unexpected aggregation and instability have prevented many of the exciting and early ideas of nanobiotechnology from reaching fruition.

We believe a key pitfall that plagues this area is the difficulty in reproducing results, where a huge amount of variability exists not only between different labs but also from day to day in the same lab. This variability is almost never reported in peer-reviewed journal papers, despite its criticality. Also, members of the nanotechnology community typically come from chemistry, physics, and materials science, and historically are not accustomed to providing stepwise protocols. Therefore, a handbook of detailed protocols and best practices for this field is critically needed.

While there have been some examples of individual protocols in protocol literature such as *Nature Methods*, *Current Protocols*, *Molecular Cloning*, and on the National Cancer Institute's Nanotechnology Characterization Laboratory website, we believe that it is time to provide a consolidated volume of step-by-step protocols in the tradition of *Methods in Molecular Biology*. *Methods in Molecular Biology* has played a major role in providing biological recipes and protocols, and in doing so has advanced biology in immeasurable ways. By creating standards for everyone and making experimental techniques more accessible, *Methods in Molecular Biology* has accelerated the discovery process. What used to be difficult experimentally now can be done by nearly anyone. These new tools have benefited the entire biology community, enabling huge leaps and bounds in scientific discovery and applications.

Thus, we present this volume with the hopes of improving the utility of nanotechnology as a tool to advance biological and medical sciences. While this volume is far from comprehensive, we hope that it will serve the new and emerging community well, and enable new capabilities and technologies that were not previously possible. The proposed protocols predominantly deal with nanomaterials, covering many of the now classic subjects, such as conjugation of nanoparticles to biomolecules and their applications in drug delivery or imaging. Additionally, some chapters are dedicated to flat surfaces because most of the methods for manipulating nanomaterials stem from those of flat surfaces, facilitated by fabrication advances to shrink the dimensions of materials.

The volume is organized into three parts: (1) protocols describing synthesis, fabrication, and construction of bio-nanomaterial interfaces, (2) characterization protocols of bio-nanomaterial interfaces, and (3) applications which utilize the bio-nanomaterial interfaces. We would like to note that we are not including significant coverage of toxicology and the unintended effects of nanomaterials. Due to the evolving scope of the toxicological studies of nanomaterials, the current state of the art is still developing, and thus is difficult to cover adequately in this volume of *Methods in Molecular Biology*. However, we hope that this collection will aid this growing field by serving as a guide.

We gratefully acknowledge all of the authors contributing to this volume, as well as our home institutions of the Department of Mechanical Engineering at MIT and the Department of Mechanical and Industrial Engineering of Università degli Studi di Brescia. This work was made possible by the UniBS-MIT-MechE faculty exchange program cosponsored by the CARIPLO Foundation, Italy under grant 2008-2290.

Brescia, Italy ***Paolo Bergese***
Cambridge, MA, USA ***Kimberly Hamad-Schifferli***

Contents

Contributors

W. Russ Algar • *Naval Research Laboratory, Washington, DC, USA*
Marie-Eve Aubin-Tam • *Delft University of Technology, Delft, The Netherlands*
Naris Barnthip • *Division of Physics, Rajamangala University of Technology, Pathum Thani, Thailand*
Irina Belozerova • *Polytechnic Institute of New York University, New York, NY, USA*
Juan B. Blanco-Canosa • *Scripps Research Institute, La Jolla, CA, USA*
Salvador Borrós • *Institut Quimic de Sarria, Universitat Ramon Llull, Barcelona, Spain*
Clemens Burda • *Case Western Reserve University, Cleveland, OH, USA*
Gabriele Candiani • *INSTM (National Interuniversity Consortium of Materials Science and Technology), Research Unit Milano Politecnico, Milan, Italy; Department of Chemistry, Materials and Chemical Engineering "Giulio Natta", Politecnico di Milano, Milan, Italy*
Xi-Jun Chen • *Department of Chemistry, University of Pennsylvania, Philadelphia, PA, USA*
Marcella Chiari • *Consiglio Nazionale delle Ricerche, Istituto di Chimica del Riconoscimento Molecolare, Milan, Italy*
Roberto Chiesa • *Department of Chemistry, Materials and Chemical Engineering "Giulio Natta", Politecnico di Milano, Milan, Italy*
Anna Cifuentes • *Institut Quimic de Sarria, Universitat Ramon Llull, Barcelona, Spain*
Joan Comenge • *Catalan Institute of Nanotechnology, Universitat Autònoma de Barcelona, Barcelona, Spain*
Núria Coronas • *Institut Quimic de Sarria, Universitat Ramon Llull, Barcelona, Spain*
Taku Cowger • *University of Georgia, Athens, GA, USA*
Marina Cretich • *Consiglio Nazionale delle Ricerche, Istituto di Chimica del Riconoscimento Molecolare, Milan, Italy*
A. Curcio • *Istituto Italiano di Tecnologia, Genoa, Italy*
Francesco Damin • *Consiglio Nazionale delle Ricerche, Istituto di Chimica del Riconoscimento Molecolare, Milan, Italy*
Kenneth A. Dawson • *Centre for BioNano Interactions, School of Chemistry and Chemical Biology, University College Dublin, Dublin, Ireland*
Philip E. Dawson • *Scripps Research Institute, La Jolla, CA, USA*
Tennyson Doane • *Case Western Reserve University, Cleveland, OH, USA*
Bradley Duncan • *University of Massachusetts, Amherst, MA, USA*
Zaki G. Estephan • *Florida State University, Tallahassee, FL, USA*
A. Falqui • *Istituto Italiano di Tecnologia, Genoa, Italy*
Francesca Frascella • *Politecnico di Torino, Torino, Italy*
Dongbiao Ge • *Polytechnic Institute of New York University, New York, NY, USA*
Christin Grabinski • *Air Force Research Laboratory, Wright-Patterson Air Force Base, Dayton, OH, USA*

Robert J. Hickey • *Department of Chemistry, University of Pennsylvania, Philadelphia, PA, USA*
Saber Hussain • *Air Force Research Laboratory, Wright-Patterson Air Force Base, Dayton, OH, USA*
Adela Isovic • *University of Illinois at Chicago, Chicago, IL, USA*
James Chen Yong Kah • *National University Singapore, Singapore*
Anna Kajaste-Rudnitski • *Division of Regenerative Medicine, Stem Cells and Gene Therapy, San Raffaele Telethon Institute for Gene Therapy, San Raffaele Scientific Institute, Milan, Italy*
Rastislav Levicky • *Polytechnic Institute of New York University, New York, NY, USA*
Luis M. Liz-Marzán • *CIC biomaGUNE, Donostia - San, Sebastián, Spain*
Iseult Lynch • *Center for BioNano Interactions, School of Chemistry and Chemical Biology, University College Dublin, Dublin, Ireland*
John Malona • *Princeton University, Princeton, NJ, USA*
Rachel L. Manthe • *Sotera Defense Solutions, University of Maryland, College Park, MD, USA*
Roberto Marotta • *Istituto Italiano di Tecnologia, Genoa, Italy*
Igor L. Medintz • *U.S. Naval Research Laboratory, Center for Bio/Molecular Science and Engineering, Washington, DC, USA*
Marco P. Monopoli • *Center for BioNano Interactions, School of Chemistry and Chemical Biology, University College Dublin, Dublin, Ireland*
Daniel Montiel • *Princeton University, Princeton, NJ, USA*
Daniel F. Moyano • *University of Massachusetts, Amherst, MA, USA*
Hyeran Noh • *Department of Optometry and Vision Science, Seoul National University and Technology, Seoul, South Korea*
Ana de Pablo • *Institut Quimic de Sarria, Universitat Ramon Llull, Barcelona, Spain*
Purnendu Parhi • *Department of Chemistry, Ravenshaw University, Cuttack, Orissa, India*
So-Jung Park • *Department of Chemistry, University of Pennsylvania, Philadelphia, PA, USA*
Sunho Park • *Dankook University, Yongin-si, Gyeonggi-do, South Korea*
Isabel Pastoriza-Santos • *Departamento de Química Física, Campus Universitario, Universidade de Vigo, Bilbao, Spain*
Teresa Pellegrino • *Istituto Italiano di Tecnologia, Genoa, Italy; National Nanotechnology Laboratory of CNR-NANO, Lecce, Italy*
Daniele Pezzoli • *INSTM (National Interuniversity Consortium of Materials Science and Technology), Research Unit Milano Politecnico, Milan, Italy*
Andrzej S. Pitek • *Center for BioNano Interactions, School of Chemistry and Chemical Biology, University College Dublin, Dublin, Ireland*
Víctor F. Puntes • *Catalan Institute of Nanotechnology, Universitat Autònoma de Barcelona, Barcelona, Spain*
A. Quarta • *Istituto Italiano di Tecnologia, Genoa, Italy; National Nanotechnology Laboratory of CNR-NANO, Lecce, Italy*
Victor Ramos-Perez • *Institut Quimic de Sarria, Universitat Ramon Llull, Barcelona, Spain*
Carlo Ricciardi • *Politecnico di Torino, Torino, Italy*
Vincent M. Rotello • *University of Massachusetts, Amherst, MA, USA*
Marissa M. Sampias • *Princeton University, Princeton, NJ, USA*

JOHN SCHLAGER • *Air Force Research Laboratory, Wright-Patterson Air Force Base, Dayton, OH, USA*
JOSEPH B. SCHLENOFF • *Department of Chemistry and Biochemistry, Florida State University, Tallahassee, FL, USA*
PRESTON T. SNEE • *University of Illinois at Chicago, Chicago, IL, USA*
LAURA SOLA • *Consiglio Nazionale delle Ricerche, Istituto di Chimica del Riconoscimento Molecolare, Milan, Italy*
ERIK J. SORENSEN • *Princeton University, Princeton, NJ, USA*
MICHAEL H. STEWART • *Division of Optical Sciences, U.S. Naval Research Laboratory, Washington, DC, USA*
LI SUN • *Princeton University, Princeton, NJ, USA*
KIMIHIRO SUSUMU • *Naval Research Laboratory, Washington, DC, USA*
CHRISTINA M. TYRAKOWSKI • *University of Illinois at Chicago, Chicago, IL, USA*
ERWIN A. VOGLER • *Department of Materials Science and Engineering, The Pennsylvania State University, University Park, PA, USA; Department of Bioengineering, The Pennsylvania State University, University Park, PA, USA*
JIN XIE • *Department of Chemistry, University of Georgia, Athens, GA, USA*
HAW YANG • *Princeton University, Princeton, NJ, USA*
EMMA V. YATES • *Princeton University, Princeton, NJ, USA*

Part I

Making Bio–Nano Interfaces

Chapter 1

Preparation of 2 nm Gold Nanoparticles for In Vitro and In Vivo Applications

Daniel F. Moyano, Bradley Duncan, and Vincent M. Rotello

Abstract

Gold nanoparticles have been a versatile tool in recent years for the exploration of biological systems. However, challenges with purification and adequate surface coverage limit the biocompatibility of gold nanoparticles. Here, we describe a detailed procedure for the synthesis, purification, and functionalization of biologically compatible gold nanoparticles for in vitro and in vivo studies.

Key words Gold nanoparticle, Biocompatibility, Purification

1 Introduction

Gold nanoparticle (AuNP) properties and applications continue to be the focus of a variety of fields since their discovery in 1857 by Michael Faraday [1]. Among these areas, avid research has been pointed towards the creation of biocompatible nanoparticles [2]. The precise control introduced by the two-phase Brust–Schiffrin method [3, 4] allowed more systematic studies of nanoparticle properties to be performed. However, the use of tetra-*n*-octylammonium bromide as the phase transfer catalyst limits biological applications due to the difficultly in separating this highly cytotoxic chemical from the nanoparticles [5, 6]. Additionally, the functionality of nanoparticle surfaces has been a prominent concern due to the direct correlation of cytotoxicity with the functionality of the surface monolayer [7]. To further improve nanoparticle utility, a fine balance between hydrophobic functionalization near the metallic core to confer monolayer stability [8] and non-fouling character provided by poly(ethylene glycol) moieties must be achieved [9]. An excellent ligand base for biological applications is 23-mercapto-3,6,9,12-tetraoxatricosan-1-ol (Fig. 1) as its components confer the desired stability and non-fouling properties [10]. This structure can be easily modified to provide *selective* binding

Paolo Bergese and Kimberly Hamad-Schifferli (eds.), *Nanomaterial Interfaces in Biology: Methods and Protocols*, Methods in Molecular Biology, vol. 1025, DOI 10.1007/978-1-62703-462-3_1,

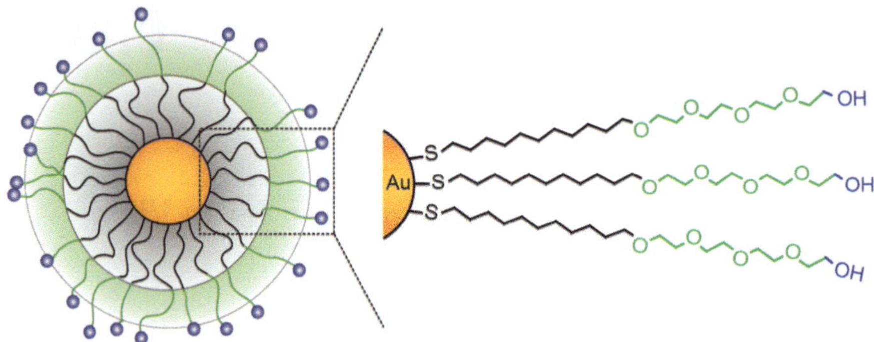

Fig. 1 Gold nanoparticle chemical structure. The ligand is composed of three parts, namely, the sulfur atom (binding with gold), an alkyl chain (nanoparticle stability), and a tetraethylene glycol spacer (biocompatibility and solubility)

making these nanoparticles a promising platform for systematic biological studies [11]. Employing these concepts, this protocol elaborates on the synthesis, purification, and functionalization of gold nanoparticles with the biocompatibility required for in vivo and in vitro studies.

2 Materials

1. Hydrogen tetrachloroaurate (III) hydrate.
2. Tetra-*n*-octylammonium bromide (TOAB).
3. 1-Pentanethiol.
4. Sodium borohydride.
5. 23-mercapto-3,6,9,12-tetraoxatricosan-1-ol, synthetized according to the reported procedure [12].
6. 200 proof ethanol, acetone, hexanes, and acetonitrile.
7. 10,000 MWCO snakeskin pleated dialysis tubing membrane.
8. 3.0 cm Grade 1 filter paper circles.

3 Methods

All procedures are done at room temperature unless otherwise specified.

3.1 Synthesis of Gold Nanoparticles (from the Brust–Schiffrin Method) [3]

1. Pour 1.0 g (2.5 mmol, 1 Eq) of hydrogen tetrachloroaurate (III) hydrate and 150 mL of Type I ultrapure water into a 1,000 mL round bottom flask. Stir slowly for 5 min using a 2 in. egg-shaped magnetic stir bar until complete dissolution. Solution should be clear yellow (*see* **Note 1**).

2. Premix 2.1 g (3.8 mmol, 1.5 Eq) of tetra-*n*-octylammonium bromide in 150 mL of toluene and add it to the gold solution in the round bottom flask. Stir at maximum speed for 15 min until the solution presents is a cloudy dark orange color (*see* **Note 2**).
3. While continuing to rapidly stir, cover the round bottom flask with a septum stopper and using a syringe add dropwise 0.7 mL (5 mmol, 2 Eq) of 1-pentanethiol over the course of 15 min. Stir until the solution presents a cloudy deep white color (*see* **Note 3**).
4. Remove the stopper and quickly add a freshly prepared solution of 2.0 g (50 mmol, 20 Eq) of sodium borohydride in 10 mL of Type I ultrapure water. Stir the solution for 5 h (*see* **Note 4**).
5. Using a separatory funnel, remove the water phase (colorless) from the toluene solution (black). Evaporate the majority of the solvent under reduced pressure at 40 °C (*see* **Note 5**).
6. Add 900 mL of 200 proof ethanol and mix until the solid is completely dispersed. Store the flask at −20 °C.

3.2 Purification of Gold Nanoparticles

1. Two days after the synthesis, the nanoparticles precipitate and a black solid is observed at the bottom of the flask. Carefully remove the ethanol solution (brown color) without dispersing the precipitate (*see* **Note 6**).
2. Add fresh 200 proof ethanol and mix the solution until the nanoparticles are dispersed. Leave the solution at −20 °C until precipitation is complete (*see* **Note** 7).
3. Repeat **steps 1** and **2** until the ethanol solution is colorless after precipitation (approximately five times) (*see* **Note 8**).
4. Remove the solvent and redisperse the precipitate in 10 mL of 200 proof ethanol. Sonicate the solution for 10 min.
5. Using a 3.0 cm grade 1 filter paper circle over a 25 mm filter holder (No. 5 stopper) attached to a vacuum line, wash the nanoparticles several times with 200 proof ethanol redispersing the precipitate in each wash (*see* **Note 8**).
6. Recover the solid from the filter paper.
7. Dissolve the nanoparticles in a minimal amount of toluene and analyze the pentanethiol-capped AuNPs using mass spectrometry (LDI) paying particular attention for the presence of a peak at 466 m/z (TOAB). If the peak is present repeat **steps 5** and **6** until it is no longer observed (*see* **Notes 9** and **10**) [13].
8. Use transmission electron microscopy (TEM) to obtain the average particle diameter and use nuclear magnetic resonance (solvent—$CDCl_3$) to confirm purity.

3.3 Functionalization of Gold Nanoparticles (from the Murray Place Exchange) [14]

1. Dissolve 30 mg of the purified pentanethiol-capped AuNPs in 10 mL of dry dichloromethane and bubble the solution with nitrogen for 5 min.
2. Separately dissolve 150 mg of 23-mercapto-3,6,9,12-tetraoxatricosan-1-ol (or the ligand of interest) in 10 mL of dichloromethane and bubble with nitrogen for 5 min.
3. Mix the two solutions in a 50 mL round bottom flask, cap with a plastic stopper, and stir (0.5″ octagonal magnetic stir bar) for 2 days.
4. Remove the solvent under reduced pressure and redissolve the now functionalized nanoparticles in 10 mL Type I ultrapure water, obtaining a brownish colored solution.
5. Purify the solution in 10,000 MWCO snakeskin pleated dialysis tubing membrane by dialyzing the system for 5 days using deionized water in a 5,000 mL polypropylene beaker (*see* **Note 11**).
6. Recover the AuNP solution from the dialysis bag and lyophilize the nanoparticles. A black solid is obtained.
7. Dissolve a small amount of the solid in D_2O (for NMR) and the rest in 2 mL Type I ultrapure water.
8. Calculate the concentration according to the reported extinction coefficient methodology using the size value obtained from TEM [15].
9. Store at 4 °C.

4 Notes

1. Weigh the gold salt using a plastic spatula, not a metallic one that can potentially reduce the gold salt. Make sure that all the gold salt is dissolved.
2. Stirring is one of the key components of the process. Maintain fast stirring for the rest of the synthesis, using the recommended stir bar.
3. Do not add more 1-pentanethiol to achieve the white color. Wait a prudent amount of time (20 min) and the color should appear, otherwise some of the components are impure and the process should be restarted with new materials.
4. A black color should appear as soon as the sodium borohydride is added. If it takes more than 3 s to appear, the process should be started again and the solution discarded as the slow appearance of the black color will indicate polydisperse product.
5. No more than 40 °C should be used, otherwise nanoparticles will grow larger and polydispersity will be observed.

6. If there is no observable precipitation (solution too dark), check with a glass pipet to observe if the solution is clear or still cloudy. In case of a cloudy solution, let sit at −20 °C for another day.
7. No sonication should be used in the first cleanings.
8. Acetone can be used for faster precipitation/filtration.
9. If TOAB is still present after several cleanings, pour the solid nanoparticles into a 1,000 mL round bottom flask, add 5 mL of toluene to dissolve (brownish color appears), and add 900 mL of ethanol [5]. Repeat the process from **step 1** (this procedure can be repeated until TOAB is removed).
10. If TOAB is still present, an alternative procedure for cleaning is the use of acetonitrile. Dissolve the AuNPs in hexanes and extract multiple times with acetonitrile (AuNPs remain in upper hexanes phase and the TOAB transfers to the lower acetonitrile layer). Collect the hexanes phase and evaporate the solvent at reduce pressure (<40 °C).
11. During the first day water should be changed every 2 h, the second and third day every 6 h, and during the last 3 days every 12 h.

Acknowledgment

This work was supported by the NIH (GM077173 and EB014277).

References

1. Faraday M (1857) The Bakerian lecture: experimental relations of gold (and other metals) to light. Philos Trans R Soc Lond 147:145–181
2. Saha K, Bajaj A, Duncan B, Rotello VM (2011) Beauty is skin deep: a surface monolayer perspective on nanoparticle interactions with cells and biomacromolecules. Small 14:1903–1918
3. Brust M, Walker M, Bethell D, Schiffrin DJ, Whyman R (1994) Synthesis of thiol-derivatised gold nanoparticles in a two-phase liquid-liquid system. J Chem Soc Chem Commun 7:801–802
4. Li Y, Zaluzhna O, Xu B, Gao Y, Modest JM, Tong YJ (2011) Mechanistic insights into the Brust-Schffrin two-phase synthesis of organo-chalcogenate-protected metal nanoparticles. J Am Chem Soc 133:2092–2095
5. Cho M-H, Niles A, Huang R, Inglese J, Austin CP, Riss T, Xia M (2008) A bioluminescent cytotoxicity assay for assessment of membrane integrity using a proteolytic biomarker. Toxicol In Vitro 22:1099–1106
6. Waters CA, Mills AJ, Johnson KA, Schiffrin DJ (2003) Purification of dodecanethiol derivatised gold nanoparticles. Chem Commun 540–541
7. Murphy CJ, Gole AM, Stone JW, Sisco PN, Alkilany AM, Goldsmith EC, Baxter SC (2008) Gold nanoparticles in biology: beyond toxicity to cellular imaging. Acc Chem Res 41:1721–1730
8. Zhu Z-J, Tang R, Yeh Y-C, Miranda OR, Rotello VM, Vachet RW (2012) Determination of intracellular stability of gold nanoparticle monolayers using mass spectrometry. Anal Chem 84:4321–4326
9. Hoffman AS (2008) The origins and evolution of "controlled" drug delivery systems. J Control Release 132:153–163
10. Kanaras AG, Kamounah FS, Schaumburg K, Kiely CJ, Brust M (2002) Thioalkylated tetraethylene glycol: a new ligand for water soluble

monolayer protected gold clusters. Chem Commun 2294–2295

11. Moyano DF, Rotello VM (2011) Nano meets biology: structure and function at the nanoparticle interface. Langmuir 27:10376–10385
12. Miranda OR, Chen H-T, You C-C, Mortenson DE, Yang X-C, Bunz UHF, Rotello VM (2010) Enzyme-amplified array sensing of proteins in solution and in biofluids. J Am Chem Soc 132:5285–5289
13. Yan B, Zhu Z-J, Miranda OR, Chompoosor A, Rotello VM (2010) Laser desorption/ionization mass spectrometry analysis of monolayer-protected gold nanoparticles. Anal Bioanal Chem 396:1025–1035
14. Hostetler MJ, Green SJ, Stokes JJ, Murray RW (1996) Monolayers in three dimensions: synthesis and electrochemistry of ω-functionalized alkanethiolate-stabilized gold cluster compounds. J Am Chem Soc 118:4212–4213
15. Liu X, Atwater M, Wang J, Huo Q (2007) Extinction coefficient of gold nanoparticles with different sizes and different capping ligands. Colloid Surf B 58:3–7

Chapter 2

DNA Conjugation to Nanoparticles

Sunho Park

Abstract

Nanoparticle–DNA (NP–DNA) conjugates have been highlighted due to their versatility in diverse science and engineering fields. The protocol of DNA conjugation to gold nanoparticles (AuNPs), which are among the most popular NPs in bio-applications, is thus described here. This protocol also includes ligand exchange of AuNP to make AuNPs suitable for conjugation process and a fluorescence technique to evaluate the average number of DNA strands attached to single AuNP.

Key words Nanoparticle, Gold nanoparticle, DNA, Conjugation, Ligand exchange, Gel electrophoresis, Lyophilization, DNA staining, Fluorescence, DNA coverage ratio

1 Introduction

DNA–nanoparticle conjugates have attracted great interest as there are numerous opportunities to utilize the properties of the nanomaterial and the DNA synergistically on small length scales [1–5]. In particular, gold nanoparticles (AuNPs) are widely used in biomedical applications such as delivery, sensing, and imaging due to their exceptional biocompatibility [6–9]. AuNPs have a relatively inert surface but can be easily conjugated to biomolecules by covalent linkage via the sulfur atom in a thiol. Therefore, thiol-capped DNA oligos have been popular for constructing AuNP–DNA conjugates.

Thiol-aided conjugation is considered feasible for the laboratory; however, some important issues are often ignored in practice. The first is the stability of the nanoparticles during the conjugation process. Conjugation via a thiol group is typically performed in concentrated ionic buffers which screen the surface charge of the particles and DNA, so that electrostatic repulsion is diminished, and therefore increasing the collision rate between the two species. Using concentrated samples also raises this collision rate. Lyophilization is a method that can elevate both the ionic strength of solution and the concentration of each species extremely high as

Paolo Bergese and Kimberly Hamad-Schifferli (eds.), *Nanomaterial Interfaces in Biology: Methods and Protocols*, Methods in Molecular Biology, vol. 1025, DOI 10.1007/978-1-62703-462-3_2,

the water content of the solution dries out. Citrate-stabilized AuNPs [10] are commonly used in laboratories but they irreversibly aggregate during lyophilization due to their low stability. A chemical ligand bis(*p*-sulfonatophenyl) phenylphosphine dihydrate dipotassium salt is thus used to replace the citrate molecules on the AuNP surface, and allows AuNPs to sustain stability under more severe ionic conditions.

Another issue is the evaluation of the exact coverage ratio of AuNP–DNA conjugates. The average number of conjugated DNA strands per each particle is lower than the number from DNA:AuNP incubation ratio in the reaction due to <100 % efficient conjugation. To quantify the coverage, the sample must be cleaned up by removing the free DNA strands from the solution, and then the conjugated DNA strands are completely liberated from the AuNPs to count their population. In this method, 6-mercapto-1-hexanol is used for the DNA displacement [11]. Fluorescence measurement of the stained DNA strands is performed afterwards to quantify the DNA concentrations. AuNPs should be excluded from the solution before the measurement to avoid fluorescence quenching [12].

This chapter includes the protocols of ligand exchange of AuNP, AuNP–DNA conjugation, and evaluation of DNA coverage in detail (Fig. 1). Purification of samples relies only on typical biochemical lab techniques such as centrifugation and agarose gel electrophoresis. The protocols provide most biochemists and nanoparticle researchers with an easier and more accurate AuNP–DNA conjugate preparation that does not require extensive effort or specialized skills.

2 Materials

All solutions are prepared with 0.5× tris-borate-EDTA buffer (0.5× TBE, pH = 8.3) unless specified otherwise.

2.1 Ligand Exchange

1. 0.5× Tris-borate-EDTA buffer (TBE): 45 mM tris, 45 mM boric acid, and 1 mM EDTA in ultrapure deionized water (resistivity of 18 MΩ cm at room temperature) (*see* **Note 1**).
2. Gold nanoparticle (AuNP): Synthesized or commercially achieved citrate-stabilized aqueous solution (*see* **Note 2**).
3. Bis(*p*-sulfonatophenyl) phenylphosphine dihydrate dipotassium salt (BPS) (*see* **Note 3**).
4. Sodium Chloride (NaCl).
5. 1 % Agarose gel: 0.5 g of agarose powder in 50 mL 0.5× TBE. Cast in an appropriate agarose gel electrophoresis device. Use 0.5× TBE as gel running buffer (*see* **Note 4**).

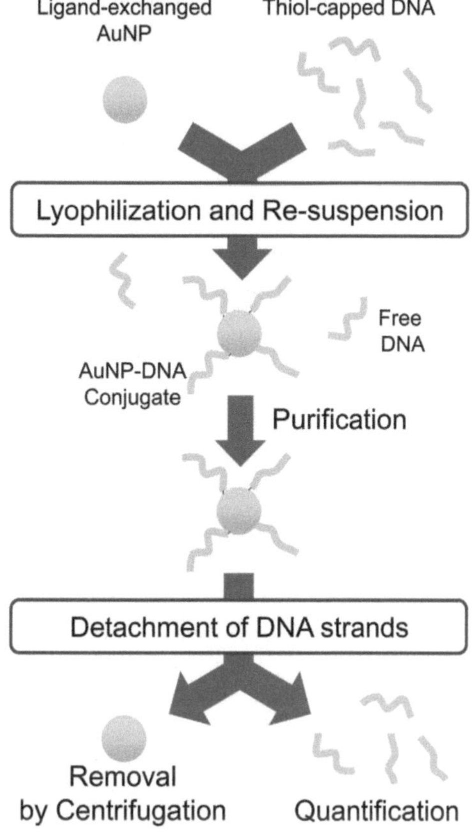

Fig. 1 An overview of the protocol of nanoparticle–DNA (NP–DNA) conjugation and coverage ratio quantification

2.2 Conjugation and Evaluation of Coverage Ratio

1. DNA oligonucleotide: commercially available thiol-capped DNA strands (this protocol works best for 10–50 mer DNA oligos). Adjust the concentration to 1 μg/μL (*see* **Note 5**).
2. 100 mM Tris(2-carboxyethyl)phosphine hydrochloride (TCEP): 0.0287 g of TCEP in 1 mL 0.5× TBE (*see* **Note 6**).
3. 10 mM 6-Mercapto-1-hexanol (MCH): 1 μL of pure MCH (~7.3 M) into 729 μL of 0.5× TBE (*see* **Note 7**).
4. DNA staining solution: 100× SYBR® Gold (Invitrogen). Dilute 1 μL of stock solution (10,000×) with 99 μL of 0.5× TBE (*see* **Note 8**).

3 Methods

3.1 Ligand Exchange and Purification of AuNPs

1. Put a pinch amount (~0.1 g) of BPS into ~100 mL of aqueous citrate stabilized AuNP solution, which is typically at ~10^{-9} M concentration and pale red. Place the solution on an orbital shaker and gently mix it for an overnight (*see* **Note 9**).

2. Put 1 g of NaCl into the AuNP–BPS solution. Stir and mix well with a disposable pipette. The color of the solution will turn dark gray in a few seconds. Separate the solution evenly into two 50 mL centrifuge tubes or eight 15 mL tubes. Place the tubes in a centrifuge and spin them for ~5 min at rcf ~1,000 × *g* (*see* **Note 10**).
3. Remove the clear supernatant from all of the tubes with a pipette and add ~100 μL of 0.5× TBE with 200 μL pipette and tip right onto the sediment at the bottom of one of the tubes. Re-disperse the particles well by pressing and releasing the pipette, take out the dark red solution, and put it right into the sediment in another one of the tubes. Repeat this step until all of the AuNP aggregates are recovered (*see* **Note 11**).
4. Place the thick red AuNP solution into a long single well in 1 % agarose gel. Apply an electric field of ~4 V/cm to the gel. The AuNP band will move from the anode (−) to cathode (+). Stop running the gel after the band shifts by ~3–4 cm (taking ~1 h) (*see* **Note 12**). Cut out the narrow AuNP band using a clean razor blade, chop it into many small pieces, and put the sliced pieces into a 15 mL tube. Pour several milliliters of 0.5× TBE into the tube until the agarose gel pieces are fully soaked. Tighten the cap and place the tubes in a 4 °C refrigerator for ~2 days (*see* **Note 13**).
5. Take out the red solution and discard the solid pieces left in the tube. Distribute the solution into 1.5 mL tubes (300 μL each) and place the tubes in a microcentrifuge. Spin the tubes for ~10 min at rcf ~10,000 × *g*. Gently remove the supernatant (pale pink) and collect the red oily layer at the bottom of the tubes (*see* **Note 14**).
6. Use a UV–VIS spectrometer to collect an absorbance spectrum. Confirm the peak absorbance is at ~520 nm and record the peak absorbance value. Determine the concentration of the stock solution using Beer–Lambert law (*see* **Note 15**).

3.2 Conjugation of DNA Oligos with AuNPs

1. Determine a target coverage ratio (average number of conjugated DNA strands per AuNP) and the amount of AuNP stock solution to be used. Calculate the necessary amount of DNA solution (*see* **Note 16**). Mix the same volume of 100 mM TCEP solution with the DNA solution. Leave the mixture on the bench for ~1 h (*see* **Note 17**).
2. Mix the TCEP + DNA solution well with AuNP solution. Separate evenly the mixture into 1.5 mL tubes (~100 μL per tube) and lyophilize the samples until they are dried out (*see* **Note 18**).

3. Re-disperse the dried sample in one of the tubes by adding ~50 μL 0.5× TBE into it, and use this solution to recover all the other samples. Place the concentrated solution in a 4 °C refrigerator for ~1 day to allow further conjugation (*see* **Note 19**).
4. Add ~1 mL of 0.5× TBE to the sample and separate the diluted solution evenly into 1.5 mL tubes (~200 μL per each tube). Spin the tubes with a microcentrifuge (rcf ~10,000 × *g*) for 10 min to collect AuNP–DNA conjugates at the bottom and leave unconjugated DNA strands in supernatant. Discard the supernatant and collect the red layer at the bottom of the tubes. Repeat this whole washing step at least three times. The red solution from the final washing is the stock solution of AuNP–DNA conjugates.
5. Take UV–VIS absorbance spectrum to achieve the concentration of the stock AuNP–DNA conjugate solution (*see* **Note 20**).

3.3 Determination of Coverage Ratio

1. Dilute the AuNP–DNA stock solution to 0.1 μM in 1 mM MCH and leave the sample undisturbed for ~1 day (*see* **Note 21**).
2. Spin the AuNP–DNA + MCH solution with a microcentrifuge (rcf ~10,000 × *g*, 10 min). Carefully transfer the clear supernatant to a new 1.5 mL tube and discard the original tube containing dark gray residue at the bottom (*see* **Note 22**). Put 10 μL of this solution into 89 μL of 0.5× TBE, and add 1 μL 100× SYBR® Gold to the mixture. Make three or four identical samples for better accuracy of coverage measurement. Do the following step (staining of reference DNA) immediately (*see* **Note 23**).
3. Dilute the stock DNA solution to 1 μM (*see* **Note 24**). Make reference DNA solutions of 1×10^{-8}, 5×10^{-8}, 1×10^{-7}, 2×10^{-7}, and 4×10^{-7} M, in 0.1 mM MCH and 1× SYBR® Gold (*see* **Note 25**). These reference samples and the samples from the previous step should be kept in the dark for ~1 h.
4. Take fluorescence spectrum of the samples from **step 2** and reference solutions from **step 3**. The excitation wavelength is at 495 nm and the emission peak is around 537 nm for SYBR® Gold (Fig. 2a). Collect the peak emission intensity of each solution. Fit the data from DNA references into a linear function, $I = a \cdot C + b$, where I is the intensity of fluorescence [cps; counts per second], C is the concentration of DNA solution [M], and a, b are constants (Fig. 2b). Calculate the DNA concentration of the samples from **step 2** by use of the measured intensity and the fitted equation. Divide this value by 1×10^{-8} M (*see* **Note 26**) to get the coverage ratio of the stock AuNP–DNA.

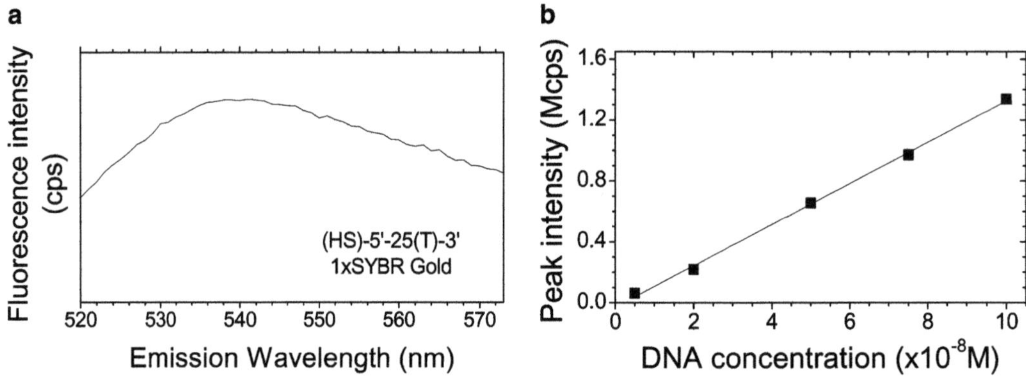

Fig. 2 Intensity of fluorescence emission from 1× SYBR® Gold-stained DNA oligo (HS-5′-TTTTT TTTTT TTTTT TTTTT TTTTT-3′) in 0.5× TBE and 0.1 mM MCH was measured. Excitation wavelength was 495 nm. (**a**) An example of emission spectrum, (**b**) A linear relationship between peak emission intensity and concentration of stained DNA

4 Notes

1. 0.5× TBE is made by either diluting commercially available concentrated TBE buffer or dissolving powdered TBE in water.
2. This protocol works for AuNP of diameter in 5–20 nm, while it is most suitable for ~10 nm AuNPs. Aqueous AuNP colloids are synthesized mostly by reducing Au^{3+} ions in sodium citrate solution; however, citrate-capped AuNPs are not stable enough to endure the conjugation process suggested here. Most failures, including irreversible aggregation of AuNPs, happen during the centrifugation or lyophilization steps.
3. BPS molecules form dipolar bonds between their own oxygen atoms and gold atoms on AuNP's surface. They are negatively charged and generate sufficiently high zeta-potential in ionized buffer.
4. Agarose gel electrophoresis is used to exclude unbound BPS molecules and residues from AuNP synthesis. Add 0.5 g agarose powder to 50 mL 0.5× TBE (or 1 g agarose to 100 mL) in 250 mL glass beaker and mix well using a disposable pipette. Pull up and push out the mixture with the pipette several times to make the agarose solution homogenized. A typical gel comb has multiple teeth to make separate wells in a gel. For this purification purpose, put some electrical tape on the teeth to cover the empty spaces between the teeth so that the comb will form a long, single well in the gel. After the gel casting device is assembled, heat up the agarose solution in the

beaker with a microwave oven. Interrupt the heating process a few times to take out and shake the beaker to re-disperse any possible sediment at the bottom. The solution is finished heating when it boils and turns clear. Take the beaker carefully out of the microwave oven, pour the hot solution quickly into the gel tray, and place the gel comb at a right position. Remove small bubbles floating on the surface level of the hot gel solution before the gel starts hardening. A 10 μL or 20 μL pipette and tip works well for pulling up the bubbles. It takes 30 min to 1 h for full hardening of the gel. Remove the gel comb and pour into the tank 0.5× TBE as gel running buffer. The gel must be fully immersed.

5. Thiol-capped DNA strands typically form dimers by making disulfide bonds, or are capped with an additional protecting group by the manufacturer. Thus, they need to be reduced before conjugation with AuNPs.
6. It is preferred to use fresh solution. Dithiothreitol (DTT) is often used for the same purpose; however, TCEP is preferable over DTT in that it is an irreversible and more powerful reducing agent.
7. It is preferred to use fresh solution. Handle the stock MCH bottle in a fume hood as it has a strong odor.
8. It is preferred to use fresh solution. This product is a DNA staining dye and stored at −20 °C. Thaw the stock in a dark place at room temperature. Put the stock bottle back into the freezer right after taking out the necessary amount.
9. The concentration of 0.1 g of BPS [$M_w = 534.6$] in 100 mL solution is ~1.9 mM, which is much higher than a typical concentration of aqueous AuNP solution. An overnight reaction is sufficient to ensure the BPS content on AuNP surface has reached an equilibrium state.
10. Excessive amounts of NaCl can make the ionic strength of the solution very high. This fully screens the surface charge of the AuNPs in solution and can result in precipitation of the AuNPs. Tune the centrifugal force and centrifugation time depending on the size and surface charge density of AuNP as the precipitated AuNPs may aggregate permanently under too strong centrifugation conditions.
11. Limit the final volume of the re-dispersed AuNP solution within a few hundred microliters so that the volume does not exceed the capacity of the gel loading well.
12. These gel running times and electric field strengths are based on lab experience. You may need to run the gel for longer if the particles are larger, ionic strength of running buffer is greater, or the electric field strength is low.

13. AuNPs in gel band will diffuse into 0.5× TBE buffer and the concentration distribution of AuNPs will reach an equilibrium state inside and outside gel. Slicing of the gel band facilitates this because it increases the surface area for diffusion flux. More AuNPs are recovered by transferring the remaining solid gel pieces into another tube filled with clean 0.5× TBE. It takes less time for diffusion for smaller AuNPs.
14. Spin the tubes for an extended time if not much of the AuNPs collect at the bottom. This is the stock solution of AuNP to be used for DNA conjugation, and the concentration is typically in the order of ~1 μM. You may adjust the concentration exactly to 1 μM using 0.5× TBE for convenience of later use.
15. $A = \varepsilon cl$, where A [OD, optical density] is the peak value of the absorbance spectrum, ε [M^{-1} cm^{-1}] is the extinction coefficient, l [cm] is the path length, and c [M] is the concentration of the solution. ε of AuNPs depends on their size and ligand types, and values can be found in literature (~10^8 M^{-1} cm^{-1} for ~10 nm AuNP) [13]. Dilute the stock AuNP solution to make the measured absorbance peak lie within a typical working range of most UV–VIS spectrometers, 0.1–10 O.D.
16. If 50 μL of 1×10^{-6} M AuNP stock solution is used to make 1:10 AuNP:DNA conjugates, for example, the exact amount of DNA strands necessary is $10 \times (50\ \mu L) \times (1 \times 10^{-6}\ M) = 500 \times 10^{-6}$ μmol. When M_w [μg/μmol] is the molecular weight of DNA oligo, this amount is equivalent to $(500 \times 10^{-6}\ \mu mol) \times M_w$ [μL] of 1 μg/μL stock DNA solution. Due to the nature of conjugation process, however, use two or three times as much of the exact amount. Not all the DNA used will be conjugated.
17. Keep the mixture in a dark place if the DNA strands have fluorescent markers.
18. An easy way of lyophilization is to put the samples in 1.5 mL tubes and place them in a vacuum chamber with mild centrifugation. Use a pushpin to make an air hole on the lid of the tubes. Centrifugation helps the samples stay at the bottom of the tube during the drying process. Spinning the samples too strongly sometimes results in irreversible aggregation. Higher success rates are usually associated with running weak centrifugation for the first ~10 min only while under vacuum.
19. This refrigeration step for further conjugation may be skipped if a higher amount of extra DNA (*see* **Note 16**) is used during the TCEP treatment step.
20. The concentration of AuNP–DNA conjugates is equivalent to the concentration of core AuNPs. DNA absorbance spectra have a peak at 260 nm and vanish in most visible light wavelength range; therefore, the AuNP absorbance peak value at 520 nm is rarely affected by conjugated DNA.

21. The sulfur atom of MCH forms a covalent bond with a gold atom on the AuNP surface, the same way of DNA conjugates to the AuNP. Due to the excessive population of MCH molecules in solution, the surface of AuNP becomes densely covered by MCH, replacing the DNA strands. The AuNP eventually becomes neutral as the negatively charged DNA and BPS are displaced in this process. If the concentration of the stock AuNP–DNA solution is 1 μM, for example, add 10 μL AuNP–DNA and 10 μL 10 mM MCH to 80 μL 0.5× TBE in a 1.5 mL tube (100 μL in total).
22. AuNPs have been aggregated and discarded by centrifugation. Remember that the DNA strands in the supernatant are in 1 mM MCH and have been detached from 0.1 μM AuNP–DNA.
23. The sample is now in 0.1 mM MCH, 1× SYBR® Gold. DNA strands in the solution were released from 1×10^{-8} M AuNP–DNA. SYBR® Gold, a staining agent of DNA, is added after all other substances have been mixed. DNA staining is fully saturated in about half an hour; however, it starts degrading after a few hours. Thus, perform the fluorescence measurement quickly in the following steps.
24. 1 μg/μL DNA stock is at molar concentration of $1/M_w$. Dilute properly with 0.5× TBE.
25. Mix together 1 μL of 1 μM DNA, 1 μL of 10 mM MCH, and 97 μL of 0.5× TBE, then put 1 μL of 100× SYBR® Gold. This makes 1×10^{-8} M DNA in 0.1 mM MCH, 1× SYBR® Gold, and 0.5× TBE. Change the amount of 1 μM DNA appropriately for each target DNA concentration, and reduce the amount of 0.5× TBE by the increased amount of 1 μM DNA to equalize the volumes of DNA reference solutions at 100 μL.
26. Note that the DNA strands have been detached from 1×10^{-8} M AuNP–DNA.

References

1. Alivisatos P (2004) The use of nanocrystals in biological detection. Nat Biotechnol 22: 47–52
2. De M, Ghosh PS, Rotello VM (2008) Applications of nanoparticles in biology. Adv Mater 20:4225–4241
3. Medintz IL, Uyeda HT, Goldman ER, Mattoussi H (2005) Quantum dot bioconjugates for imaging, labelling and sensing. Nat Mater 4:435–446
4. Nel AE, Madler L, Velegol D, Xia T, Hoek EMV, Somasundaran P, Klaessig F, Castranova V, Thompson M (2009) Understanding biophysicochemical interactions at the nano-bio interface. Nat Mater 8:543–557
5. Choi HS, Liu W, Liu F, Nasr K, Misra P, Bawendi MG, Frangioni JV (2010) Design considerations for tumor-targeted nanoparticles. Nat Nanotechnol 5:42–47
6. Chen C-C, Lin Y-P, Wang C-W, Tzeng H-C, Wu C-H, Chen Y-C, Chen C-P, Chen L-C, Wu Y-C (2006) DNA-gold nanorod conjugates for remote control of localized gene expression by near infrared irradiation. J Am Chem Soc 128:3709–3715
7. Ghosh PS, Kim C-K, Han G, Forbes NS, Rotello VM (2008) Efficient gene delivery vectors by tuning the surface charge density of amino acid-functionalized gold nanoparticles. ACS Nano 2:2213–2218

8. Thomas M, Klibanov AM (2003) Conjugation to gold nanoparticles enhances polyethylenimine's transfer of plasmid DNA into mammalian cells. Proc Natl Acad Sci U S A 100: 9138–9143
9. Macfarlane RJ, Lee B, Jones MR, Harris N, Schatz GC, Mirkin CA (2011) Nanoparticle superlattice engineering with DNA. Science 334:204–208
10. Frens G (1973) Controlled nucleation for the regulation of the particle size in monodisperse gold suspensions. Nat Phys Sci 241:20–22
11. Park S, Brown KA, Hamad-Schifferli K (2004) Changes in oligonucleotide conformation on nanoparticle surfaces by modification with mercaptohexanol. Nano Lett 4:1925–1929
12. Dulkeith E, Ringler M, Klar TA, Feldmann J (2005) Gold nanoparticles quench fluorescence by phase induced radiative rate suppression. Nano Lett 5:585–589
13. Liu X, Atwater M, Wang J, Qun H (2007) Extinction coefficient of gold nanoparticles with different sizes and different capping ligands. Colloids Surf B 58:3–7

Chapter 3

Conjugation of Nanoparticles to Proteins

Marie-Eve Aubin-Tam

Abstract

Nanoparticle–protein conjugates hold great promise in biomedical applications. Diverse strategies have been developed to link nanoparticles to proteins. This chapter describes a method to assemble and purify nanoparticle–protein conjugates. First, stable and biocompatible 1.5 nm gold nanoparticles are synthesized. Conjugation of the nanoparticle to the protein is then achieved via two different approaches that do not require heavy chemical modifications or cloning: cysteine–gold covalent bonding, or electrostatic attachment of the nanoparticle to charged groups of the protein. Co-functionalization of the nanoparticle with PEG thiols is recommended to help protein folding. Finally, structural characterization is performed with circular dichroism, as this spectroscopy technique has proven to be effective at examining protein secondary structure in nanoparticle–protein conjugates.

Key words Nanoparticles, Proteins, Bioconjugates, Agarose gel electrophoresis, Circular dichroism

1 Introduction

The conjugation of nanoparticles (NPs) to proteins is of increasing interest in biomedicine. NP–protein conjugates hold great promise in sensing/diagnostics [1–3], in targeted delivery [4], in the control of protein activity [5, 6] and in imaging [7]. To this end, efforts have been made to synthesize biocompatible NPs that are nontoxic and stable at physiological pH [8].

Several strategies have been developed to assemble NP–protein conjugates. A commonly used approach is to attach the NP on the protein via non-covalent interactions, mostly through electrostatic adsorption. In the 1970s, negatively charged gold NPs were already linked non-covalently to various proteins, such as antibodies, horseradish peroxidase, and bovine serum albumin, to serve as electron dense labels for electron microscopy (EM) of tissue sections [9, 10]. Although this approach is relatively straightforward and does not require cloning or chemical cross-linkage, the NP can dissociate from the protein, especially at pH values above the protein isoelectric point (p*I*) [10] or at high salt concentrations [11].

Paolo Bergese and Kimberly Hamad-Schifferli (eds.), *Nanomaterial Interfaces in Biology: Methods and Protocols*, Methods in Molecular Biology, vol. 1025, DOI 10.1007/978-1-62703-462-3_3, © Springer Science+Business Media New York 2013

In contrast, the formation of a covalent bond with either the NP core atoms or the NP ligand molecules creates a more stable NP–protein complex and allows to control the site of attachment on the protein more easily. Such site-specific labeling is especially advantageous in imaging and sensing applications. In EM imaging applications, for instance, site-specific NP attachment on the protein enables one to go beyond the traditional histology applications and to localize specific structures within large proteins [12, 13].

Various purification schemes are reported for isolating NP–protein conjugates. Size exclusion chromatography [7, 14] and gel electrophoresis [14–17] have been used to assess conjugation yields and to purify NP–DNA and NP–protein conjugates of desired stoichiometry. Agarose gel electrophoresis is particularly suitable because it allows sample to migrate in the gel towards both electrodes. Both positively charged and negatively charged species can be viewed simultaneously. This is especially useful for conjugating proteins to NPs of opposite charge. Furthermore, since no denaturant such as sodium dodecyl sulfate (SDS) is used, the conjugate can be directly extracted from the agarose gel with spin centrifugation columns.

Unfortunately, NP conjugation may result in protein unfolding, as previously reported for a number of proteins [11, 17–20]. Therefore, prior to any application, it is advised to verify that the protein maintains its native fold and activity when conjugated to the NP. In particular, proteins are more prone to being denatured or destabilized if covalently attached to the NP [17]. Co-functionalization of the NP with polyethylene glycol (PEG) molecules can help to circumvent this issue by preventing unfolding. As thiolated PEG molecules replace BPS ligands on the surface of the gold NP, the detrimental electrostatic interactions between the NP and its associated protein can be reduced [17].

This chapter describes the process of purifying and characterizing proteins labeled with gold NPs. In the first section (Subheading 3.1), a method of synthesizing soluble biocompatible 1.5 nm gold NPs is presented. Then, labeling (Subheading 3.2), purification (Subheading 3.3), and structural characterization (Subheading 3.4) of NP–protein conjugates is described—these protocols are easily adapted to protein labeling with NPs of different sizes, materials, or surface ligands.

2 Materials

All solutions should be prepared with deionized (DI) water. A Millipore water purification system can be used to reach a resistivity of 18 MΩ·cm at 25 °C.

2.1 Nanoparticle Synthesis

1. 125 mL conical flask and Teflon stir bar.
2. Separatory funnel of 50 mL capacity.

3. 250 mg hydrogen tetrachloroaurate ($HAuCl_4$) and 400 mg of tetraoctylammonium bromide dissolved in 12.5 mL DI water and 16.25 mL toluene.
4. 580 mg triphenylphosphine.
5. 352.5 mg of sodium borohydride.
6. 2.14 g of bis(*p*-sulphonatophenyl) phenylphosphine dihydrate (BPS) dissolved in 20 mL of DI water.
7. High-performance liquid chromatography (HPLC) system.
8. HPLC gel filtration column (e.g.: TSK-gel G4000PW, TOSOH bioscience).

2.2 NP Conjugation

1. PBS 1× or 10 mM NaPi buffer (pH 7.4).
2. Dithiothreitol (DTT).
3. Hexa(ethylene glycol) thiol (Polypure).
4. Dialysis cassette (e.g., Slide-A-Lyzer 2K MWCO, Pierce).

2.3 Agarose Gel Purification

1. 0.5× TBE buffer (Tris-Borate-EDTA): 44.5 mM Tris Base, 44.5 mM boric acid, 1 mM EDTA. Dissolve 10× from 5× TBE buffer stock (*see* **Note 1**).
2. 3 g ultrapure low melting point agarose dissolved in 100 mL 0.5× TBE buffer.
3. 500 mL beaker or conical flask.
4. Horizontal gel electrophoresis system.
5. SimplyBlue SafeStain (Invitrogen).
6. Centrifugal filter with 0.2 μm pore size membrane (Nanosep MF Centrifugal Devices with Bio-Inert Membrane, PALL).

2.4 Structural Characterization Equipment

1. UV–Vis spectrophotometer.
2. Circular dichroism spectrometer.

3 Methods

All procedures should be carried at room temperature unless specified otherwise.

3.1 Synthesis and Purification of Water Soluble 1.5 nm NPs

The first eight steps are adapted from Weare et al. [21].

1. Mix 12.50 mL DI water with 16.25 mL toluene in a 125 mL conical flask.
2. Weigh 250 mg of $HAuCl_4$ and 400 mg of tetraoctylammonium bromide. Add to the water–toluene mixture and mix with magnetic stir bar until the yellow color is all transferred in the toluene phase.

3. Weigh 580 mg of triphenylphosphine and add to the solution. Stir vigorously for at least 10 min or until the toluene phase is white and cloudy.
4. Weigh 352.5 mg of sodium borohydride. Dissolve in 2.5 mL of DI water and immediately add to the $HAuCl_4$ and triphenylphosphine solution in one step as quickly as possible while stirring.
5. Stir for 2 h.
6. To wash the NPs, separate the organic phase from the water phase using a separatory funnel. Discard the water phase.
7. Add 25 mL of DI water to the solution of NPs in toluene. Separate the organic phase from the water with the separatory funnel. Discard the water phase.
8. Repeat **step 7** two additional times.
9. Weigh 2.14 g of bis(*p*-sulphonatophenyl) phenylphosphine dihydrate (BPS) and dissolve in 20 mL of DI water to make a 0.2 M solution.
10. Mix the solution of NPs in toluene with the BPS aqueous solution in a closed bottle. Stir gently with a magnetic bar for 48 h. The NP will transfer to the water phase upon ligand exchange (*see* **Note 2**). Using a separatory funnel, separate the organic phase from the aqueous solution of BPS NPs. Dispose properly of the organic phase. Store the NP solution at 4 °C.
11. (Optional) Purify the water soluble NPs by size exclusion HPLC to obtain a narrower NP size distribution and to remove excess BPS ligand. A size exclusion HPLC column such as the TSK-gel G4000PW (TOSOH bioscience) can be used with DI water flowing at a rate of 0.5 mL/min. Collect 500 μL fractions. Keep the fractions that show broad UV–Vis absorbance in the 200–800 nm range with no plasmon peak at ~520 nm (*see* Fig. 1). Store at 4 °C. After HPLC purification, it is possible to measure the stoichiometry of ligands on the surface of the NPs with analytical techniques like ICP-OES (*see* **Note 3**).
12. Determine the NP concentration from the absorbance at 420 nm using an extinction coefficient of 110 mM^{-1} cm^{-1}.

3.2 NP Conjugation to Protein

This protocol can be used either with 1.5 nm gold NPs covered with negatively charged BPS ligands synthesized as previously described, or with NPs of other sizes, materials, or surface ligands (*see* **Note 4** for protocols to synthesize larger NPs).

NPs can be attached via electrostatic interactions. For example, a protein with a net positive charge (e.g., cytochrome *c*, ribonuclease A) would readily attach to a negatively charged BPS NP or citrate NP. Alternatively, a gold NP can also be attached to an exposed cysteine on the surface of a protein via gold-thiol chemistry.

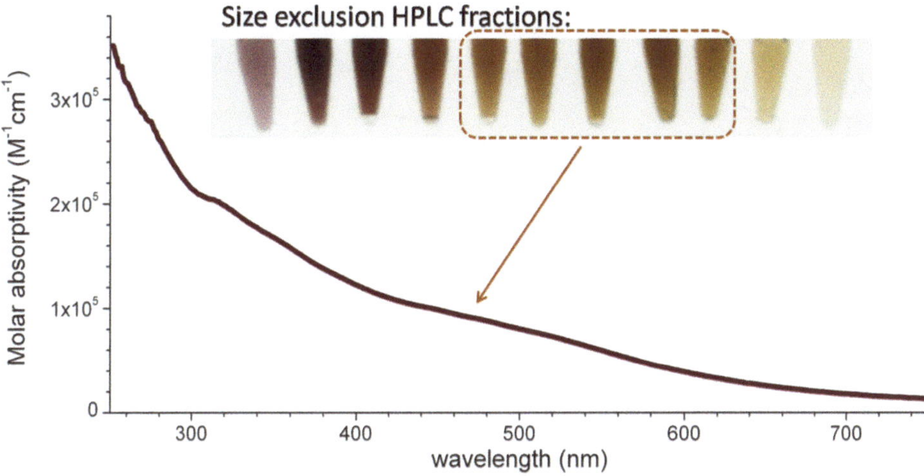

Fig. 1 Size exclusion HPLC purification of gold NPs. The picture *above* shows fractions collected consecutively. The fractions that contain solutions of NPs of ~1.5 nm diameter (*encircled* by *dashed line*) do not show any plasmon absorption peak at ~520 nm in the UV–Vis spectra. Larger NPs elute from the column earlier and are pink/purple because of the plasmon absorption. Free BPS molecules elute later

1. (Optional) If the protein attaches to the NP via a surface cysteine using gold-thiol chemistry, incubate the protein with 10 mM DTT for 15 min. Remove excess DTT by dialyzing against conjugation buffer (*see* **Note 5**) in dialysis cassettes, or by four successive concentration–dilution cycles with 5 kDa centrifugal filters using a refrigerated microcentrifuge at 4 °C.
2. Mix NPs and protein solution (both in conjugation buffer) in the desired NP–protein molar incubation ratio. If using 1.5 nm gold NPs, NP concentration should be ~0.1 mM to permit gel visualization.
3. (Optional) Simultaneous incubation of the gold NPs and proteins with poly(ethylene glycol) thiol can prevent protein unfolding (Fig. 2). If working with 1.5 nm BPS gold NPs, a 5×–25× molar excess of hexa(ethylene glycol) thiol (Polypure) relative to NP concentration can be used.
4. Incubate overnight at 4 °C.

3.3 Agarose Gel Electrophoresis Purification

1. Prepare a 3 % agarose gel by mixing 3 g of ultrapure agarose with 100 mL of 0.5× TBE in a 500 mL beaker or conical flask. Microwave the solution 20 s at HIGH intensity. Swirl the solution gently. Continue microwaving in 20 s increments and swirling the solution until it boils and the agarose is fully dissolved. Pour the gel into a horizontal gel electrophoresis system using a pipettor to prevent bubbles to set into the gel. Insert a comb in the middle of the gel with wells at least 1 cm wide. When the gel is set, fill the gel box with 0.5× TBE and remove comb.

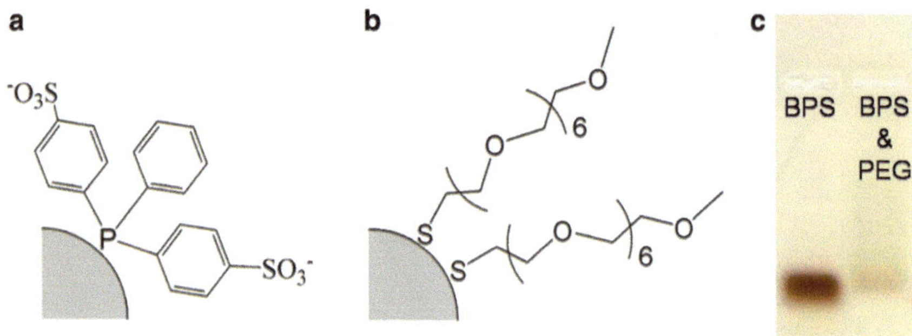

Fig. 2 Schematics of BPS ligand (**a**) and PEG ligand (**b**) at the surface of a NP. (**c**) Agarose gel electrophoresis of NP covered with BPS ligand (*left lane*) and co-functionalized with mPEG-thiol (*right lane*) in a 25× molar excess relative to the NP concentration

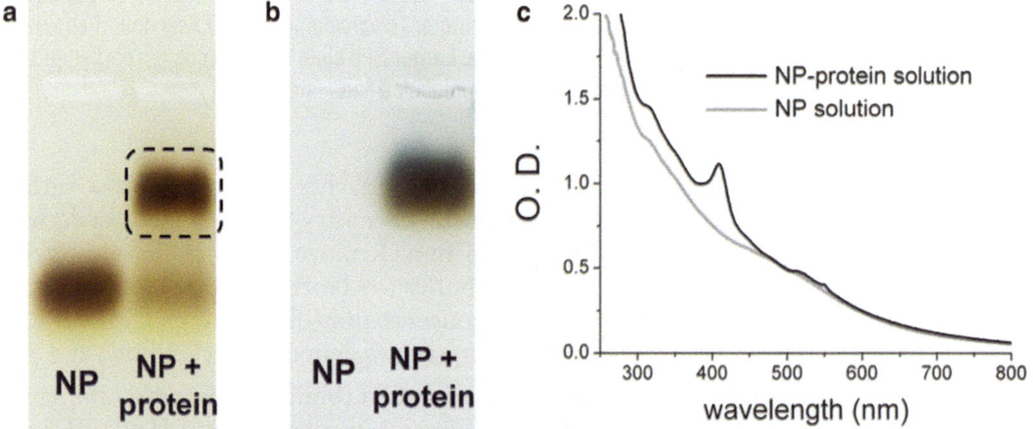

Fig. 3 (**a**) NP conjugation is confirmed by agarose gel electrophoresis. The 1.5 nm gold NPs with BPS ligand run as a single band towards the positive electrode (*left lane*). The NPs migrate more slowly in the agarose gel when bound covalently to *Saccharomyces cerevisiae* cytochrome *c* via cysteine C102. (**b**) *Blue* staining for protein confirms the presence of protein in the slower band. Cytochrome *c* is positively charged and would run in the opposite direction towards the negative electrode. (**c**) UV–Vis absorption spectra of solutions of NPs (*grey line*) and NP–protein conjugates (*black line*)

2. Load the NP, protein and NP–protein solutions into successive wells. Run the gel with field strength of 10 V/cm until the NPs have entered into the gel matrix and you get the desired separation between the bands (Fig. 3).
3. To blue stain the gel for proteins, immerse the gel in SimplyBlue SafeStain (Invitrogen). Rock the gel gently for 2 h. Discard staining solution. Rinse with DI water. Immerse the gel in DI water. Rock the gel gently overnight.
4. If the protein or the NP band position overlaps with the NP–protein band, run the gel for longer times and/or repeat the

gel electrophoresis with different concentrations of agarose (in the range of 1–4 %) until the bands run at distinct positions as shown in Fig. 3.

5. When appropriate gel electrophoresis conditions have been identified, run another agarose gel without blue staining.
6. Using a clean razor blade, cut the NP–protein band and place gel into a centrifugal filter with 0.2 μm pore size membrane (PALL). Cap the filter.
7. Spin at 14,000 × *g* at 4 °C for 30 min or until the NP–protein solution has been extracted from the agarose gel.

3.4 Structural Characterization of Conjugate

Circular dichroism (CD) spectroscopy in the far-UV is an effective and widely used technique of probing protein structure in NP–protein conjugates [11, 17–20, 22]. CD measurements are performed in solution. Clean quartz CD cuvettes with 5 M nitric acid and rinse them thoroughly with DI water.

Buffers that absorb strongly at low wavelengths (195–250 nm) should be avoided if possible when performing CD in the far-UV. NP ligands that strongly absorb in the UV can make CD measurements noisier (*see* **Note 6**). If large NPs or aggregates of NPs are present, correction for absorption flattening might be necessary (*see* **Note 7**).

1. Measure UV–Vis absorption spectra of a 200 μL solution of NPs and a 200 μL solution of NP–protein (there can be excess NPs, *see* **Note 8**). Both should be in the same buffer. Using the NP extinction coefficient, calculate the NP concentration. If needed, add some buffer to have same NP concentration in both solutions. If the protein absorbs in the visible (e.g., Heme containing proteins), deconvolute the NPs and proteins contribution to the UV–Vis absorption spectra to calculate NP and protein concentration.
2. Measure CD spectra of the NPs and NP–protein solutions (with same NP concentrations) in the 195–250 nm. Subtract the CD spectrum of NP solution from the CD spectrum of NP–protein solution.
3. Secondary structure algorithms such as CDSSTR [23] can be used to deconvolute the proportion of α-helices, β-strands, turns, and unordered structures.

4 Notes

1. To make 5× TBE (445 mM Tris Base, 445 mM boric acid, 10 mM EDTA), weight 54 g of Tris base and 27.5 g of boric acid. Add 20 mL of 0.5 M EDTA (pH 8.0) and complete to 1 L with DI water. The pH of this stock solution will be around pH 8.3.

2. During ligand exchange, the two phases (organic and aqueous) might not be distinguishable at some point. Eventually, the NPs will transfer to the aqueous solution of BPS and the organic phase will become clear.
3. The stochiometry of ligand on the NPs can be measured with inductively coupled plasma optical emission spectrometry (ICP-OES). For that purpose, use a solution of HPLC purified NPs in DI water, with aqua regia if needed.
4. Larger gold NPs can be synthesized as follow. 30 mg $HAuCl_4$ is dissolved in 240 mL DI water and heated to 60 °C. A 60 mL aqueous solution that contains 120 mg sodium citrate with tannic acid and sodium carbonate is also heated to 60 °C. NPs size can be tuned by the amount of tannic acid and sodium carbonate. For 10 nm NPs, use 250 mg/L tannic acid and 22 mg/L sodium carbonate. For 16 nm NPs, use 8.33 mg/L tannic acid and 0.73 mg/L sodium carbonate. Once the temperature in both solutions reaches 60 °C, they are rapidly mixed together and stirred for 10 min. Let cool to room temperature. NPs are functionalized with BPS by adding 0.1 g of BPS to the solution and incubating overnight. The NPs are then precipitated by adding NaCl to the solution. The supernatant is discarded and the NPs are resuspended in a low salt buffer. The final step is to run the NP sample in a 2 % agarose gel and extract the NPs from the gel with centrifugal filters as described in (Subheading 3.3).
5. Several buffers are appropriate for conjugation. Examples of suitable buffers include 10 mM NaPi buffer (pH 7.4) or 1× PBS. In high salt concentration, NPs tend to precipitate. DTT in buffers can bind to gold atoms on the NP surface. In high enough concentration, DTT can induce NP aggregation.
6. Some NP ligands, like BPS, might have a strong optical absorption at short wavelengths (~200 nm), which renders the CD measurements very noisy unless averaging over long times.
7. Due to their small diameter, 1.5 nm NPs do not cause considerable scattering of the UV light. However, if large aggregates are formed or large NPs are used, correction for absorption flattening might be necessary [19, 24].
8. If conjugation conditions are such that all proteins are labeled with NPs (as in Fig. 3a,b right lane), it is not necessary to purify the protein–NP conjugates with agarose gel electrophoresis. It is possible to directly measure the CD signal of the solution of NPs and proteins, since it does not contain any unlabelled proteins.

References

1. Chah S, Hammond MR, Zare RN (2005) Gold nanoparticles as a colorimetric sensor for protein conformational changes. Chem Biol 12:323–328
2. De M et al (2009) Sensing of proteins in human serum using conjugates of nanoparticles and green fluorescent protein. Nat Chem 1:461–465
3. Willner I, Baron R, Willner B (2007) Integrated nanoparticle-biomolecule systems for biosensing and bioelectronics. Biosens Bioelectron 22:1841–1852
4. Bhattacharyya S et al (2011) Efficient delivery of gold nanoparticles by dual receptor targeting. Adv Mater 23:5034–5038
5. Park S, Kimberly H-S (2010) Enhancement of in vitro translation by gold nanoparticle-DNA conjugates. ACS Nano 4:2555–2560
6. Aubin M-E, Morales DG, Hamad-Schifferli K (2005) Labeling Ribonuclease S with a 3 nm Au nanoparticle by two-step assembly. Nano Lett 5:519–522
7. Hu M et al (2008) Gold nanoparticle-protein arrays improve resolution for cryo-electron microscopy. J Struct Biol 161:83–91
8. Zheng M, Li Z, Huang X (2004) Ethylene glycol monolayer protected nanoparticles: synthesis, characterization, and interactions with biological molecules. Langmuir 20:4226–4235
9. Faulk WP, Taylor GM (1971) An immunocolloid method for the electron microscope. Immunochemistry 8:1081–1083
10. Geoghegan WD, Ackerman GA (1977) Adsorption of horseradish peroxidase, ovomucoid and anti-immunoglobulin to colloidal gold for the indirect detection of concanavalin A, wheat germ agglutinin and goat anti-human immunoglobulin G on cell surfaces at the electron microscopic level: a new method, theory and application. J Histochem Cytochem 25:1187–1200
11. Aubin-Tam M-E, Hwang W, Hamad-Schifferl K (2009) Site-directed nanoparticle labeling of cytochrome c. Proc Natl Acad Sci USA 106:4095–4100
12. Montesano-Roditis L et al (2001) Cryo-electron microscopic localization of protein L7/L12 within the Escherichia coli 70 S ribosome by difference mapping and Nanogold labeling. J Biol Chem 276:14117–14123
13. Buechel C et al (2001) Localisation of the PsbH subunit in photosystem II: a new approach using labelling of His-tags with a Ni^{2+}-NTA gold cluster and single particle analysis. J Mol Biol 312:371–379
14. Ackerson CJ et al (2006) Rigid, specific, and discrete gold nanoparticle/antibody conjugates. J Am Chem Soc 128:2635–2640
15. Zanchet D et al (2001) Electrophoretic isolation of discrete Au nanocrystal/DNA conjugates. Nano Lett 1:32–35
16. Mamedova NN et al (2001) Albumin-CdTe nanoparticle bioconjugates: preparation, structure, and interunit energy transfer with antenna effect. Nano Lett 1:281–286
17. Aubin-Tam M-E, Hamad-Schifferli K (2005) Gold nanoparticle-cytochrome c complexes: the effect of nanoparticle ligand charge on protein structure. Langmuir 21: 12080–12084
18. Tellechea E et al (2012) Engineering the interface between glucose oxidase and nanoparticles. Langmuir 28:5190–5200
19. Vertegel A, Siegel RW, Dordick JS (2004) Silica nanoparticle size influences the structure and enzymatic activity of adsorbed lysozyme. Langmuir 20:6800–6807
20. Srivastava S et al (2005) Controlled assembly of protein-nanoparticle composites through protein surface recognition. Adv Mater 17:617–621
21. Weare WW et al (2000) Improved synthesis of small (dcore 1.5 nm) phosphine-stabilized gold nanoparticles. J Am Chem Soc 122:12890–12891
22. Lundqvist M et al (2006) Induction of structure and function in a designed peptide upon adsorption on a silica nanoparticle. Angew Chem Int Ed 45:8169–8173
23. Sreerama N, Woody RW (2004) Computation and analysis of protein circular dichroism spectra. Methods Enzymol 383:318–351
24. Gerdova A, Kelly SM, Halling P (2011) Experimental test of absorption flattening correction for circular dichroism of particle suspensions. Chirality 23:574–579

Chapter 4

Water-Solubilization and Functionalization of Semiconductor Quantum Dots

Christina M. Tyrakowski, Adela Isovic, and Preston T. Snee

Abstract

Semiconductor quantum dots (QDs) are highly fluorescent nanocrystals that have abundant potential for uses in biological imaging and sensing. However, the best materials are synthesized in hydrophobic surfactants that prevent direct aqueous solubilization. While several methods have been developed to impart water-solubility, an aqueous QD dispersion has no inherent useful purpose and must be functionalized further. Due to the colloidal nature of QD dispersions, traditional methods of chemical conjugation in water either have low yields or cause irreversible precipitation of the sample. Here, we describe several methods to water-solubilize QDs and further functionalize the materials with chemical and/or biological vectors.

Key words Quantum dot, Semiconductor nanocrystal, Cap-exchange, Encapsulation, Functionalization, Carbodiimide coupling

1 Introduction

Semiconductor quantum dots (QDs, or nanocrystals) have gained a significant level of interest since their discovery in the mid-1980s due to their size-dependent optical properties [1–3]. Over the next ~25 years of study, many fundamental physical properties have been characterized; applications were also demonstrated for the use of QDs in lasers [4], indoor lighting [5], and energy generation [6] to name a few. Although several physical phenomena are still the subject of investigation, the development of very high quantum yield core/shell water-soluble QD systems [7, 8] created inroads for biological imaging applications. These initial reports also demonstrated dramatic enhancement of photostability and brightness of QD-based bioimaging agents compared to organic

Paolo Bergese and Kimberly Hamad-Schifferli (eds.), *Nanomaterial Interfaces in Biology: Methods and Protocols*, Methods in Molecular Biology, vol. 1025, DOI 10.1007/978-1-62703-462-3_4, © Springer Science+Business Media New York 2013

fluorescent dyes. Chemical and biological sensing has also been reported using mixed QD-organic coupled fluorophores [9, 10].

Significant challenges still exist concerning the use of colloidal QD systems for imaging and sensing applications. Biological targeting requires the use of functionalized fluorophores, and chemical and biological sensing capability can be imparted by conjugating QDs with energy-accepting organic dyes or other compounds. Furthermore, the research scientist must have control over all the steps in the water-solubilization and subsequent functionalization processes to guarantee reproducibility and the development of transferable technologies. We have written this chapter with the intention of aiding the development of QD-based bioimaging technology starting with as-prepared CdSe/CdZnS nanocrystals dispersed in hydrophobic solvents. We detail three water-solubilization methods to control size and chemical compatibility, as well as strategies, that depend on the method of water-solubilization employed, for functionalizing the dispersions with chemical and biological vectors.

Described in this chapter are procedures for purifying as-prepared QDs as well as the recently developed zinc-metathesis cap-exchange method that creates small water-soluble QDs with enhanced aqueous stability [11]. We also detail the highly robust method of polymer encapsulation with 40 % octylamine-modified poly(acrylic acid) [12]. As is well-known, carboxylic-acid coated cap-exchanged or polymer-encapsulated QDs cannot be easily functionalized with amine-functional chemical or biological vectors using the common reagent 1-ethyl-3-(3-dimethylaminopropyl) carbodiimide hydrochloride (EDC) [13], which we have recently demonstrated is due to the cationic character of EDC [14]. Neutral reagents such as methoxypolyethylene glycol carbodiimide (MPEG CD) are highly effective for QD functionalization [14], yet the synthesis is challenging due to the use of highly specific chemical transformation techniques. As such, we have detailed the synthesis and usage of this reagent as a part of this chapter. A method for functionalizing silica-coated QDs is also detailed, which relies on the presence of thiols on the surface of the QD arising from the thiol-silica precursor used in the water-solubilization process. The effect of these water-solubilization methods on the emission of CdSe/CdZnS QDs is shown in Fig. 1. A sample of hydrophobic QDs was water-solubilized using the three methods described in this chapter. The quantum yields, relative to Rhodamine 6G in ethanol, are 81 % for hydrophobic, 50 % for DHLA-capped, 50 % for silica-coated, and 36 % for polymer-encapsulated QDs.

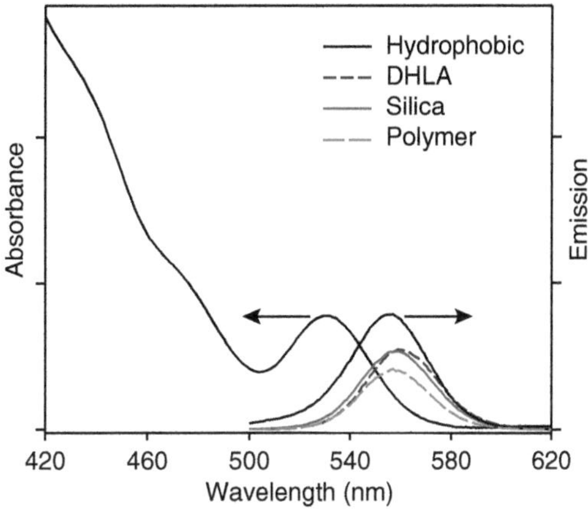

Fig. 1 The absorption spectrum of hydrophobic QDs and the emission spectra relative to quantum yield of hydrophobic, DHLA-capped, silica-coated, and polymer-encapsulated QDs

2 Materials

There are several bench-top solvents and equipment that are used in most of the procedures; they are listed below, with the specific solvents and equipment needed in an individual procedure listed in that procedure's materials section. We store most precursors under ambient conditions grouped by chemical compatibility unless specified otherwise by the manufacturer. Note that some procedures use toxic and explosive sodium azide; not only is the use of this chemical an inherent hazard, extreme caution must also be taken in the disposal of excess azide reagent.

Common Solvents

1. Chloroform (≥99.5 %), dichloromethane (≥99.5 %), dimethylformamide (99.8 %, dried over molecular sieves), ethyl acetate (≥99.5 %), ethyl ether (99.9 %), hexanes (≥98.5 %, mixture of isomers), methanol (≥99.8 %), 2-propanol (≥99.5 %), tetrahydrofuran, toluene (99.9 %), silicone oil for a heat bath, and 10 MΩ-cm deionized water (D.I. water) (Milli-Q Water System, Millipore, Billerica, MA); however, we find that house deionized water is sufficient.
2. 0.1 M sodium hydroxide solution (0.1 M NaOH): Take an aluminum weigh boat or a glass vial and record its mass. Add ~4.0 g of NaOH (97 %) to the vessel, dry in an oven to a constant weight, and transfer with small portions of D.I. water to a 1 L glass container on a kitchen scale. Next, add an additional amount of D.I. water (~1,000 g) as measured using the kitchen scale to make a 0.1 M solution; store under ambient conditions.

Common Equipment

1. Vacuum gas manifold (Schlenk line) with a 5.9 cfm pump that generally delivers 80–200 mTorr of low pressure as measured with a vacuum gauge and nitrogen gas that is purified by passing over activated BASF catalyst (R3-11G) and Drierite (*see* **Note 1**).
2. Centrifuge with rotor capable of achieving an RCF (relative centrifugal force) of 2,000 × *g*.
3. Rotary evaporator with a cold-water condenser and hot water bath.
4. Electronic scale (110 g capacity, 0.0001 g accuracy) and kitchen scale (3 kg capacity, 1 g accuracy).

2.1 Precipitation and Purification of As-Prepared CdSe/CdZnS QDs

1. Hydrophobic QDs: We use emissive CdSe/CdZnS QDs prepared according to the procedures outlined in ref. 15. We believe that the following procedures will also be useful when handling commercially available QD materials when following the manufacturer's guidelines for purification.
2. Solvents: Hexane, 2-propanol, and methanol.
3. Equipment: 7 mL glass vials with caps, turnover septum stopper, 21 gauge needle, Schlenk line, sonicator, and centrifuge.

2.2 Water-Soluble Silica-Coated QDs

1. Precipitated and purified CdSe/CdZnS QDs, prepared as described in Subheading 3.1.
2. (3-Mercaptopropyl)trimethoxysilane (95 %).
3. Cesium carbonate (99.5 %).
4. Either zinc chloride (≥98.0 %) or zinc nitrate hexahydrate (98.0 %).
5. Solvents: dry dichloromethane, D.I. water, 0.1 M NaOH, and hexane.
6. Equipment: 7 mL glass vial with cap, Teflon flea micro stir bar (10 × 3 mm), Teflon polygon stir bar (38 × 9.5 mm), turnover septum stopper, 21 gauge needle, 10 mL disposable syringe fitted with a 100 or 200 nm filter, 10 mL dialysis tube (catalog number G235071, Float-A-Lyzer, Type: CE, MWCO: 100 kDa, Spectrum Labs, Rancho Dominguez, CA, USA), large 1 or 2 L flask, pH indicator strips, Schlenk line, stir plate, and centrifuge.

2.3 Water-Soluble Dihydrolipoic Acid Cap-Exchanged QDs

1. Precipitated and purified CdSe/CdZnS QDs, prepared as described in Subheading 3.1.
2. Dihydrolipoic acid (DHLA) (*see* **Note 2**).
3. Sodium hydroxide (99.5 %).
4. Zinc nitrate hexahydrate (98.0 %).

5. Solvents: D.I. water, and 0.1 M NaOH, methanol.
6. Equipment: 7 mL glass vial with cap, Teflon flea micro stir bar (10 × 3 mm), Teflon polygon stir bar (38 × 9.5 mm), turnover septum stopper, 21 gauge needle, centrifuge, Schlenk line, 10 mL disposable syringe fitted with a 100 or 200 nm filter, 10 mL dialysis tube (catalog number G235071, Float-A-Lyzer, Type: CE, MWCO: 100 kDa, Spectrum Labs, Rancho Dominguez, CA, USA), large 1 or 2 L flask, stirring hot plate, and oil bath.

2.4 Water-Soluble 40 % Octylamine-Modified Poly(acrylic acid) Polymer-Encapsulated QDs

1. Precipitated and purified CdSe/CdZnS QDs, prepared as described in Subheading 3.1.
2. Poly(acrylic acid), 1,800 MW.
3. Octylamine (98 %).
4. 1-ethyl-3-(3-dimethylaminopropyl)carbodiimide hydrochloride.
5. Concentrated HCl (37 %) and diluted HCl: Take an appropriately sized glass flask and fill to 100 mL with D.I. water. Add 4 mL of concentrated HCl and mix with a glass stir rod or pipet.
6. 0.33 M sodium hydroxide solution (0.33 M NaOH): Take an aluminum weigh boat or a glass vial and record its mass. Add ~1.33 g of NaOH (97 %), dry in an oven to a constant weight, and transfer with small portions of D.I. water to a glass container. Next, add an appropriate amount of D.I. water (~100 g) using a kitchen scale to make a 0.33 M solution.
7. Solvents: chloroform, ethyl acetate, D.I. water, 0.1 M NaOH, dimethylformamide, and methanol.
8. Equipment: Glass 1-neck 150 and 200 mL round bottom flasks fitted with glass hose adapters, glass 500 mL separatory funnel with a Teflon stopcock and glass stopper, Teflon egg-shaped stir bars (19 × 9.5 mm), 5 mL disposable syringe, Schlenk line, stir plate, sonicator, 50 mL polypropylene centrifuge tubes, turnover septum stoppers, 21 gauge needle, 10 mL disposable syringe fitted with a 100 or 200 nm filter, Amicon Ultra-15 100K Centrifugal Filter (Millipore, Billerica, MA, USA), pH indicator strips, and centrifuge.

2.5 Methoxypolyethylene Glycol Carbodiimide Reagent Preparation

1. Methoxypolyethylene glycol 350, Average Molecular Weight: 350.
2. Thionyl chloride (≥99 %).
3. Sodium azide (99 %).
4. Triphenylphosphine (99 %).
5. Ethyl isothiocyanate (≥97.0 %).

6. Mercury (II) oxide (HgO), yellow (98 %): Add 6.5 g of HgO to a 50 mL polypropylene centrifuge tube, and then add 30 mL of D.I. water. Sonicate for 30 min, centrifuge at 2,000 × *g*, and discard the supernatant. Repeat this D.I. water wash three times. Transfer the wet HgO to a 20 mL glass vial (*see* **Note 3**) fitted with a screw cap with septa (*see* **Note 1**), connect to a Schlenk line, and remove the D.I. water by vacuum, which may take upwards of 2 days.
7. Solvents: dichloromethane, dimethylformamide, D.I. water, ethyl ether, tetrahydrofuran, and toluene.
8. Equipment: Glass 3-neck 50 mL round bottom flasks fitted with glass hose adapters, glass stoppers, and turnover septum stoppers. Glass 500 and 250 mL 1-neck round bottom flasks fitted with glass hose adapters, Büchner funnel with a fritted glass filter, glass 500 mL separatory funnel with a Teflon stopcock and glass stopper, Teflon egg-shaped stir bars (19 × 9.5 mm), 3 mL disposable syringe, aluminum foil, 50 mL polypropylene centrifuge tube, small glass flask, grade 1 filter paper, Schlenk line, sonicator, stir plate, and centrifuge.

2.6 Functionalizing Aqueous Carboxylic-Acid Functional QDs Using MPEG CD

1. Aqueous dihydrolipoic acid cap-exchanged QDs or 40 % octylamine-modified PAA polymer-encapsulated QDs, prepared as described in Subheadings 3.3 or 3.4.
2. Rhodamine B piperazine dye, as prepared by ref. 16 (or other amine-functional compounds such as a protein).
3. Methoxypolyethylene glycol carbodiimide (MPEG CD), prepared as described in Subheading 3.5.
4. pH 8 phosphate buffer: To a 1 L glass bottle on a kitchen balance, add 6.81 g of potassium phosphate monobasic (99.5 %), 467 mL of 0.1 M NaOH, and dilute to 1 L with D.I. water. Cap bottle and dissolve by stirring or sonication.
5. Solvents: D.I. water, and 0.1 M NaOH.
6. Equipment: Teflon flea micro stir bar (10 × 3 mm), Teflon polygon stir bar (38 × 9.5 mm), 5 mL disposable syringe fitted with a 100 or 200 nm filter, Amicon Ultra-15 100K Centrifugal Filter (Millipore, Billerica, MA, USA) or 10 mL dialysis tube (catalog number G235071, Float-A-Lyzer, Type: CE, MWCO: 100 kDa, Spectrum Labs, Rancho Dominguez, CA, USA), pH indicator strips, stir plate, and centrifuge.

2.7 Functionalizing Aqueous Silica-Coated QDs Using Sulfo-SMCC

1. Water-soluble silica-coated QDs, prepared as described in Subheading 3.2.
2. Rhodamine B piperazine dye, as prepared by ref. 16 (or other amine-functional compound, such as a protein).
3. Sulfosuccinimidyl-4-(*N*-maleimidomethyl)cyclohexane-1-carboxylate (Sulfo-SMCC).

4. pH 6 phosphate buffer: To a 1 L glass bottle on a kitchen balance, add 6.81 g of potassium phosphate monobasic (99.5 %), 56 mL of 0.1 M NaOH, and dilute to 1 L with D.I. water. Cap bottle, and dissolve by stirring or sonication.
5. pH 8 phosphate buffer: To a 1 L glass bottle on a kitchen balance, add 6.81 g of potassium phosphate monobasic (99.5 %), 467 mL of 0.1 M NaOH, and dilute to 1 L with D.I. water. Cap bottle, and dissolve by stirring or sonication.
6. Solvents: D. I. water, and 0.1 M NaOH.
7. Equipment: 7 mL glass vial with cap, Teflon flea micro stir bar (10 × 3 mm), Teflon polygon stir bar (38 × 9.5 mm), 50 mL polypropylene centrifuge tube, 5 mL disposable syringe fitted with a 100 or 200 nm filter, spin desalting column (catalog number 89893, Zeba™, MWCO: 7 kDa, 10 mL, Thermo Scientific, Rockford, IL, USA), 1 mL dialysis tube (catalog number G235035, Float-A-Lyzer, Type: CE, MWCO: 100 kDa, Spectrum Labs, Rancho Dominguez, CA, USA), large flask, pH indicator strips, stir plate, and centrifuge.

3 Methods

All reactions should be performed at room temperature in a fume hood unless stated otherwise.

3.1 Precipitation and Purification of As-Prepared CdSe/CdZnS QDs

1. Add ~1×10^{-8} mol of QDs in growth solution to a 7 mL glass vial; this is typically 0.5 g of growth solution for QDs synthesized according to the published protocols in ref. 15. Next, add 1 mL of hexane, 0.5 mL of 2-propanol, and fill to the top with methanol. Vigorously shake the vial to induce flocculation. Centrifuge the sample until ~1 mL of a gelatinous paste is coating the bottom of the flask, typically ~4 min at 1,700 × *g*. Discard the clear, noncolored supernatant.
2. Add 1–2 mL of hexane to the gel; strongly shake the vial to break up the gel to wash it thoroughly with hexane. Sonication may be necessary. Centrifuge the cloudy mixture at 1,700 × *g* until the hexane supernatant appears to be a clear, colored solution that is free of any particulates (typically 10 min). Carefully transfer the QD-hexane extract to another pre-weighed 7 mL glass vial using a pipet as it is essential that the gel material is removed at this step. As some QDs may remain in the gel layer, repeat the hexane wash and extraction if desired and combine all extracts into the 7 mL glass vial, but do not fill over half the volume of the vial. Add 0.5 mL of 2-propanol to the QD-hexane extract and fill to the top with methanol. Vigorously shake the vial to induce flocculation and then centrifuge the cloudy mixture at 1,700 × *g* for a minimum of

10 min until the QDs have coated the walls of the glass vial. Discard the supernatant.

3. Remove the cap, attach a turnover septum stopper, and pierce the top with a 21 gauge needle. Dry under vacuum with a Schlenk line (*see* **Note 1**). The sample is now ready for use in any of the water-solubilization procedures described below; however, processed samples should be water-solubilized on the same day as the phase-transfer efficiency suffers with time significantly.

3.2 Water-Soluble Silica-Coated QDs

1. Purified, dried CdSe/CdZnS QDs in a 7 mL glass vial from Subheading 3.1 are solubilized by addition of ~2 mL dichloromethane. The solution should be optically clear; if not, the sample should be discarded and a new QD sample should be prepared.
2. 50 μL of (3-mercaptopropyl)trimethoxysilane (0.27 mmol), 88 mg of cesium carbonate (0.27 mmol), and either 9 mg of zinc chloride (0.067 mmol) or 20 mg of zinc nitrate hexahydrate (0.067 mmol) are added with a flea micro stir bar to the QD dispersion in dichloromethane.
3. Stir the solution 1–2 days. Next, remove the stir bar and centrifuge the vial at 1,700 × *g* until the cesium carbonate is pelleted to the bottom of the vial and the solution is optically clear, typically 5 min. Transfer the clear supernatant using a glass pipet into a new 7 mL glass vial. Add hexane to the top of the vial to induce precipitation (*see* **Note 4**).
4. Once the QDs have flocculated, centrifuge the vial at 1,700 × *g* until the QDs become pelleted to the bottom of the vial. Uncap, remove the supernatant, and let dry under ambient conditions for ~1 h. The gel of QDs should noticeably shrink and crack during this time.
5. Add two to three drops of methanol and a Teflon flea micro stir bar, and break up the dried gel until most of the material appears “wet” with methanol, normally after ~1 min stir. Next, fill the vial with 0.1 M NaOH and stir (*see* **Note 5**).
6. Filter the sample with a 100 or 200 nm filter into a 10 mL dialysis tube; fill the dialysis tube to the top with D.I. water and cap tightly. Fill a 1 L or 2 L glass flask containing a polygon stir bar with D.I. water to allow the dialysis tube to float fully extended. Stir 24 h while replacing the D.I. water regularly (*see* **Note 6**) or until the pH of the sample matches the same of the D.I. water. Transfer to a capped glass vial. Samples have been found to be stable for months under ambient conditions; regardless, storage at 4 °C is advised.

3.3 Water-Soluble Dihydrolipoic Acid Cap-Exchanged QDs

1. Add 60–80 mg of sodium hydroxide and a flea micro stir bar to a 7 mL glass vial. Dissolve the sodium hydroxide in ~2 mL of methanol by stirring.
2. Add 100 mg of DHLA and 70 mg of zinc nitrate hexahydrate to the sodium hydroxide solution. Place a turnover septum stopper on the vial, pierce the top with a 21 gauge needle, and put the solution under N_2 via a Schlenk line. Using an oil bath to heat the solution to ~50 °C (*see* **Note 7**), stir until the solution is clear (*see* **Note 8**) and then cool to room temperature.
3. Once cool, add this solution and its stir bar to purified and dried CdSe/CdZnS QDs. Add a turnover septum stopper and stir overnight under N_2 using a Schlenk line.
4. The next day, remove the stir bar and centrifuge the vial at 1,700 × *g* for 10 min. Drain the clear supernatant. Place a turnover septum stopper on the vial, pierce the top with a 21 gauge needle and dry the sample very briefly (1–2 min) by vacuum via a Schlenk line. Add ~6 mL D.I. water and a few drops of 0.1 M NaOH to the sample and shake vigorously until the QDs have dissolved completely.
5. Filter the samples with a 200 nm or 100 nm filter into a 10 mL dialysis tube. Fill the dialysis tube to the top with D.I. water and cap tightly. Fill a 1 L or 2 L glass flask containing a polygon stir bar with D.I. water and place the dialysis tube in the water, allowing the tube to float fully extended. Stir 24 h while replacing the D.I. water regularly (*see* **Note 6**). Transfer to a capped glass vial and store at 4 °C.

3.4 Water-Soluble 40 % Octylamine-Modified Poly(acrylic acid) Polymer-Encapsulated QDs

1. Add 5.00 g of poly(acrylic acid) (PAA) with a Teflon egg-shaped stir bar to a glass 1-neck 150 mL round bottom flask followed by addition of 95 mL of dimethylformamide (*see* **Note 9**). Cap with a hose adapter and connect to a Schlenk line under N_2 while stirring PAA until completely dissolved to yield a clear solution, typically 5 min.
2. Add 5.33 g EDC to the polymer solution and stir under N_2 until the EDC has completely dissolved, typically 30 min. Add additional solvent if the EDC doesn't fully dissolve. Next, load a 5 mL syringe with 4.6 mL of octylamine. Briefly remove the hose adapter and quickly inject the octylamine while the solution is being stirred vigorously. Reattach the host adapter and stir the solution under N_2 overnight.
3. The next day, the system is placed under vacuum to reduce the solvent volume, typically for 5 h; this yields a very viscous, slightly discolored solution that is transferred in equal parts to three 50 mL centrifuge tubes (*see* **Note 10**). Next, add D.I. water to the centrifuge tubes to capacity to precipitate the

polymer. Shake the tubes vigorously, and then centrifuge them for 10 min at 2,000 × *g*. There should be a white or near-white precipitate at the bottom of the centrifuge tube and an aqueous top layer. Discard the supernatant.

4. Add 33 mL of 0.33 M NaOH to each centrifuge tube. Shake and sonicate until all polymer is dissolved, or add additional quantities of 0.1 M NaOH dropwise if the solids do not appear to be dissolving beyond a certain point (*see* **Note 11**).
5. Transfer the basic solution of octylamine-modified PAA to a separatory funnel and wash three times with ~75 mL aliquots of ethyl acetate. Pour the bottom aqueous layer in equal parts into three 50 mL centrifuge tubes and add dilute HCl dropwise, vigorously shaking the tubes every few drops, during which time the polymer precipitates out as a white solid. Dilute HCl should be added until the pH of the solution is below 5.
6. Centrifuge the samples for 5 min at 2,000 × *g* to collect all the 40 % octylamine-modified PAA at the bottom of the centrifuge tube. Discard the supernatant.
7. Add D.I. water to the centrifuge tubes and shake them vigorously to break up the polymer solids. Centrifuge the tubes, and decant the supernatant after measuring its pH. This process is repeated until the pH of the supernatant is the same as D.I. water, typically three times.
8. Transfer the 40 % octylamine-modified PAA samples to a 200 mL round bottom flask (*see* **Note 12**) and dry with a rotary evaporator. Transfer to a Schlenk line and dry further with vacuum. The sample should be broken up into a coarse powder and put in a centrifuge tube or capped bottle and stored at 4 °C. These polymers appear stable on the order of several years.
9. Calculate the mass of the purified and dried CdSe/CdZnS QDs from Subheading 3.1. Multiply this mass by 5× and add at least this weight of 40 % octylamine-modified PAA to the vial containing QDs.
10. Add ~3 mL chloroform to the vial, shake vigorously, and sonicate. If the solution is not optically clear, add two to three drops of methanol to dissolve the polymer; the sample should become clear after sonicating for several minutes.
11. Uncap the vial and add a turnover septum stopper. Pierce the stopper with a 21 gauge needle, and remove the solvent under vacuum with a Schlenk line. It is imperative that the sample be thoroughly dried at this step.
12. Fill the vial with 0.1 M NaOH and stir or sonicate until completely dissolved. Filter with a 100 or 200 nm filter into a 15 mL 100K MWCO concentrating centrifugal filter; fill to

the top via dilution with D.I. water, and centrifuge at the lowest speed possible to concentrate the sample to ~3 mL. Discard the rinse in the bottom of the tube (*see* **Note 13**), re-dilute the sample with D.I. water, and repeat as above a minimum of five times to remove the excess polymer. Transfer to a capped glass vial and store under ambient conditions. Samples are stable on the order of years; however, bacterial infections may cause the formation of a cloudy "plume" that is removable with filtration. This can be suppressed with storage at 4 °C and/or by addition of sodium azide.

3.5 Methoxypolyethylene Glycol Carbodiimide Reagent Preparation

1. Add 10 g (28.6 mmol) of methoxypolyethylene glycol 350 (MPEG) and a Teflon egg-shaped stir bar to a 3-neck 50 mL round bottom flask fitted with a glass stopper, a turnover septum stopper, and a hose adapter. Degas and dry the MPEG by stirring at 80 °C in an oil bath under vacuum via a Schlenk line for several hours (*see* **Notes 1** and 7). Flush the flask with dry N_2 gas from the Schlenk line.
2. Under N_2, cool the solution in an ice bath, and then slowly drip in 3 mL (41 mmol) of thionyl chloride using a syringe, equipped with a 21 gauge needle, through the septum while stirring. Allow the solution to warm to room temperature, and let the solution stir overnight under N_2.
3. Remove the thionyl chloride from the solution by vacuum via a Schlenk line, which may take several hours. Add 10 mL of dimethylformamide (DMF) using a graduated cylinder to dilute the solution in the round bottom, and subsequently remove the DMF by vacuum. Repeat this addition and removal of DMF for a total of three times to ensure removal of all thionyl chloride; this may take several days.
4. Add 75 mL of DMF to the chlorinated MPEG and transfer the solution with the stir bar to a 1-neck 250 mL round bottom flask. Add 2.83 g (43.5 mmol) of sodium azide to the solution, and cover the flask with aluminum foil to protect from light. Stir the solution overnight under N_2 at 85 °C using an oil bath (*see* **Note 7**). The next day, cool the solution to room temperature, and reduce the DMF by vacuum, which may take about 2 days.
5. Add 60 mL of dichloromethane (DCM), and remove undissolved solids using a Büchner funnel with a fritted glass filter (*see* **Note 14**) while collecting the clear solution in a second 1-neck 250 mL round bottom flask. Add two 4 mL portions of DCM to rinse out the first round bottom, pour over the precipitate in funnel, and collect the wash in the second round bottom flask.

6. Add a Teflon egg-shaped stir bar to the 250 mL round bottom, and remove the DCM by vacuum.
7. Add 90 mL of tetrahydrofuran (THF) to the solution, and then add 8.26 g (31.5 mmol) of triphenylphosphine. Stir this for 4 h under N_2, and then add 1.2 mL of D.I. water by graduated cylinder and stir overnight under N_2.
8. Remove the THF by vacuum. Next, add 200 mL of D.I. water to dilute the solution in the round bottom and transfer to a separatory funnel. Rinse the round bottom with two 4 mL portions of D.I. water and transfer to the separatory funnel. Add 150 mL of toluene to the separatory funnel, shake, and let the layers separate. Remove the top organic layer, and repeat the toluene wash of the bottom aqueous layer two more times.
9. Transfer the bottom aqueous layer to a 1-neck 500 mL round bottom flask, and remove the D.I. water by rotary evaporation. Dry further using a Schlenk line while stirring using a Teflon egg-shaped stir bar until the vacuum pressure stabilizes. Store the light yellow oil product MPEG 350 amine in an airtight glass vial at 4 °C.
10. Add 3.0 g (8.6 mmol) of MPEG 350 amine with a Teflon egg-shaped stir bar to a 3-neck round bottom flask fitted with a glass stopper, a turnover septum stopper, and a hose adapter. Degas and dry by stirring under vacuum via a Schlenk line at 70 °C using an oil bath for several hours (*see* **Note 7**). After cooling to room temperature under vacuum, purge the round bottom with N_2, add 5 mL of DMF by graduated cylinder, and re-purge the round bottom with N_2. While stirring under N_2, slowly drip in 1.1 mL (12.6 mmol) of ethyl isothiocyanate by syringe through the septum, and then let the solution stir overnight under N_2.
11. Remove the DMF under vacuum, and then add 10 mL of DCM. Next, add 4.65 g (21.5 mmol) of washed and dried HgO, and stir the solution under N_2 overnight. The next day, the solution should be a dark brown-green color due to the formation of HgS. Add 1.0 g more of washed and dried HgO and stir overnight under N_2 (*see* **Note 15**).
12. Transfer the solution to a 50 mL centrifuge tube with three 3 mL portions of dichloromethane (DCM) to rinse the product out of the round bottom flask. Centrifuge the tube at 2,000 × *g* for 10 min and filter the supernatant through a filter paper-lined glass funnel into a small glass flask. Add 10 mL DCM to the precipitate in the tube and sonicate for 15 min. Centrifuge the tube at 2,000 × *g* for 10 min and again filter the supernatant through the filter paper. Rinse the filter paper with a 5 mL portion of DCM. Using new filter paper, refilter the collected supernatant through a filter paper-lined glass funnel into a 1-neck 50 mL round bottom (*see* **Note 16**).

Rinse filter paper with a 5 mL portion of DCM into the round bottom, add a stir bar to the round bottom, and remove the DCM by vacuum.

13. Put 50 mL of ethyl ether in a small glass bottle, cap loosely, and chill in a container of ice in a ventilation hood. Once cold, add 20 mL ether to the round bottom to dissolve the product and filter through new filter paper in a glass funnel into a new 1-neck 50 mL round bottom flask. Rinse the round bottom with three 5 mL portions of cold ether and filter into the new round bottom. Remove the ether by vacuum until the pressure gauge is stable at a low pressure. Store the dark yellow oil product MPEG CD in an airtight glass vial at 4 °C or at −80 °C for a longer shelf life.

3.6 Functionalizing Aqueous Carboxylic-Acid Functional QDs Using MPEG CD

1. Add 1 mL of aqueous DHLA cap-exchanged or 40 % octylamine-modified PAA polymer-encapsulated QDs (typically 1 μM concentration) to a 7 mL glass vial with a cap (*see* **Note 17**). Add 1–10 mg of MPEG CD (*see* **Note 18**) and a flea micro stir bar. Cap the vial and stir for 10 min.
2. Prepare a solution (*see* **Note 19**) of rhodamine B piperazine (RBpip) dye or other amine-functional compound in pH 8 phosphate buffer.
3. Once the QD solution has stirred for 10 min, drip the desired amount of prepared RBpip dye solution or other amine-functional compound solution into the activated QD dispersion. Check the pH of the QD solution; if the solution is not at pH 8, add more pH 8 phosphate buffer. Recap and stir the solution overnight.
4. The next day, if necessary, filter the QDs with a 5 mL disposable syringe fitted with a 100 or 200 nm filter. In the case of functionalized 40 % octylamine-modified PPA polymer-encapsulated QDs, transfer the sample to a 100K MWCO concentrating centrifugal filter. Fill to the top via dilution with D.I. water, and centrifuge at the lowest speed possible to concentrate the sample to ~3 mL. Discard the rinse liquid (*see* **Note 13**) in the bottom portion of the centrifuge tube. Repeat dilution with D.I. water and concentration of sample for a minimum of five times to ensure that all unreacted RBpip dye (or other amine-functional compound) is removed. Functionalized DHLA cap-exchanged QDs should be transferred to a 1 mL or 10 mL dialysis tube (*see* **Note 20**); fill tube to the top with D.I. water, and cap tightly. Fill a 1 L or 2 L glass flask containing a polygon stir bar with D.I. water to allow the dialysis tube to float fully extended. Stir 24 h while replacing the D.I. water regularly (*see* **Note 6**). Transfer the dialyzed solution to a glass vial and store under ambient conditions in the case of functionalized polymer-encapsulated QDs or at 4 °C in the case of functionalized DHLA cap-exchanged QDs.

3.7 Functionalizing Aqueous Silica-Coated QDs Using Sulfo-SMCC

1. Add 1 mg of sulfo-SMCC to a 7 mL glass vial. Add 0.5 mL of pH 6 phosphate buffer to dissolve the reagent. If the pH is lower than pH 6, add more pH 6 phosphate buffer.
2. Add 1 mL of silica-coated water-solubilized QDs to the sulfo-SMCC solution. Add 0.1 M NaOH dropwise until the pH is close to or at pH 7 (*see* **Note 21**). Add a flea micro stir bar to the vial, cap, and stir for 30 min.
3. Prepare a spin desalting column by opening the bottom, placing in a 50 mL centrifuge tube, and removing the storage solution by centrifugation at 1,000 ×*g* for 2 min. Drain the liquid in the bottom of the centrifuge tube. Add ~8 mL of pH 8 phosphate buffer solution to the top of the desalting column, centrifuge at 1,000 ×*g* for 2 min, and drain the liquid from the bottom of the tube. Repeat this addition and centrifugation of pH 8 phosphate buffer through the desalting column two more times (*see* **Note 22**).
4. Once the QD solution has stirred for 30 min, add the QD solution to the top of the desalting column followed by ~0.2 mL of pH 8 phosphate buffer. Centrifuge at 1,000 ×*g* for 2 min. Transfer the QD solution at the bottom of the centrifuge tube to a new 7 mL glass vial and add a flea micro stir bar.
5. Prepare a solution (*see* **Note 19**) of rhodamine B piperazine (RBpip) dye or other amine-functional compound in pH 8 phosphate buffer. Drip desired amount of prepared RBpip dye solution or other amine-functional compound solution into the activated QD dispersion. Cap and stir the QD solution overnight.
6. The next day, filter the QD solution if necessary with a 5 mL disposable syringe fitted with a 100 or 200 nm filter, and add to a 1 mL dialysis tube (*see* **Note 20**). Fill dialysis tube to the top with D.I. water, and cap tightly. Fill a large glass flask containing a polygon stir bar with D.I. water to allow the dialysis tube to float fully extended. Stir 24 h while replacing the D.I. water regularly (*see* **Note 6**). Transfer the solution to a glass vial and store at 4 °C.

4 Notes

1. When drying a chemical to remove a solvent using a Schlenk line, monitor the pressure using a vacuum gauge to determine when the solute is dry. To dry materials in a vial with a turnover septum stopper, insert a 1 mL plastic syringe with a luer slip fitting into the end of the Schlenk line vacuum rubber tubing connected to the vacuum manifold. The syringe is held

in place with a ring clamp. A needle can be attached to the luer slip end to preserve the integrity of the vacuum while drying.

2. While DHLA is commercially available, due to the expense we prepared the material using the methods outlined in either ref. 17 or 18.
3. The wet HgO is very hard to transfer.
4. Sometimes the cap-exchanged QDs do not precipitate vigorously upon addition of hexane. In this case, transfer the contents of the vial to a larger vessel and add more hexane. Chilling the solution and waiting for a few hours will increase the efficiency of precipitation. Repeating the process using a greater amount of silica precursors (zinc and carbonate as well) can prevent problems with poor precipitation yield.
5. In some samples, the QDs will become instantly solubilized in the base water while others may require overnight stirring. Regardless, it is recommended that the sample stir overnight. Low transfer yields may be the result of water in the dichloromethane solvent.
6. The rinse water in the flask should be monitored for QD emission to ensure that the dialysis tube has not broken. Transfer the remaining sample to a new dialysis tube if there is QD emission in the rinse.
7. A stirring hot plate with oil bath must have the oil stirred (a paper clip is sufficient); the temperature of the oil must be monitored with a thermometer.
8. The solution may have a very slight yellow tint. Sometimes very fine white particles are present in the solution even after stirring with heat for a very long time. This does not seem to affect the cap-exchange of the QDs.
9. The solution might have a slight yellow-brown color; this does not affect the polymer effectiveness or yield of the product.
10. The round bottom flask should be turned over to allow the product on the walls of the round bottom flask to drip into the third centrifuge tube.
11. The 40 % octylamine-modified PAA polymer precipitates due to the acidic pH of the water. The subsequent addition of 0.08 Eq of NaOH is added to bring the pH over 7.
12. The sample will be very viscous; scrape as much as possible into the round bottom.
13. Monitor the rinse liquid for QD emission to ensure that the centrifugal filter has not broken; if rinse liquid has strong QD emission, switch to a new centrifugal filter. Very minor QD

leakage sometimes occurs when the solution is basic and generally cannot be avoided.

14. While using a fritted filter is best to remove the excess sodium azide and sodium chloride, the solid precipitate can also be removed by centrifugation in a 50 mL centrifuge tube followed by transferring the supernatant to a round bottom. This needs to be performed in portions.
15. Some samples may need a second 1.0 g of HgO. The solution should be centrifuged and the supernatant filtered (as in **step 12** of Subheading 3.5) before this addition. Add washed and dried HgO until no more green byproduct is produced.
16. The filter paper typically tears and leaks some of the green HgS byproduct when filtering the first time. After the second filtering, some green color may still remain in the solution.
17. The pH of the QD solution needs to be ~pH 6 for activation with MPEG CD. The QD solution should be at this pH if it was dialyzed in D.I. water after water solubilization.
18. 10 mg of MPEG CD will be less than one drop by pipet. The amount of MPEG CD necessary to observe a good conjugation efficiency depends on the nature of the QDs and amine-functional substrate. We have found yields as high as 95 % using the procedures reported here.
19. If a certain ratio of amine-functional compound to QD is desired, first calculate how many QDs are in the 1 mL solution. Then, determine how many amine-functional compounds are needed to achieve the ratio. Prepare a solution of the amine-functional compound, and then determine how much of this solution should be added to the QD solution to obtain the desired ratio.
20. If the sample is under 2 mL, the entire sample will fit in the 1 mL dialysis tube.
21. Sulfo-SMCC needs to be at pH 6.5–7.5 in order for the maleimide to react with the thiols on the QDs.
22. The pH of the liquid in the bottom of the centrifuge tube should be monitored each time. Once it has reached pH 8, no further additions of pH 8 phosphate buffer are needed.

Acknowledgments

This work was supported by the ACS PRF (50859-ND10) and by funds from the University of Illinois at Chicago. We would like to thank Prof. Hedi Mattoussi for helpful discussions concerning the synthesis of cap-exchanged CdSe/CdZnS QDs.

References

1. Brus LE (1983) A simple model for the ionization potential, electron affinity, and aqueous redox potentials of small semiconductor crystallites. J Chem Phys 79:5566–5571
2. Rossetti R, Nakahara S, Brus LE (1983) Quantum size effects in the redox potentials, resonance Raman-spectra, and electronic-spectra of CdS crystallites in aqueous-solution. J Chem Phys 79:1086–1088
3. Ekimov AI, Onushchenko AA (1984) Size quantization of the electron energy spectrum in a microscopic semiconductor crystal. JETP Lett 40:1136
4. Klimov VI et al (2000) Optical gain and stimulated emission in nanocrystal quantum dots. Science 290:314–317
5. Colvin VL, Schlamp MC, Alivisatos AP (1994) Light-emitting-diodes made from cadmium selenide nanocrystals and a semiconducting polymer. Nature 370:354–357
6. Huynh WU, Dittmer JJ, Alivisatos AP (2002) Hybrid nanorod-polymer solar cells. Science 295:2425–2427
7. Hines MA, Guyot-Sionnest P (1996) Synthesis and characterization of strongly luminescing ZnS-capped CdSe nanocrystals. J Phys Chem 100:468–471
8. Michalet X et al (2005) Quantum dots for live cells, in vivo imaging, and diagnostics. Science 307:538–544
9. Snee PT et al (2006) A ratiometric CdSe/ZnS nanocrystal pH sensor. J Am Chem Soc 128:13320–13321
10. Medintz IL et al (2003) Self-assembled nanoscale biosensors based on quantum dot FRET donors. Nat Mater 2:630–638
11. Liu D, Snee PT (2011) Water soluble semiconductor nanocrystals cap exchanged with metallated ligands. ACS Nano 5:546–550
12. Wu XY et al (2003) Immunofluorescent labeling of cancer marker Her2 and other cellular targets with semiconductor quantum dots. Nat Biotechnol 21:41–46
13. Mattoussi H et al (2000) Self-assembly of CdSe-ZnS quantum dot bioconjugates using an engineered recombinant protein. J Am Chem Soc 122:12142–12150
14. Shen HY, Jawaid AM, Snee PT (2009) Poly(ethylene glycol) carbodiimide coupling reagents for the biological and chemical functionalization of water-soluble nanoparticles. ACS Nano 3:915–923
15. Snee PT et al (2011) Quantifying quantum dots with Förster resonant energy transfer. J Phys Chem C 115:19578–19582
16. Nguyen T, Francis MB (2003) Practical synthetic route to functionalized rhodamine dyes. Org Lett 5:3245–3248
17. Gunsalus IC, Barton LS, Gruber W (1956) Biosynthesis and structure of lipoic acid derivatives. J Am Chem Soc 78:1763–1768
18. Susumu K, Mei BC, Mattoussi H (2009) Multifunctional ligands based on dihydrolipoic acid and polyethylene glycol to promote biocompatibility of quantum dots. Nat Protoc 4:424–436

References

Chapter 5

Synthesizing and Modifying Peptides for Chemoselective Ligation and Assembly into Quantum Dot-Peptide Bioconjugates

W. Russ Algar, Juan B. Blanco-Canosa, Rachel L. Manthe, Kimihiro Susumu, Michael H. Stewart, Philip E. Dawson, and Igor L. Medintz

Abstract

Quantum dots (QDs) are well-established as photoluminescent nanoparticle probes for in vitro or in vivo imaging, sensing, and even drug delivery. A critical component of this research is the need to reliably conjugate peptides, proteins, oligonucleotides, and other biomolecules to QDs in a controlled manner. In this chapter, we describe the conjugation of peptides to CdSe/ZnS QDs using a combination of polyhistidine self-assembly and hydrazone ligation. The former is a high-affinity interaction with the inorganic surface of the QD; the latter is a highly efficient and chemoselective reaction that occurs between 4-formylbenzoyl (4FB) and 2-hydrazinonicotinoyl (HYNIC) moieties. Two methods are presented for modifying peptides with these functional groups: (1) solid phase peptide synthesis; and (2) solution phase modification of pre-synthesized, commercial peptides. We further describe the aniline-catalyzed ligation of 4FB- and HYNIC-modified peptides, in the presence of a fluorescent label on the latter peptide, as well as subsequent assembly of the ligated peptide to water-soluble QDs. Many technical elements of these protocols can be extended to labeling peptides with other small molecule reagents. Overall, the bioconjugate chemistry is robust, selective, and modular, thereby potentiating the controlled conjugation of QDs with a diverse array of biomolecules for various applications.

Key words Quantum dot, Peptide, Bioconjugation, Polyhistidine, Self-assembly, Chemoselective ligation, Hydrazone

1 Introduction

1.1 Quantum Dot Bioconjugates

Nanotechnology continues to provide new tools for studying biochemical and biological systems. In particular, nanoparticle (NP) materials have much to offer due to their small size, large surface area-to-volume ratio, and interesting electronic/optical/magnetic properties that are not accessible at the bulk scale. NPs are being widely explored as *in vitro* and *in vivo* probes for imaging and sensing, as well as platforms for diagnostics and drug delivery

Paolo Bergese and Kimberly Hamad-Schifferli (eds.), *Nanomaterial Interfaces in Biology: Methods and Protocols*, Methods in Molecular Biology, vol. 1025, DOI 10.1007/978-1-62703-462-3_5, © Springer Science+Business Media New York 2013

[1–3]. Colloidal semiconductor nanocrystals, or quantum dots (QDs), are currently among the most widely utilized NPs for these applications. The unique, size-dependent electronic properties of QDs result in bright luminescence that offers many advantages over conventional molecular fluorophores: strong, broad absorption characterized by molar absorption coefficients in the range of 10^5–10^6 M^{-1} cm^{-1}; large two-photon absorption cross-sections, typically 10^3–10^4 GM; narrow photoluminescence (PL) characterized by full-widths-at-half-maxima in the range 25–40 nm; peak PL wavelengths that can be continuously tuned across a broad spectral range through control of nanocrystal size and semiconductor composition; and superior resistance to photochemical degradation [4, 5]. As a consequence of these properties, QDs greatly facilitate multiplexed/multicolor experiments, are excellent emitters for single molecule (particle) fluorescence imaging/tracking, and are ideal donors in Förster resonance energy transfer (FRET) [6–8]. The latter has been exploited to develop a diverse array of biosensing configurations based on QDs [9, 10].

High quality QDs are typically synthesized in hydrophobic media as a type I core/shell structure [11]. The core size and material determines the optical properties of the QD, while the growth of a thin shell of a higher band-gap semiconductor protects and enhances those optical properties. The most well-established example of a core/shell QD is CdSe/ZnS, which provides bright, size-tunable PL across the visible spectrum. However, CdSe/ZnS and other QD materials are not biocompatible or biofunctional as synthesized. Utility in biological applications is predicated on applying a hydrophilic coating to the QD (e.g., a monolayer of bifunctional ligands, or an amphiphilic polymer), which imbues biocompatibility, as well as conjugating biomolecules (e.g., proteins, peptides, oligonucleotides) to provide biofunctionality. The bioconjugation of QDs enables targeting of tissues and biomarkers, cellular uptake, and selective recognition for sensing, among other capabilities [1, 4]. The challenge is not necessarily in identifying functional biomolecules, but rather in reliably and controllably conjugating them to the QDs. The notion of control includes the number of biomolecules per QD, their attachment point and orientation, and the homogeneity of those properties—all of which depend on the bioconjugate chemistry utilized and critically impact the activity of the final QD-bioconjugate [1].

1.2 Self-Assembly of Polyhistidine Appended Biomolecules

One of the most effective means through which to conjugate biomolecules to QDs is via metal-affinity coordination between polyhistidine motifs and the ZnS shell of CdSe/ZnS QDs. This association occurs spontaneously, rapidly ($k_{on} \sim 10^4$ M^{-1} s^{-1}) and with high-affinity ($K_d \sim 1$ nM) [12]. The conjugate valence (i.e., number of biomolecules per QD) can also be controlled on the basis of mixing stoichiometry, albeit subject to a Poisson distribution [13]. We, along with several other groups, have used polyhistidine motifs to assemble recombinantly expressed proteins, synthetic peptides, and even chemically modified oligonucleotides to QDs [12, 14–19]. Up to 50 ± 10 peptides (<3 kDa) or 12 ± 2 proteins (~44 kDa) can be assembled per QD [20]. We have also developed the concept of a "starter peptide," which has both a polyhistidine motif and a chemical handle for ligation to an oligonucleotide, another peptide, or a protein modified with a cognate chemical handle. The starter peptide strategy has the advantage of being modular to allow ligation with functionally diverse biomolecules [14–16], and/or introducing a chemical bond that, while generally stable, is selectively labile under certain conditions (e.g., disulfide, hydrazone) [21–24].

1.3 Chemoselective Hydrazone Ligation

The reaction between 4-formylbenzoyl and 2-hydrazinonicotinoyl moieties to yield a hydrazone bond is chemically orthogonal to most biological functionalities, and has been employed for the chemoselective modification of proteins and peptides [25–29]. Other forms of imine chemistry have been widely used for protein and peptide modification [30–42]; however, the above reaction is particularly promising. At slightly acidic or neutral pH, conjugation rates are typically 10^1–10^2 M^{-1} s^{-1} with 10 μM reactants and 100 mM aniline as a catalyst, going to completion in ca. 30 min [25]. The aniline reacts with the 4-formylbenzoyl group to form a highly reactive iminium intermediate that will react with the 2-hydrazinonicotinoyl group to yield a hydrazone. This reaction has an equilibrium constant of $K_{eq} = 2.3 \times 10^6$ M^{-1} at pH 7.0, and the resulting hydrazone bond has slow hydrolytic kinetics ($k_{hydrolysis} = 0.08 \times 10^{-3}$ s^{-1}) in the absence of aniline [25, 43, 44]. Although slow, the hydrolysis of similar hydrazone bonds is sufficiently accelerated under the moderately acidic conditions of the endolysomal system or cancerous tissue (e.g., pH 5) to be useful for drug delivery [21, 22]. Recently, we have shown that this chemistry can be used to ligate a starter polyhistidine peptide to another peptide or oligonucleotide, both in bulk solution and when the starter peptide is preassembled to QDs [14, 16]. The starter peptides were ligated with a cell-penetrating peptide for cellular delivery, substrate peptides for protease activity, or oligonucleotide probes for hybridization on microarrays.

Here, we describe two general routes to modify a pair of peptides with 4-formylbenzoyl (4FB) and 2-hydrazinonicotinoyl (HYNIC) moieties, respectively, for use as cognate reactants in chemoselective hydrazone ligations. The first peptide includes a polyhistidine motif for assembly to CdSe/ZnS QDs, and functions as a starter peptide. The second peptide is ligated to the starter peptide and is also labeled with a fluorescent dye. Upon assembly to a QD, this configuration provides the opportunity for FRET and utility in biosensing experiments targeting, for example, various proteases [9, 10]. The first route to these peptides that we describe is 9-fluorenylmethoxycarbonyl (Fmoc)-solid phase peptide synthesis (SPPS) and in-situ modification with 4FB and/or HYNIC. This method is suited to well-equipped chemistry laboratories and yields the largest amount of modified peptide. The second route is based on the solution-phase modification of pre-synthesized peptides, purchased from commercial sources, with 4FB and HYNIC reagents. Although its scale is typically limited to smaller quantities, the second method can be carried out in almost any laboratory by chemists and non-chemists alike. The products of the SPPS and solution-phase routes are functionally equivalent and can both be chemoselectively ligated using a protocol that we describe here. Some elements of the two methods can be interchanged and, cumulatively, the various steps we describe can be adapted and applied to labeling peptides with almost any small molecule reagent.

2 Materials

2.1 General

1. Personal safety equipment: safety glasses, laboratory coat, nitrile gloves, and fumehood.
2. Chemical waste receptacles.
3. Ultrapure water with a resistivity of 18.2 MΩ cm, for example, from a Milli-Q purification system (Millipore, Billerica, MA).
4. Retort stand and adjustable clamps.
5. UV–visible spectrophotometer and quartz cuvette. The cuvette should be transparent to wavelengths longer than 250 nm. A microcuvette with a sub-milliliter volume is strongly recommended.
6. Spectrofluorimeter.
7. Gel imaging system with ultraviolet transillumination.

2.2 Solid Phase Peptide Synthesis of 4FB/HYNIC Modified Peptides

2.2.1 Equipment

1. A polypropylene syringe fitted with a frit and stopcock (teflon/polypropylene). This is referred to as the "reaction vessel." The reaction vessel should have a 10 mL capacity for the use of 200–250 mg of resin.
2. Erlenmeyer flask with side-arm (125 mL capacity or larger).
3. Vacuum source and vacuum trap. The vacuum source may be a stand-alone pump, central/house vacuum, or even a Chapman water aspirator pump. Connect the reaction vessel to the vacuum as shown in Fig. 1. The reaction vessel is connected to the Erlenmeyer flask, which is connected to the vacuum source via the vacuum trap.
4. Teflon rod.

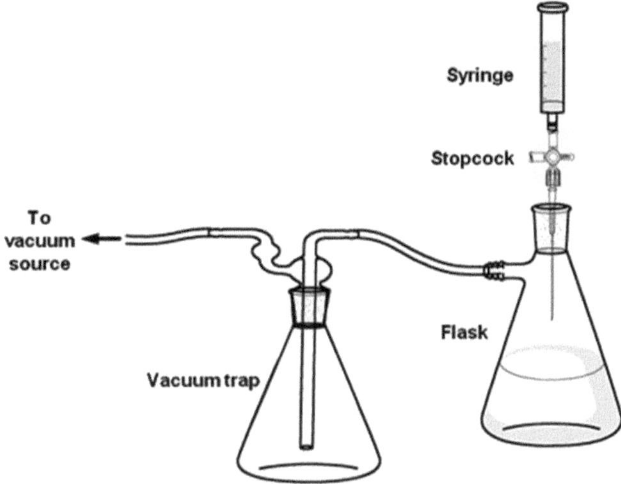

Fig. 1 Apparatus for solid phase peptide synthesis. The syringe needle is punched through a septum on the flask to maintain a vacuum seal

5. Glass vials (e.g., scintillation vials, 10 mL or three dram vials).
6. Glass tubes, disposable (i.e., culture tubes/test tubes).
7. Centrifuge tubes, polypropylene with screw cap: 50 mL capacity.
8. Centrifuge (with rotor to spin 50 mL centrifuge tubes at 10,000 rcf).
9. Rotary evaporator.
10. Lyophilizer.
11. Block heater (and block to hold glass tubes).
12. Semi-preparative, reverse phase RP-HPLC system. Column: C-8 or C-12 or C-18, 5 μm, 90–110 Å, 150 × 21.2 mm.
13. Analytical RP-HPLC system. Column: C-18, 5 μm, 90–110 Å, 150 × 4.5 mm.
14. Mass-spectrometer: electrospray ionization mass spectrometer (ESI-MS; e.g., hyphenated with the analytical RP-HPLC system) or matrix assisted laser desorption ionization-time of flight-mass spectrometer (MALDI-TOF-MS).

2.2.2 Reagents

1. Rink amide aminomethyl resin: 100–200 mesh, loading 0.45 mmol g^{-1}. Approximately 220 mg of resin is useful for a 0.1 mmol scale synthesis (Novabiochem-EMD Millipore, San Diego, CA).
2. Fmoc-protected amino acids (*see* **Note 1**). As an example, we use the Fmoc-protected amino acids necessary to synthesize the peptides listed in Table 1.

Table 1
Peptides made using SPPS

Name	Sequence (N- to C-terminal)	Modifications	
		Synthetic	Post-synthetic
Peptide-1a	GSGAAAGLSHHHHHH	N-terminal 4FB C-terminal amidation	–
Peptide-2	GLYRGSGEGC	N-terminal HYNIC	Cys(C)-TMR

3. Peptide synthesis grade (or better) solvents: dimethylformamide (DMF), dichloromethane (DCM).
4. Acetonitrile (MeCN), HPLC grade.
5. *N,N*-diisopropylethylamine (DIEA).
6. 2-(1*H*-benzotriazol-1-yl)-1,1,3,3-tetramethyluronium hexafluorophosphate) (HBTU): prepare 100 mL of a 0.4 M solution in DMF (Solution A; *see* **Note 2**). *Caution*: HBTU and HATU (see below) can potentially illicit strong allergic reactions.
7. 2-(7-aza-1*H*-benzotriazol-1-yl)-1,1,3,3-tetramethyluronium hexafluorophosphate (HATU).
8. Piperidine: prepare 500 mL of a 20 % v/v solution in DMF (Solution B).
9. Trifluoroacetic acid (TFA), neat. *Caution*: highly corrosive.
10. Trifluoroacetic acid, 0.1 % v/v (aq): prepare 100 mL of solution by adding 0.1 mL of neat TFA to 100 mL of ultrapure water.
11. Triisopropylsilane (TIS).
12. Ethanedithiol (EDT). *Caution*: toxic, stench.
13. *tert*-butylmethylether (TBME).
14. 4-formylbenzoic acid (4FB).
15. 6-Boc-hydrazinonicotinic acid (6-Boc-HYNIC) (Solulink, San Diego, CA).
16. Ninhydrin (*a.k.a.* Kaiser) test solutions (Applied Biosystems by Life Technologies, Foster City, CA): phenol in ethanol (M1), potassium cyanide/pyridine (M2), ninhydrin in ethanol (M3).
17. Maleimide mono-reactive fluorescent dye. We use tetramethylrhodamine-5-maleimide (TMR-mal; Invitrogen by Life Technologies, Carlsbad, CA) as an example.
18. Tris buffer: 50 mM, pH 7.5. The buffer can be degassed by sparging with nitrogen.

2.3 Solution Phase Modification of Commercial Peptides with 4FB/HYNIC

2.3.1 Equipment

1. Disposable polypropylene syringes: 1, 3, 5, and 10 mL.
2. Polypropylene microcentrifuge tubes, ≥1.5 mL capacity.
3. Vacuum concentrator (with rotor to fit microcentrifuge tubes).
4. Snap-tight plastic cartridges with frits and Luer-Lok/Slip fittings on each end; ca. 0.4 mL capacity (*see* **Note 3**).
5. Column with frit and stopcock: 2.5 cm diameter × 20 cm.
6. Apparatus/cell for running polyacrylamide gel electrophoresis (PAGE).
7. Electrophoresis power supply.

2.3.2 Reagents

1. Sodium tetraborate buffer: 100 mM, pH 8.5.
2. Phosphate buffered saline (10× PBS): 100 mM phosphate, 1.37 M NaCl, 30 mM KCl, pH 7.4–7.8.
3. Phosphate buffered saline (PBS): 10 mM phosphate, 0.137 M NaCl, 3 mM KCl, pH 7.4. Prepare 100 mL of solution by diluting 10 mL of 10× PBS with 90 mL of ultrapure water.
4. Ethanol (EtOH), 50 % (aq): prepare 30 mL of a 1:1 dilution with PBS.
5. Imidazole. Prepare a 300 mM solution by dissolving 0.20 g in 10 mL of PBS.
6. MeCN, neat.
7. MeCN, 70 % v/v (aq): prepare 10 mL of a 7:3 dilution with ultrapure water.
8. Triethylamine acetate (TEAA) buffer: 2.0 M, pH 7.0 (Applied Biosystems).
9. TEAA buffer, 0.2 M: prepare 100 mL of a 1:9 dilution with ultrapure water.
10. Two peptides from a commercial supplier. There are many companies that offer custom synthetic peptides. We use the peptides listed in Table 2 (Biosynthesis Inc., Lewisville, TX) as an example. One peptide should have a polyhistidine sequence at one terminus and an amine group at the opposite terminus (e.g., Peptide-1b); the second peptide should have an amine group at one terminus and a thiol group at the opposite terminus (e.g., Peptide-3). The amino acid sequence between the termini is selected to suit the application.
11. Sulfo-*N*-succinimidyl-4-formylbenzamide (sNHS-4FB) and sulfo-succinimidyl-6-hydrazino-nicotinamide (sNHS-HYNIC) (Solulink; *see* **Note 4**).
12. Maleimide mono-reactive fluorescent dye. As an example, we use Cy3 and Cy5 maleimide mono-reactive dye (Cy3/5-mal), which are sold dried in tubes as "enough to label 1.0 mg of antibody" (GE Healthcare, Piscataway, NJ).

Table 2
Peptides purchased commercially

Name	Sequence (N- to C-terminal)	Modifications	
		Synthetic	Post-synthetic
Peptide-1b	GSGAAAGLSHHHHHH	C-terminal amidation	N-terminal 4FB
Peptide-3	KSSTGGQNGEDTGTSC	N-terminal acetylation C-terminal amidation	Lys(K)-HYNIC Cys(C)-Cy3/5

13. Nickel(II)-nitrilotriacetic acid (Ni-NTA)-agarose (Qiagen, Valencia, CA).
14. Oligonucleotide purification cartridges (OPC; Applied Biosystems).
15. Permeation gel for size-exclusion chromatography. The fractionation range should be suitable for the size of the peptides. As an example, we use ca. 15 g of BioGel P4 (fractionation range 800–4000 Da; BioRad, Hercules, CA). The permeation gel is prepared as a slurry in PBS and loaded into the column according to the manufacturer's instructions.
16. Reagents for casting a polyacrylamide gel (e.g., acrylamide, bisacrylamide, ammonium persulfate, tetramethylethylenediame (TEMED), and running buffer) or, alternatively, precast PAGE gels obtained commercially. We use 8–25 % gradient polyacrylamide Pharmacia PhastGel SDS buffer strips (GE Healthcare) for an automated Pharmacia LKB PhastSystem.

2.4 Chemoselective Hydrazone Ligation

1. **Items 1–4**, **6**, and 7 from Subheading 2.3.1.
2. **Items 2–9**, **13**, and **14** from Subheading 2.3.2.
3. Aniline, neat (*see* **Note 5**).
4. Ammonium acetate buffer (NH_4OAc): 100 mM, pH 5.5 (optional).
5. HYNIC modified peptide, as prepared and purified in Subheading 3.
6. 4FB modified peptide, as prepared and purified in Subheading 3.

2.5 Peptide Assembly to Quantum Dots

1. Water soluble QDs coated with a ligand such as dihydrolipoic acid (DHLA), DHLA-PEG, or a zwitterionic dithiol (*see* **Note 6**).
2. Cell for agarose gel electrophoresis.
3. Running buffer: 90 mM tris-borate-ethylenediaminetetraacetic acid (EDTA; 2 mM), pH 8.0–8.5 (TBE buffer).
4. Agarose: prepare a 1.5 % w/w solution in running buffer.

3 Methods

3.1 Solid Phase Synthesis of Modified Peptides

Fmoc solid phase peptide synthesis (Fmoc-SPPS) can be done either manually or using automated methods. Here, we detail the procedure for manual synthesis. Stepwise SPPS is a cyclic process that comprises (1) deprotection of the amine group of an amino acid (or initial linker) attached to the solid support; (2) chemical coupling of the Fmoc-protected amino acid in the peptide sequence being synthesized; and (3) repetition of the previous steps with the next amino acid. As an example, we synthesize Peptide-1a and Peptide-2 (*see* Table 1) and modify their N-termini with 4FB and HYNIC, respectively. Peptide-2 is further modified with a fluorescent dye in a solution phase labeling step. Figures 2 and 3 illustrate the synthesis and modification of peptides with 4FB and HYNIC/TMR-mal, respectively.

3.1.1 Fmoc-SPPS

1. Initial deprotection of the rink amide resin. Transfer the resin (0.1 mmol, ca. 220 mg at a loading of 0.45 mmol/g resin) to a polypropylene reaction vessel. Swell the resin in DMF for 5 min and then drain the cartridge by applying vacuum.
2. Add 5 mL of Solution B (piperidine) to the resin and stir the suspension occasionally with a teflon rod. Drain the reaction vessel after 5 min, add another 5 mL of Solution B, and let stand 5 min further.

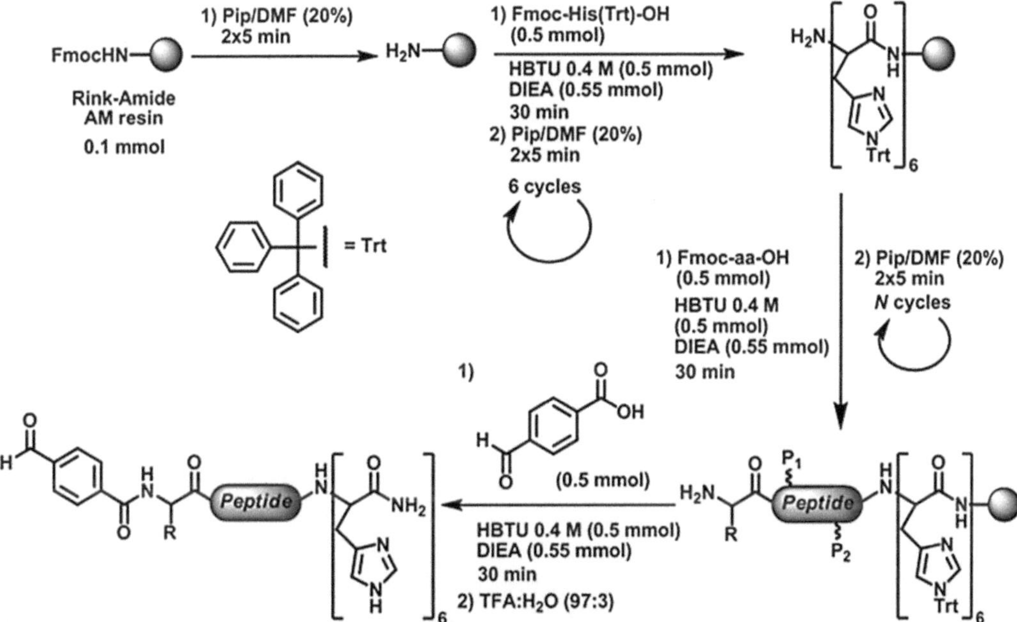

Fig. 2 SPPS scheme for a C-terminal polyhistidine-appended peptide-1a with a N-terminal 4FB modification

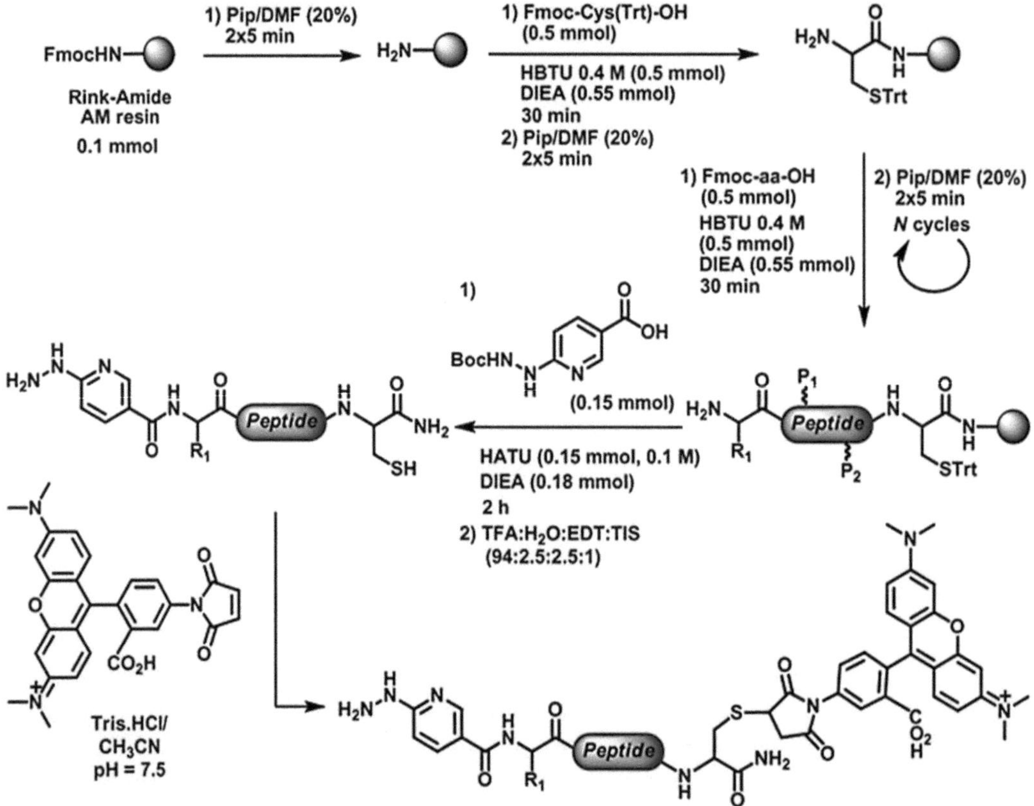

Fig. 3 SPPS scheme for a peptide-2 with a N-terminal HYNIC modification and C-terminal cysteine residue for labeling with TMR-mal

3. Drain and wash with 5 × 5 mL of DMF followed by 3 × 5 mL of DCM; each wash should take ca. 30 s (*see* **Note 7**).
4. Take ca. 1 mg of the resin and perform the ninhydrin test (*see* Subheading 3.1.2) in a small glass tube: a blue color should appear. Parallel treatment of a control resin with the Fmoc group intact is useful for comparison (*see* **Note 8**).
5. Swell the resin in DMF for 2 min and then drain. The next step is coupling the first amino acid residue.
6. Dissolve Fmoc-aa$_1$-OH (0.5 mmol; "aa$_1$" is the first amino acid in the peptide sequence attached to the resin) in 1.25 mL of Solution A (HBTU) in a glass vial (*see* **Note 9**). Add neat DIEA (96 μL, 0.55 mmol) to the solution, shake for 30 s, and then add to the resin.
7. Stir the mixture periodically for 30 min.
8. Drain the reaction vessel and wash the resin with 5 × 5 mL of DMF and 3 × 5 mL of DCM in succession; each wash should take ca. 30 s.

9. Take ca. 1 mg of the resin and perform the ninhydrin test. If a positive result is obtained (i.e., blue color) repeat **steps 6–8**; if a negative result is obtained, proceed to **step 10**.
10. Couple the remaining amino acid residues in the peptide sequence by repeating **steps 2–9**, which correspond to deprotection using Solution B and coupling Fmoc-aa$_n$-OH using Solution A and DIEA (where $n > 1$ is the cycle number and corresponding residue from the C-terminus of the peptide).
11. Proceed to Subheading 3.1.3 for modification with 4FB or Subheading 3.1.4 for modification with HYNIC.

3.1.2 The Ninhydrin Test

The ninhydrin or Kaiser test is a colorimetric chemical test that is used for quality control during SPPS. Primary amines react with ninhydrin to form a chromophore with a deep blue/violet or "Ruhemann's Purple" color. A positive test is desired following deprotection steps in SPPS since this indicates that primary amines are available for coupling the next amino acid in the peptide sequence. Conversely, a negative test is desired following coupling steps since this indicates that the reaction went to completion. We use the ninhydrin test qualitatively to minimize the amount of time needed to synthesize the peptide; however, the test can also be used quantitatively if desired [45].

1. *Caution*: treat the reagents and spent glass tubes as hazardous materials.
2. Transfer ca. 1 mg of dry resin to a glass tube.
3. Add two drops each of the M1, M2, and M3 solutions and heat at 100 °C for 5 min.
4. An intense blue color will appear in the presence of free amine, which may correspond to successful deprotection (**step 4** in Subheading 3.1.1) or incomplete coupling of an amino acid (**step 9** in Subheading 3.1.1).

3.1.3 Solid Phase Modification with 4FB

1. Dissolve 4FB (75 mg, 0.5 mmol) in 1.25 mL of Solution A (HBTU). Add neat DIEA (96 μL, 0.55 mmol) and shake the resulting mixture for 30–60 s. Add this solution to the resin and stir occasionally for 30 min.
2. Drain and wash the resin with 5 × 5 mL of DMF and 3 × 5 mL of DCM in succession; each wash should take ca. 30 s.
3. Confirm coupling using the ninhydrin test (*see* Subheading 3.1.2). Repeat **step 1** if a positive result is obtained (i.e., blue color).
4. Dry the resin under vacuum for 2 h.
5. Proceed to Subheading 3.1.5.

3.1.4 Solid Phase Modification with HYNIC

1. Dissolve 6-Boc-HYNIC (38 mg, 0.15 mmol) and HATU (57 mg, 0.15 mmol) in 1 mL of DMF. Add neat DIEA (31 μL, 0.18 mmol) and shake the solution for 30–60 s.
2. Add the solution from **step 1** to the resin and stir periodically for 2 h.
3. Drain and wash the resin with 5 × 5 mL of DMF and 3 × 5 mL of DCM in succession; each wash should take ca. 30 s.
4. Confirm coupling using the ninhydrin test (*see* Subheading 3.1.2). Repeat **steps 1** and **2** if a positive test result is obtained (i.e., blue color).
5. Dry the resin under vacuum for 2 h.
6. Proceed to Subheading 3.1.5.

3.1.5 Cleavage of Peptides from Rink Resin and Removal of Protecting Groups

The following steps are used to cleave the SPPS-grown peptides from the rink amide resin and remove the protecting groups from the amino acid side chains. Pay close attention to the protocol as there are variations depending on the nature of the peptide synthesized.

1. *Important*: the 4FB is potentially sensitive to EDT and TIS (*see* **Note 10**). If cleaving and deprotecting a 4FB-modified peptide (e.g., Peptide-1a), skip **step 2** and proceed with **step 3**. If the peptide is HYNIC-modified (e.g., Peptide-2), proceed with **step 2**.
2. Prepare a 10 mL solution of TFA:EDT:H_2O:TIS (94:2.5:2.5:1 v/v) and cool to ca. −8 °C in a freezer or on dry ice (*see* **Notes 11** and **12**). Skip **step 3** and proceed with **step 4**.
3. Prepare 10 mL of 97 % v/v TFA (aq) and cool to ca. −8 °C in a freezer or on dry ice. Proceed to **step 4**.
4. Add the cleavage/deprotection solution to the resin. Close the reaction vessel and gently agitate for 90 min at room temperature (*see* **Note 13**).
5. Drain the resin and collect the solution, which contains peptide, into a round bottom flask. Wash the resin with 2 × 5 mL of neat TFA and 2 × 5 mL of DCM in succession, and collect the washes. Each wash should take ca. 30 s.
6. *Important*: if the peptide contains a polyhistidine sequence (e.g., Peptide-1a) proceed to **step 7** (*see* **Note 14**). Otherwise, skip **steps 7–9** and proceed to **step 10**.
7. Concentrate the peptide containing TFA solution to dryness in a rotary evaporator. In the absence of a rotary evaporator, this can be done under a stream of nitrogen.
8. Add 20 mL of TBME and 20 mL of 0.1 % TFA (aq) to the peptide residue. Transfer this mixture to a 50 mL centrifuge tube or extraction funnel. Mix the two phases vigorously and

let them separate. Retain the aqueous phase, which contains peptide, and extract the TBME phase further with 3 × 20 mL portions of 0.1 % TFA (aq).

9. Combine and lyophilize the aqueous phase extracts. Proceed to Subheading 3.1.6.
10. Concentrate the peptide containing solution to ca. 1 mL in a rotary evaporator.
11. Add the concentrated peptide/TFA solution to 30 mL of cold TBME in a 50 mL centrifuge tube. A white precipitate will form.
12. Isolate the precipitate by centrifugation at ~1,000 rcf for 6 min and discard the supernatant. Proceed to Subheading 3.1.6.

3.1.6 Peptide Purification

1. The peptide can be purified by semipreparative RP-HPLC, and the purity confirmed by analytical HPLC and mass spectrometric analysis (e.g., ESI-MS, MALDI-TOF MS).
2. Set up the analytical/semipreparative RP-HPLC. The details of both the instrument setup and operation will vary between manufacturers, models, and software. A two-component gradient elution will be used: Eluent A is 0.1 % v/v TFA (aq); Eluent B is 0.05 % v/v TFA in MeCN.
3. For semipreparative RP-HPLC, the elution program is from 0 to 5 % Eluent B over 5 min and 5–70 % Eluent B over 80 min at a flow rate of 15 mL/min.
4. For analytical RP-HPLC, the elution program is from 0 to 70 % Eluent B over 30 min at a flow rate of 1 mL/min.
5. Lyophilize and store the peptide as a TFA salt at –20 °C. Protect from light.

3.1.7 Fluorescent Labeling

Any peptide with a cysteine residue can be labeled with a maleimide derivative of a fluorescent dye of interest. The maleimide is selective for thiols such as those on the cysteine side chain, at near neutral pH; labeling can therefore proceed in the presence of amine groups or HYNIC. We label the terminal cysteine residue of HYNIC-modified Peptide-2 with TMR-mal as an example (*see* Fig. 3).

1. Dissolve TMR-mal (4.1 μmol) in 2.0 mL of degassed 1:1 Tris buffer/MeCN (*see* **Note 15**) and add HYNIC-modified Peptide-2 (6.2 μmol).
2. Agitate the reaction at room temperature until completion, which can be monitored by analytical RP-HPLC (*see* Subheading 3.1.6).
3. Purify the product by semi-preparative RP-HPLC (*see* Subheading 3.1.6). In the special case that the peptide has a polyhistidine sequence, the purification steps in Subheading 3.2.3 may be used as an alternative.

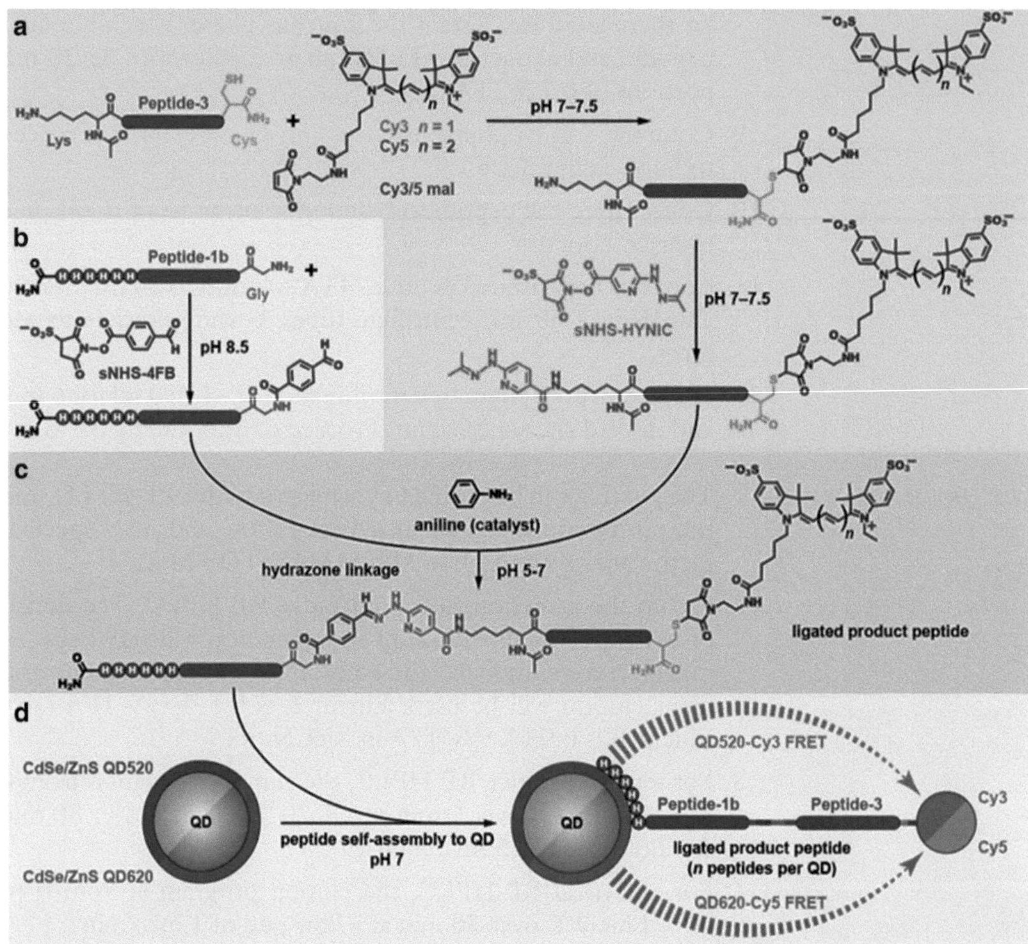

Fig. 4 Solution phase modification and labeling scheme: (**a**) modification of peptide-3 with HYNIC and Cy3/5; (**b**) modification of peptide-1b with 4FB; (**c**) chemoselective ligation; and (**d**) assembly to a QD (peak PL 520 or 620 nm) and FRET

3.2 Solution Phase Modification of Commercial Peptides

As an alternative to SPPS, peptides can be purchased commercially from several suppliers and modified with 4FB or HYNIC using commercial reagents and solution phase labeling techniques. While SPPS is typically limited to organic chemistry laboratories, the solution phase protocol in this section is more widely accessible to non-specialists. As an example, we modify commercially obtained Peptide-1b and Peptide-3 (*see* Table 2) with 4FB and HYNIC, respectively. The latter is also modified with a fluorescent dye in a one-pot, two-step method. The full sequence of steps is illustrated in Fig. 4, including chemoselective hydrazone ligation (Subheading 3.3) and assembly to a QD (Subheading 3.4).

3.2.1 Solution Phase Modification with 4FB

The following steps are used to modify a peptide at an available amine group (e.g., N-terminus or lysine side chain) using sNHS-4FB. We modify the α-amine at the N-terminus of Peptide-1b.

These steps can be extended as a general method for labeling any peptide with any NHS-ester activated reagent (*see* **Note 16**), assuming that the peptide has a reactive amine group (e.g., unprotected N-terminus or Lys residue).

1. Dissolve ca. 1 mg of peptide in 10 μL of DMSO (*see* **Note 17**) and add 0.5 mL of borate buffer. Other non-amine containing buffers with pH ≥7 are also suitable.
2. Dissolve 3.0 mg of sNHS-4FB in 30 μL of DMSO. Add the peptide solution from **step 1** and dilute to 1.0 mL with borate buffer (*see* **Note 18**). Mix well. The final peptide concentration should generally be in the range of 0.1–1 mM (ca. 10- to 100-fold excess of sNHS-4FB).
3. Agitate the reaction for 4–8 h at room temperature. For convenience, the reaction mixture can be stored overnight at 4 °C before proceeding with the next steps.
4. Complete the steps in Subheading 3.2.3 and then proceed with **step 5** (*see* **Note 19**).
5. Pre-weigh a microcentrifuge tube.
6. Complete the steps in Subheading 3.2.4 and elute the desalted peptide into the pre-weighed tube.
7. Evaporate the solvent to dryness in a vacuum concentrator and re-weigh the centrifuge tube. The difference in mass corresponds to the quantity of peptide. The result should only be used as a rough estimate of the amount of material. Peptides with tryptophan or tyrosine residues can be more accurately quantitated prior to drying by measuring their absorbance at 280 nm.

3.2.2 Solution Phase Modification with HYNIC and Fluorescent Labeling

The following steps are used to sequentially modify a peptide with a fluorescent dye and HYNIC at available thiol (e.g., cysteine side chain) and amine (e.g., N-terminus or lysine side chain) groups, respectively, in a one-pot reaction. As an example, we modify Peptide-3 (*see* Table 2) with sNHS-HYNIC and Cy3/5-mal at its N-terminal lysine residue and C-terminal cysteine residue, respectively. The dual modification is enabled by the selective reaction of the Cy3/5-mal with the cysteine residues at near neutral pH. Following the conversion of these residues, the sNHS-HYNIC can react exclusively with the lysine residues. The order of these steps cannot be interchanged since NHS esters react efficiently with both amine and thiol groups.

1. Dissolve ca. 1 mg of Peptide-3 in 10 μL of 50 % MeCN (*see* **Note 17**). Add 200 μL of 10× PBS.
2. Obtain two tubes of Cy3/5 maleimide mono-reactive dye and dissolve each in 10 μL of DMSO. Add the peptide solution from **step 1** to each vial in succession. Dilute with 200 μL of ultrapure water (*see* **Note 18**).

3. Agitate the reaction for 4–8 h at room temperature. For convenience, the reaction mixture can be stored overnight at 4 °C before proceeding with the next steps.
4. Dissolve 3.0 mg sNHS-HYNIC in 30 μL of DMSO (*see* **Note 18**).
5. Transfer the Cy3/5 maleimide reaction mixture to the sNHS-HYNIC. Add 200 μL of ultrapure H_2O and 200 μL of 10× PBS. Mix well.
6. Agitate the reaction for 4 h at room temperature. For convenience, the reaction mixture can be stored overnight at 4 °C before proceeding with the next steps.
7. Load the reaction mixture onto a gel permeation column and elute using PBS. Collect 2–3 mL fractions in small glass tubes. The first dye-colored band that elutes is the product; combine the corresponding fractions (*see* **Note 20**). Alternatively, the peptide can be purified by RP-HPLC.
8. Proceed to Subheading 3.2.4.
9. Quantitate the peptide using the absorbance of the dye. The molar absorption coefficient (Cy3, $\varepsilon_{552}=150{,}000\ M^{-1}\ cm^{-1}$; Cy5, $\varepsilon_{650}=250{,}000\ M^{-1}\ cm^{-1}$) is generally included in the product literature or available online from the manufacturer (*see* **Note 21**).

3.2.3 Purification of Polyhistidine Appended Peptides

The following steps are used to quickly and conveniently purify polyhistidine appended peptides from excess labeling reagents. The purified peptide is obtained in buffer with high concentrations of imidazole, and typically requires subsequent desalting (*see* Subheading 3.2.4). As an example, we purify Peptide-1b from excess sNHS-4FB after modification using the steps in Subheading 3.2.1.

1. Load the Ni-NTA-agarose into a plastic cartridge (*see* **Note 3**). Wash the Ni-NTA-agarose with 5 mL of PBS. Repeat this procedure with a second purification cartridge (*see* **Note 22**). If not proceeding to **step 2** immediately, the Ni-NTA-agarose should be kept hydrated.
2. Flush the reaction mixture through the first Ni-NTA-agarose cartridge using a double-syringe technique for ca. 2–3 min (*see* **Note 23**). The polyhistidine sequence of the peptide will bind to the Ni-NTA-agarose. Retain the reaction mixture.
3. Wash the Ni-NTA-agarose with 10 mL of PBS, 10 mL of 1:1 EtOH:PBS (*see* **Note 24**), and 2 × 10 mL of PBS in succession (*see* **Note 25**).
4. Flush the reaction mixture through the second Ni-NTA-agarose cartridge using a double-syringe technique. Repeat **step 3** with this cartridge.

5. Elute the bound peptide from one of the Ni-NTA-agarose cartridges using successive 3 × 0.5 mL portions of 300 mM imidazole in PBS (*see* **Note 26**).
6. Repeat **step 5** with the second Ni-NTA-agarose cartridge. It is preferable to use ≤3 mL of total eluent between both cartridges (*see* **Note 23**).
7. Proceed to Subheading 3.2.4.

3.2.4 Peptide Desalting

1. Condition a fresh OPC by flushing with 3 mL of MeCN followed by 3 mL of 2 M TEAA buffer.
2. Flush the peptide containing solution through the OPC using a double-syringe technique (*see* **Note 23**). The peptide will bind to the resin.
3. Wash the OPC with 4 × 10 mL of 0.2 M TEAA buffer (*see* **Note 25**).
4. Elute the peptide from the resin with 2 × 0.3–0.5 mL of 70 % MeCN (aq) (*see* **Note 26**).
5. Regenerate the OPC by repeating **step 1**.
6. Repeat **steps 2–4** and combine the eluents. If the peptide is dye/chromophore labeled, or has Try, Trp, or Cys residues, its concentration can be determined by UV–visible spectrophotometry.
7. Vacuum concentrate to dryness.
8. Store at −20 °C until needed.

3.3 Chemoselective Hydrazone Ligation

Since reaction between the 4FB and HYNIC moieties has an equilibrium constant of $K_{eq} = 2.3 \times 10^6\ M^{-1}$ at pH 7.0 [25], a 1:1 stoichiometric reaction does not go to completion at low millimolar concentrations (i.e., 0.01–0.05 mM), and the hydrazone product is in equilibrium with some 4FB and HYNIC. In order to achieve full conversion to the hydrazone ligated product, a small excess (≥1.5-fold) of one of the reagents is required. The following steps assume that one of the peptides has a polyhistidine sequence. As examples, we ligated 4FB-modified Peptide-1a and HYNIC/TMR-modified Peptide 2 from Subheading 3.1, as well as 4FB-modified Peptide-1b and HYNIC/Cy3/5-modified Peptide 3 from Subheading 3.2 (*see* Fig. 4).

1. Dissolve ca. 0.1–1 μmol of 4FB-modified polyhistidine peptide (*see* **Note 27**) in 0.5 mL of either 10× PBS or NH_4OAc buffer (0.2–2 mM peptide).
2. Dissolve ca. 2–3 Eq of the HYNIC-modified peptide in 0.5 mL of the same buffer used in **step 1** (*see* **Note 28**).
3. Mix the two peptide solutions from **steps 1** and **2** together.
4. Add 1–10 μL of aniline (*see* **Note 29**), which corresponds to a final concentration of 10–100 mM in a 1 mL reaction volume.

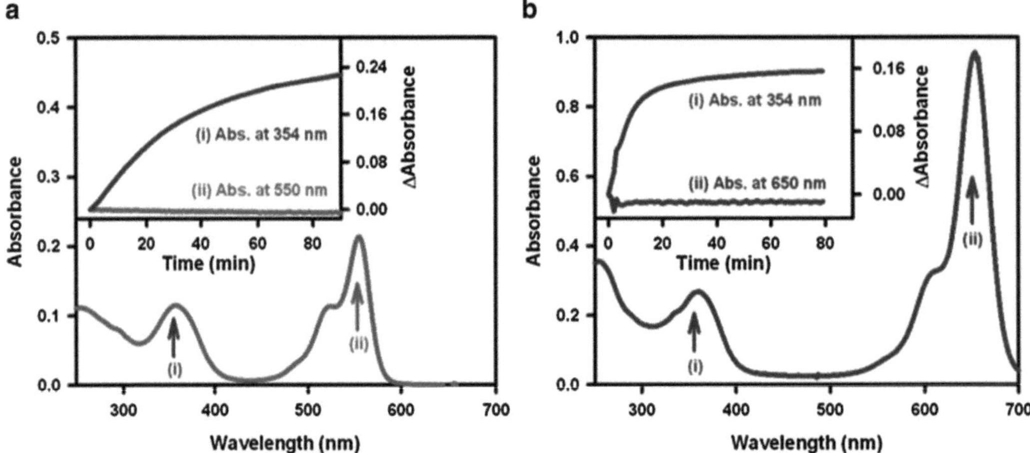

Fig. 5 UV–visible absorption characterization of the chemoselective hydrazone reaction between (**a**) HYNIC-Peptide-3-Cy3 and 4FB-Peptide-1b, and (**b**) HYNIC-Peptide-3-Cy5 and 4FB-Peptide-1b. The *insets* show reaction progress curves from monitoring the change in absorption at 354 nm. In (**a**), the aliquot extracted for this purpose was catalyzed by 1 mM aniline, whereas in (**b**) the reaction was catalyzed by 10 mM aniline. The full absorption spectra shown are for the final ligated and purified products, highlighting the (i) hydrazone and (ii) Cy3/5 absorption bands

5. To monitor the progress of the reaction, remove an aliquot of the reaction mixture and dilute by a factor of 1/100 (1 mM peptide) or 1/10 (0.1 mM peptide) into a UV-transparent cuvette (so that $A_{354} < 1$ throughout the reaction). Aniline can be (optionally) added to the same final concentration as in **step 4**. Track the change in absorbance at 354 nm with time using a UV–visible spectrophotometer (*see* Fig. 5). This dilution will have a slower reaction rate than the reaction mixture since the peptide concentration is lower.
6. The reaction should go to completion in ca. 30–120 min depending on concentrations and conditions. For convenience, the reaction mixture can be stored overnight at 4 °C before proceeding with the next steps.
7. Complete the steps in Subheading 3.2.3.
8. Complete the steps in Subheading 3.2.4.
9. The peptide can be quantitated using the absorption of the hydrazone chromophore at 354 nm ($\varepsilon_{354} = 29{,}000\ M^{-1}\ cm^{-1}$; *see* Fig. 5). Dilute an aliquot of the desalted peptide in 70 % MeCN with ultrapure water to an optical density <1 for the measurement. The amount of peptide is calculated using the Beer-Lambert Law.
10. The purity of the peptide can be confirmed by RP-HPLC (*see* Subheading 3.1.6) or using polyacrylamide gel electrophoresis (*see* Fig. 6). If the peptides were prepared using the protocol in Subheading 3.2 it is useful to set aside and save

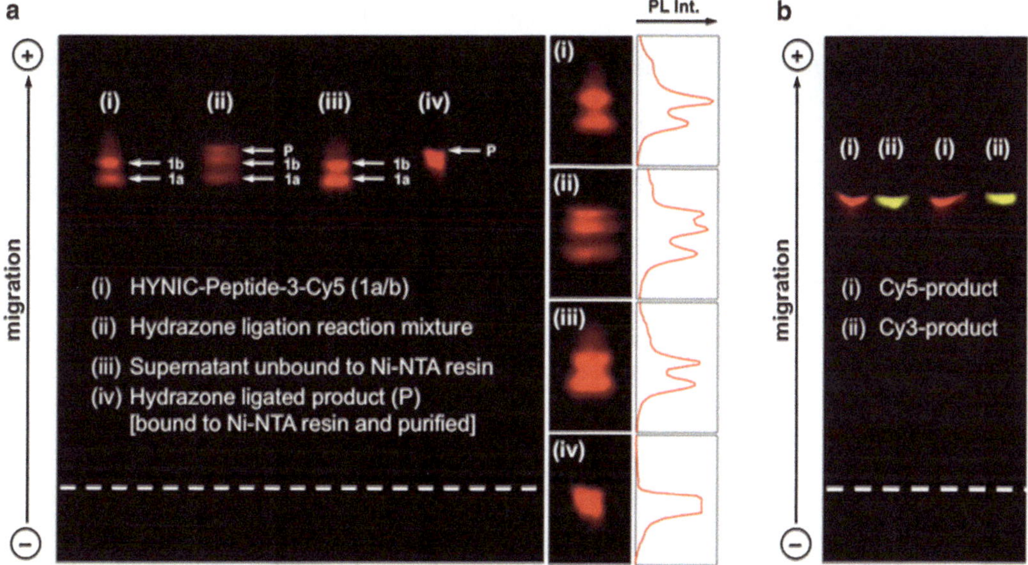

Fig. 6 (**a**) Pseudo-color image of a UV-illuminated polyacrylamide gel confirming solution-phase peptide modification and chemoselective hydrazone ligation: (i) HYNIC-Peptide-3-Cy5 reactant, (ii) the hydrazone ligation reaction mixture, (iii) material remaining in the supernatant following Ni-NTA-agarose purification, and (iv) the ligated and purified reaction product. (**b**) Pseudo-color image of a UV-illuminated polyacrylamide gel showing the purity and equivalence of (i) Cy5 and (ii) Cy3 labeled hydrazone ligation products following purification over Ni-NTA-agarose and desalting. PL collected using an optical filter for Cy3 is colored *yellow*; PL collected using an optical filter for Cy5 is colored *red*

small aliquots (e.g., 10 μL) of dye labeled peptide at different stages of the process. These aliquots can be run on a PAGE gel alongside the purified ligated peptide (*see* Fig. 6).

11. Dry the peptide using a vacuum concentrator (without applying heat) and store at −20 °C (*see* **Note 30**).

3.4 Assembly of Ligated Peptides to Quantum Dots

The ligated peptides can be assembled with QDs simply by mixing these two components at the desired stoichiometry, provided the concentration of both is sufficiently high (≥100 nM).

1. If the ligated peptide was dried prior to use, dissolve by adding 10–20 μL of 50 % MeCN (aq) (*see* **Note 17**). Add the desired buffer (e.g., PBS, borate) to a peptide concentration between five- to tenfold larger than the stock solution of QDs.
2. Select the (average) number of peptides to be assembled per QD.
3. Select the QD concentration to be used. A concentration in the range of 0.1–0.5 μM will provide ample signal for most spectrofluorimeters.
4. The QD-peptide conjugates are assembled by mixing the peptides and QDs in buffer (*see* **Note 31**). Calculate the volumes of the peptide and QD stock solutions needed (*see* **Note 32**) based on **steps 1** and **2**. Calculate the volume of buffer needed to bring the sample up to the desired concentration.

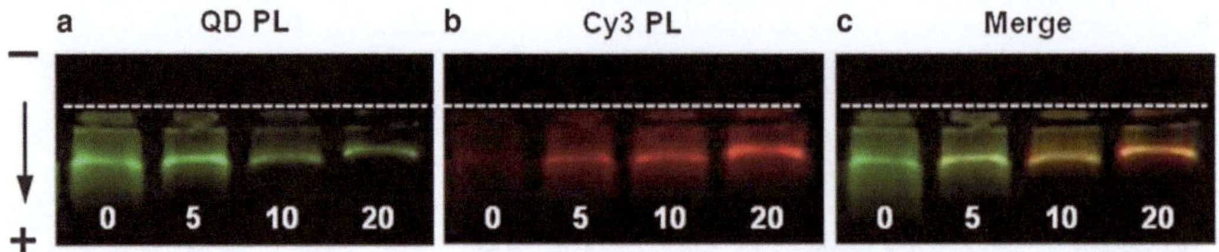

Fig. 7 Pseudo-color images of a UV-illuminated agarose gel (1.5 %) showing the assembly of Cy3 labeled hydrazone ligation product with CL4-coated QD520 (*see* **Note 6**). The numbers below each band are the average number of peptides added per QD. PL collected using an optical filter (**a**) for the QD520 is shown in *green*; (**b**) for the Cy3 is shown in *red*; and (**c**) a merge of the two images showing colocalization. Each well contained 6 pmol of QD, loaded in 15 μL of buffer. The run time as ca. 5 min at a field strength of 10 V/cm

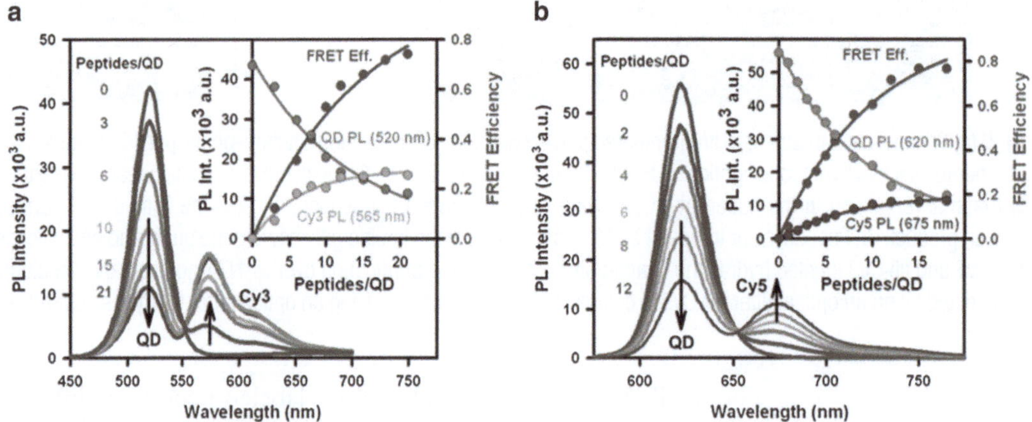

Fig. 8 Confirmation of assembly between (**a**) QD520 and Cy3 labeled hydrazone ligation product, and (**b**) QD620 and Cy5 labeled hydrazone ligation product. Each plot shows progressive quenching of QD PL and sensitization of Cy3/5 PL as more QDs are added per QD. Cy3/5 PL from direct excitation has been subtracted from the spectra. The *insets* show the change in peak PL intensities and calculated FRET efficiencies

5. Add the peptide and buffer to a microcentrifuge tube, followed by the QDs. Mix immediately and let stand at room temperature for ≥30 min. The sample is then ready for measurement or other use.
6. Assembly can be confirmed by taking an aliquot of the sample (or preparing a second analogous sample in parallel) to run alongside a sample of only QDs on a 1.5 % agarose gel (*see* **Note 33**; *see* Fig. 7).
7. If the QD and dye label on the peptide form a donor–acceptor pair for Förster resonance energy transfer (FRET; *see* **Note 34**), then quenching of the QD (donor) luminescence and sensitization of the dye (acceptor) luminescence can provide confirmation of peptide assembly (*see* Fig. 8). Separate controls with the equivalent amount of QD and peptide *alone* should be measured in parallel.

4 Notes

1. Amino acid side chain protecting groups (in parenthesis) in Fmoc-SPPS: Arg(Pbf), Asn(Trt), Asp(^{t}Bu), Cys(Trt), Gln(Trt), Glu(^{t}Bu), His(Trt), Lys(Boc), Ser(^{t}Bu), Thr(^{t}Bu), Trp(Boc), and Tyr(^{t}Bu), where Pbf is 2,2,4,6,7-pentamethyl-2,3-dihydrobenzofuran-5-sulfonyl, Trt is trityl, ^{t}Bu is *tert-butyl*, and Boc is *tert*-butyloxycarbonyl.
2. The HBTU solution is generally stable for 2 weeks at room temperature when protected from light; however, storage at 4 °C and under inert atmosphere is recommended when possible.
3. Empty oligonucleotide purification cartridges (OPC; Applied Biosystems) work well for this purpose. The cartridges can be snapped apart and the resin removed. Cartridges can be cleaned by storage in a 50 % aqueous solution of a polar organic solvent such as methanol, ethanol, or acetonitrile, and then reused over several peptide purifications.
4. The sNHS-4FB and sNHS-HYNIC are sold by Solulink under the abbreviated names "sulfo-S-4FB" and "sulfo-S-HYNIC."
5. Aniline will gradually oxidize in air and change from colorless to red/brown. Although a partially oxidized aniline solution may retain sufficient catalytic activity for chemoselective hydrazone ligation, we recommend the use of fresh reagent to obtain the best results.
6. CdSe/ZnS core-shell QDs were synthesized with a coating of hydrophobic ligands including trioctylphosphine (TOP), trioctylphosphine oxide (TOPO), and hexadecylamine (HDA) as described previously [46, 47]. The QDs were made water soluble by exchanging the TOP/TOPO/HDA with polyethylene glycol-appended dihydrolipoic acid (DHLA-PEG) or a compact, zwitterionic dithiol ligand (CL4) using published procedures [47, 48]. The ligand exchange procedure can be applied to hydrophobic QDs purchased from commercial sources. The details of these methods are beyond our current scope but can be found in the cited references. Other thiol-based ligands can be used in preference to DHLA-PEG or CL4. QDs with peak PL at 520 nm are referred to as QD520, and those with peak PL at 620 nm are referred to as QD620.
7. It is critical to remove all the piperidine from the resin; traces of piperidine will interfere with the subsequent coupling reaction.
8. The ninhydrin test should strictly be performed following every deprotection step and coupling reaction; however, in practice, it is generally only necessary to perform a ninhydrin test after each coupling reaction. Fmoc deprotection with Solution B

typically goes to completion without difficulty. As an alternative to the ninhydrin test, Fmoc removal can be monitored using the absorbance of the piperidine–dibenzofulvene adduct at 301 nm.

9. Complete dissolution of the Fmoc-protected amino acid can be assisted by low power ultrasonication.
10. TIS and EDT can react with 4FB to yield a corresponding alcohol and thioacetal, respectively. Many other scavengers are also reactive and, as a consequence, we recommend that only water be added to the TFA cleavage solution. The success of the 4FB-modified peptide synthesis largely depends on the success of this cleavage step.
11. *Caution*: EDT has a stench and is toxic; handle with care. Glassware or plasticware contaminated with EDT should be cleaned with bleach solution.
12. The addition of 2.5 % of acetone to the cleavage mixture will yield the corresponding 6-hydrazinonicotinate acetone hydrazone, which is the commercial HYNIC derivative used in the solution phase modification protocol (*see* Subheading 3.2.2).
13. It is possible that a fraction of the HYNIC is trifluoroacetylated in this step; longer cleavage times lead to higher trifluoroacetylation yields of the HYNIC. The HYNIC can be de-trifluoroacetylated in 0.1 M NaOH buffer for 30 min [49].
14. Polyhistidine peptides tend to precipitate from TBME very poorly and must be recovered using a different method. Protecting groups (e.g., trityl) are removed by extraction into the TBME and the peptide is retained in the aqueous phase. Extraction of the lipophilic protecting groups also facilitates purification of the peptide by RP-HPLC.
15. Other buffers that do not contain thiols may be used for this reaction; however, the solution should be kept between pH 7.0–7.5. Although higher pH yields a faster reaction, it also increases the hydrolysis rate of the succinimide product.
16. While NHS-esters are generally selective for amines over hydroxyls, they do react efficiently with thiols to yield thioesters. Cysteine residues should thus be avoided in the peptide sequence (or protected as a disulfide) for specific labeling of an amine group.
17. The most reliable approach to dissolving peptides of almost any amino acid sequence in buffer is to first add a small amount of an organic solvent such as DMSO or 50 % MeCN (aq).
18. Solulink conveniently sells the sNHS-4FB and sNHS-HYNIC as 1.0 mg quantities in individual vials. The contents of each vial may be dissolved in 10 μL of DMSO and the 0.5 mL peptide solution added to each in succession. We use three vials in

our suggested protocol, totaling 3.0 mg dissolved in 30 μL of DMSO. To help maximize the transfer of material, water/buffer may be used in two portions to rinse each vial in succession with addition to the reaction mixture. This strategy is also effective with the premeasured tubes of Cy3/5 maleimide mono-reactive dye sold by GE Healthcare. While the pre-weighed packages are convenient, they are not essential to the protocol.

19. This step assumes that the 4FB-modified peptide has a polyhistidine sequence, as in the case of our example. If the peptide does not, skip this step and proceed with **step** 7 in Subheading 3.2.2 instead.

20. Excess Cy3/5 and HYNIC have larger retention volumes than the HYNIC/Cy3/5-modified peptide. If the resolution of two dye-colored bands is poor, use polyacrylamide gel electrophoresis to determine which fractions contain the modified peptide. These fractions can then be combined, concentrated using the desalting protocol in Subheading 3.2.4, and purified using a gel permeation column of greater length and/or more optimal fractionation range.

21. The accuracy of the dye absorption coefficients when a peptide is also modified with the HYNIC (or hydrazone) chromophore is uncertain. Our experience suggests that the use of the Cy3/5 absorption coefficients underestimates the amount of peptide. Figure 5 shows how the ratio between the hydrazone and Cy3/5 absorption bands does not correspond to the ratio of their reported absorption coefficients. In practice, we use the hydrazone absorption band to quantitate the purified product from Subheading 3.3.

22. We have found that each 0.4 mL plastic cartridge (e.g., empty OPC cartridge; *see* **Note 3**) filled with Ni-NTA-agarose is sufficient for each 0.5 mg of peptide (MW ≥1.5 kDa) used as starting material in the labeling reaction.

23. Our recommended double-syringe technique is to connect a 3 mL syringe to one Luer fitting of the plastic/OPC cartridge and another 3 mL syringe (without plunger) to the other Luer fitting. The reaction mixture can be added to the latter and the plunger carefully inserted into the syringe barrel. The opposing syringes can be pushed back and forth to flush the reaction mixture over the Ni-NTA-agarose/OPC resin. The similar diameters of the cartridge and 3 mL syringe minimize backpressure. This helps avoid leakage from any defects in the snap-tight joint between the two pieces forming the cartridge or, worse, breaking a seal at one of the Luer fittings with spillage or spray of solution. If 3 mL syringes provide insufficient volume, then use of a syringe with the smallest possible diameter for the needed volume is recommended.

24. We have noted that frits often nonspecifically adsorb fluorescent dyes and other molecules with hydrophobic character. This is largely ameliorated by the washing step with the 50 % buffered solution of ethanol (or another polar organic solvent, e.g., MeCN).
25. Washes can be done with syringe sizes that match the wash volume with minimal concern for backpressure. It is advisable to use air to expel residual solution from the Ni-NTA-agarose/OPC resin between washes. This is particularly true on the final washes so as to avoid diluting the imidazole/70 % MeCN (aq) solutions used to elute peptide.
26. To help maximize the efficiency of peptide elution from the media, the eluent can be gently and slowly pulsated by pushing the syringe barrel 5 mm, pulling back 2–3 mm, and repeating this cycle through the full volume of eluent.
27. If the number of moles of the peptide is determined by weight, and the peptide has not gone through any desalting steps (i.e., prepared via the SPPS method in Subheading 3.1), it is important to take into account the mass of the TFA counter-ions associated with basic residues. When ligating two peptides modified similarly those in Subheading 3.2, and at the same reaction scales, it is most convenient to use the full amount of HYNIC-Cy3/5 modified peptide prepared and half the amount 4FB-modified peptide (i.e., an approximately 2:1 stoichiometry).
28. Although the HYNIC-acetone hydrazone is stable in aqueous buffer at pH 7.0, it rapidly equilibrates with the HYNIC form in presence of aniline and reacts with 4FB yielding a more thermodynamically stable hydrazone.
29. Increasing the amount of aniline will increase the reaction rate. The reaction will also proceed faster at moderately acidic pH (e.g., NH_4OAc buffer at pH 5.5). We typically use 100 mM aniline with buffers at neutral pH (e.g., PBS buffer at pH 7), and 10 mM aniline with acidic buffers; however, there is no strict rule in this regard. Further, if it is inconvenient to measure ≤10 μL of aniline, 0.1–1.0 mL of aniline can be diluted to 10 mL with buffer and 0.1 mL added to the 1 mL reaction mixture; however, these stock solutions should always be prepared fresh.
30. We typically divide the total amount of peptide between several microcentrifuge tubes with ca. 5–20 nmol of peptide per tube. These aliquots are concentrated individually and allow the ligated peptide to be dissolved and used in experiments only as needed.
31. The polyhistidine appended peptides will assemble to the inorganic shell of QDs coated with thiol-based ligands (e.g., DHLA, DHLA-PEG, and CL4). There is also evidence that suggests these peptides will also assemble to the organic corona of commercially available QDs with carboxylated polymer

(Invitrogen) or phospholipid (eBioscience, San Diego, CA) coatings [50].

32. As always, the equation $V_o = V_s(C_s/C_o) = n_s/C_o$ is convenient to use, where V_o and C_o are the volume and concentration (respectively) of the stock solution to be added to the sample, V_s is the final sample volume, and n_s is the number of moles per sample.
33. In general, agarose gels should be run using negatively charged QDs (e.g., a coating of DHLA ligands). A decrease in electrophoretic mobility with increases in the amount of assembled peptide is typically expected. In the special case of highly charged peptides, neutral QDs (e.g., a coating of DHLA-PEG ligands) can be used and assembly is indicated by an increase in mobility. In either case, and with appropriate optical filters for the gel imager, it is often possible to observe colocalization of the different wavelengths of fluorescence from the QD and dye labeled peptide. This serves as additional confirmation of assembly—particularly if the mobility shifts are small. Finally, we note that the hydrazone bond sometimes degrades during gel electrophoresis, and that the running time should be minimized.
34. The details of a FRET analysis are beyond our current scope but can be found in both a recent review [10] and a previous method published in this series [51].

Acknowledgments

The authors acknowledge NRL, NRL NSI, ONR, and DTRA-JSTO MIPR # B112582M for financial support. W.R.A. is grateful to the Natural Sciences and Engineering Research Council of Canada (NSERC) for support through a postdoctoral fellowship. J.B.B.-C. acknowledges a Marie Curie IOF. RLM thanks National Science Foundation Graduate Research Fellowship (Grant no. DGE-0750616) for support.

References

1. Algar WR, Prasuhn DE, Stewart MH, Jennings TL, Blanco-Canosa JB, Dawson PE, Medintz IL (2011) The controlled display of biomolecules on nanoparticles: a challenge suited to bioorthogonal chemistry. Bioconjug Chem 22:825–858
2. Rosi NL, Mirkin CA (2005) Nanostructures in biodiagnostics. Chem Rev 105:1547–1562
3. Niemeyer CM (2001) Nanoparticles, proteins, and nucleic acids: biotechnology meets materials science. Angew Chem Int Ed 40: 4128–4158
4. Algar WR, Susumu K, Delehanty JB, Medintz IL (2011) Semiconductor quantum dots in bioanalysis: crossing the valley of death. Anal Chem 83:8826–8837
5. Rosenthal SJ, Chang JC, Kovtun O, McBride JR, Tomlinson ID (2011) Biocompatible quantum dots for biological applications. Chem Biol 18:10–24
6. Algar WR, Krull UJ (2010) New opportunities in multiplexed bioanalyses using quantum dots and donor-acceptor interactions. Anal Bioanal Chem 398:2439–2449

7. Bruchez MP (2011) Quantum dots find their stride in single molecule tracking. Curr Opin Chem Biol 15:775–780
8. Chan WCW, Maxwell DJ, Gao XH, Bailey RE, Han MY, Nie SM (2002) Luminescent quantum dots for multiplexed biological detection and imaging. Curr Opin Biotechnol 13: 40–46
9. Algar WR, Tavares AJ, Krull UJ (2010) Beyond labels: a review of the application of quantum dots as integrated components of assays, bioprobes, and biosensors utilizing optical transduction. Anal Chim Acta 673: 1–25
10. Medintz IL, Mattoussi H (2009) Quantum dot-based resonance energy transfer and its growing application in biology. Phys Chem Chem Phys 11:17–45
11. Biju V, Itoh T, Anas A, Sujith A, Ishikawa M (2008) Semiconductor quantum dots and metal nanoparticles: syntheses, optical properties, and biological applications. Anal Bioanal Chem 391:2469–2495
12. Sapsford KE, Pons T, Medintz IL, Higashiya S, Brunel FM, Dawson PE, Mattoussi H (2007) Kinetics of metal-affinity driven self-assembly between proteins or peptides and CdSe-ZnS quantum dots. J Phys Chem C 111:11528–11538
13. Pons T, Medintz IL, Wang X, English DS, Mattoussi H (2006) Solution-phase single quantum dot fluorescence resonance energy transfer. J Am Chem Soc 128:15324–15331
14. Prasuhn DE, Blanco-Canosa JB, Vora GJ, Delehanty JB, Susumu K, Mei BC, Dawson PE, Medintz IL (2010) Combining chemoselective ligation with polyhistidine-driven self-assembly for the modular display of biomolecules on quantum dots. ACS Nano 4:267–278
15. Medintz IL, Berti L, Pons T, Grimes AF, English DS, Alessandrini A, Facci P, Mattoussi H (2007) A reactive peptidic linker for self-assembling hybrid quantum dot-DNA bioconjugates. Nano Lett 7:1741–1748
16. Blanco-Canosa JB, Medintz IL, Farrell D, Mattoussi H, Dawson PE (2010) Rapid covalent ligation of fluorescent peptides to water solubilized quatnum dots. J Am Chem Soc 132:10027–10033
17. Dennis AM, Bao G (2008) Quantum dot-fluorescent protein pairs as novel fluorescence resonance energy transfer probes. Nano Lett 8:1439–1445
18. Zhang Y, So MK, Loening AM, Yao HQ, Gambhir SS, Rao JH (2006) HaloTag protein-mediated site-specific conjugation of bioluminescent proteins to quantum dots. Angew Chem Int Ed 45:4936–4940
19. Wang JH, Xia J (2011) Preferential binding of a novel polyhistidine peptide dendrimer ligand on quantum dots probed by capillary electrophoresis. Anal Chem 83:6323–6329
20. Prasuhn DE, Deschamps JR, Susumu K, Stewart MH, Boeneman K, Blanco-Canosa JB, Dawson PE, Medintz IL (2010) Polyvalent display and packing of peptides and proteins on semiconductor quantum dots: predicted versus experimental results. Small 6:555–564
21. Sawant RM, Hurley JP, Salmaso S, Kale A, Tolcheva E, Levchenko TS, Torchilin VP (2006) “SMART” drug delivery systems: double-targeted pH-responsive pharmaceutical nanocarriers. Bioconjug Chem 17: 943–949
22. Savla R, Taratula O, Garbuzenko O, Minko T (2011) Tumor targeted quantum dot-mucin 1 aptamer-doxorubicin conjugate for imaging and treatment of cancer. J Control Release 153:16–22
23. Kam NWS, Liu Z, Dai HJ (2005) Functionalization of carbon nanotubes via cleavable disulfide bonds for efficient intracellular delivery of siRNA and potent gene silencing. J Am Chem Soc 127:12492–12493
24. Takae S, Miyata K, Oba M, Ishii T, Nishiyama N, Itaka K, Yamasaki Y, Koyama H, Kataoka K (2008) PEG-detachable polyplex micelles based on disulfide-linked block catiomers as bioresponsive nonviral gene vectors. J Am Chem Soc 130:6001–6009
25. Dirksen A, Dawson PE (2008) Rapid oxime and hydrazone ligations with aromatic aldehydes for biomolecular labeling. Bioconjug Chem 19:2543–2548
26. Chen Y, Aveyard J, Wilson R (2004) Gold and silver nanoparticles functionalized with known numbers of oligonucleotides per particle for DNA detection. Chem Commun 2804–2805
27. Zhong XB, Reynolds R, Kidd JR, Kidd KK, Jenison R, Marlar RA, Ward DC (2003) Single-nucleotide polymorphism genotyping on optical thin-film biosensor chips. Proc Natl Acad Sci USA 100:11559–11564
28. Chang YS, Jeong JM, Lee YS, Kim HW, Rai GB, Lee SJ, Lee DS, Chung JK, Lee MC (2005) Preparation of F-18-human serum albumin: a simple and efficient protein labeling method with F-18 using a hydrazone-formation method. Bioconjug Chem 16:1329–1333
29. Steinberg-Tatman G, Huynh M, Barker D, Zhao CF (2006) Synthetic modification of silica beads that allows for sequential attachment of two oligonucleotides. Bioconjug Chem 17:841–848
30. King PT, Zhao SW, Lam L (1986) Preparation of protein conjugates via intermolecular hydrazone linkage biochemistry. Biochemistry 25:5774–5779
31. Gaertner HF, Rose K, Cotton R, Timms D, Camble R, Offord RE (1992) Construction of

protein analogues by site-specific condensation of unprotected fragments. Bioconjug Chem 3:262–268

32. Rose K (1994) Facile synthesis of homogeneous artificial proteins. J Am Chem Soc 116:30–33
33. Weden RC, Lankinen M, Rose K, Blakey D, Shuttleworth H, Melton R, Offord RE (1994) Site-specific conjugation of an enzyme and an antibody fragment. Bioconjug Chem 5:411–417
34. Canne LE, Ferré-D'Amaré AR, Burley SK, Kent SBH (1995) Total chemical synthesis of a unique transcription factor-related protein: cMyc-Max. J Am Chem Soc 117:2998–3007
35. Shao J, Tam JP (1995) Unprotected peptides as building blocks for the synthesis of peptide dendrimers with oxime, hydrazone, and thiazolidine linkages. J Am Chem Soc 117: 3893–3899
36. Cervigini SE, Dumy P, Mutter M (1996) Synthesis of glycopeptides and lipopeptides by chemoselective ligation. Angew Chem Int Ed 35:1230–1232
37. Mahal LK, Yarema KJ, Bertozzi CR (1997) Engineering chemical reactivity on cell surfaces through oligosaccharide biosynthesis. Science 276:1125–1128
38. Chen I, Howarth M, Lin W, Ting AY (2005) Site-specific labeling of cell surface proteins with biophysical probes using biotin ligase. Nat Methods 2:99–104
39. Singh Y, Renaudet O, Defrancq E, Dumy P (2005) Preparation of a multitopic glycopeptide-oligonucleotide conjugate. Org Lett 7:1359–1362
40. Larsen K, Thygesen MB, Guillaumie F, Willats WGT, Jensen K (2006) Solid-phase chemical tools for glycobiology. Carbohydr Res 341:1209–1234
41. Scheck RA, Francis MB (2007) Regioselective labeling of antibodies through N-terminal transamination. ACS Chem Biol 4:247–251
42. Carrico IS, Carlson LB, Bertozzi CR (2007) Introducing genetically encoded aldehydes into proteins. Nat Chem Biol 3:321–322
43. Dirksen A, Dirksen S, Hackeng TM, Dawson PE (2006) Nucleophilic catalysis of hydrazone formation and transamination: implications for dynamic covalent chemistry. J Am Chem Soc 128:15602–15603
44. Dirksen A, Hackeng TM, Dawson PE (2006) Nucleophilic catalysis of oxime ligation. Angew Chem Int Ed 118:7743–7746
45. Sarin VK, Kent SB, Tam JP, Merrifield RB (1981) Quantitative monitoring of solid-phase peptide synthesis by the ninhydrin reaction. Anal Biochem 117:147–157
46. Dabbousi BO, Rodriguez-Viejo J, Mikulec FV, Heine JR, Mattoussi H, Ober R, Jensen KF, Bawendi MG (1997) (CdSe)ZnS core-shell quantum dots: synthesis and optical and structural characterization of a size series of highly luminescent materials. J Phys Chem B 101:9463–9475
47. Susumu K, Oh E, Delehanty JB, Blanco-Canosa JB, Johnson BJ, Jain V, Hervey WJ, Algar WR, Boeneman K, Dawson PE, Medintz IL (2011) Multifunctional compact zwitterionic ligands for preparing robust biocompatible semiconductor quatnum dots and gold nanoparticles. J Am Chem Soc 133:9480–9496
48. Mei BC, Susumu K, Medintz IL, Mattoussi H (2009) Polyethylene glycol-based bidentate ligands to enhance quantum dot and gold nanoparticle stability in biological media. Nat Protoc 4:412–423
49. Surfraz MBU, King R, Mather SJ, Biagini SCG, Blower PJ (2007) Trifluoroacetyl-HYNIC peptides: synthesis and 99mTc radiolabeling. J Med Chem 50:1418–1422
50. Dennis AM, Sotto DC, Mei BC, Medintz IL, Mattoussi H, Bao G (2010) Surface ligand effects on metal-affinity coordination to quantum dots: implications for nanoprobe self-assembly. Bioconjug Chem 21:1160–1170
51. Prasuhn DE, Susumu K, Medintz IL (2011) Multivalent conjugation of peptides, proteins, and DNA to semiconductor quantum dots. In: Hurst SJ (ed) Biomedical nanotechnology: methods and protocols. Humana Press-Springer, New York, pp 95–110

Chapter 6

Reliable Methods for Silica Coating of Au Nanoparticles

Isabel Pastoriza-Santos and Luis M. Liz-Marzán

Abstract

The inherent properties of silica, such as optical transparency, high biocompatibility, chemical and colloidal stability, controllable porosity, and easy surface modification, provide silica materials with a tremendous potential in biomedicine. Therefore, the coating of Au nanoparticles with silica largely contributes to enhance the important applications of metal nanoparticles in biomedicine. We describe in this chapter a number of reliable strategies that have been reported for silica coating of different types of Au nanoparticles. All descriptions are based on tested protocols and are expected to provide a reference for scientists with an interest in this field.

Key words Au nanoparticles, Silica coating, Core-shell, Surface modification, Optical properties

1 Introduction

The intense research and the great progress that have been achieved over the past two decades in the synthesis, characterization and surface modification of gold nanoparticles have demonstrated that their properties go beyond their beauty [1]. The unique electronic properties, different from their bulk counterpart, arise from the collective oscillation of conduction electrons in resonance with an incident electromagnetic radiation [2, 3]. The phenomenon of localized surface plasmon resonance (LSPR) constitutes the basis of numerous and important applications of metal nanoparticles in a wide range of fields (biomedicine, optoelectronics, photovoltaics, (bio)sensing, catalysis, etc.) [1]. Importantly, LSPR frequency is strongly dependent on a number of factors, including particle size, shape and composition, as well as the dielectric properties of the surrounding medium and interparticle distance [4, 5].

In the fields of biology and medicine, Au nanoparticles present a wide range of potential applications, derived not only from their optical properties (enhanced radiative properties) but also from their low inherent toxicity, large surface area, and easy surface

Paolo Bergese and Kimberly Hamad-Schifferli (eds.), *Nanomaterial Interfaces in Biology: Methods and Protocols*, Methods in Molecular Biology, vol. 1025, DOI 10.1007/978-1-62703-462-3_6, © Springer Science+Business Media New York 2013

functionalization (which is important in terms of biological compatibility and specificity). Murphy, El-Sayed, and others have recently published a critical review entitled "The golden age: gold nanoparticles for biomedicine" [6]. In this review article, apart from summarizing the most promising achievements in the field, they express their belief in that "a new Golden Age of biomedical applications is truly upon us." Gold nanoparticles have been used to fabricate highly sensitive assays for in vitro diagnostics [7, 8], devices that are mainly based on surface-enhanced Raman scattering (SERS) [9] and LSPR [10] (the shift in the extinction spectra or even simply a color change, in response to changes in the local refractive index surrounding the nanoparticle provides a direct readout of target binding) assays. Their strongly enhanced radiative properties also render Au nanoparticles useful probes for bioimaging, either in vitro or in vivo [6, 11]. Moreover, they are promising scaffolds for gene and drug delivery applications—either as delivery platforms or as stabilizers—or even to trigger drug release [6, 10, 12]. Even more fascinating is the fact that gold colloids themselves can act as therapeutic agents. The absorption of light when the nanoparticle is irradiated with a laser beam produces local heating that has been proven capable of killing cancer cells, viruses, etc. [6]. The immediate future for Au nanoparticles in biomedicine is very promising and one of the most important challenges will be assessing the toxicity and long-term effects on humans and the environment, as well as the development of new strategies for nanoparticles surface functionalization.

Another interesting material for biomedicine is silica. Apart from exhibiting optical transparency, high biocompatibility and strong chemical and colloidal stability, silica particles can be easily labeled (e.g., via encapsulation of molecular tags) and surface modified (with amino-, mercapto-, carboxy-, aldehyde-terminated silanes, polyelectrolytes, etc.), while their porosity is highly tunable (useful for drug loading and delivery) [13]. Additionally, the coating of metal nanoparticles with silica has been demonstrated not just to facilitate bioconjugation of such Au nanoparticles but also to increase their biocompatibility or to fabricate multifunctional systems. Thus, silica-coated Au nanoparticles (Au@SiO_2) show a great potential for designing colorimetric [14] and SERS biosensors, for in vivo and in vitro bioimaging [15–17], for drug delivery [18], in therapy [19, 20], or in theranostics [21]. In this chapter we describe in detail various protocols that have been successfully applied for coating Au nanoparticles with silica.

2 Materials

Some rules of general application should be indicated. Colloid synthesis in general is extremely sensitive to the presence of potential nuclei in the reaction medium and therefore all glassware and solvents must be extremely clean. In the case of metal nanoparticles in general (and gold in particular), it is standard to require that all glassware is thoroughly washed with aqua regia (*see* **Note 1**) prior to use and Milli-Q grade water should be used in all preparations. Although all chemicals can be used as received, in some cases the quality of the reagents or even the specific supplier may also critically affect the quality of the resulting nanoparticles. Therefore, those critical reagents have been listed below.

2.1 Reagents

1. Cetyltrimethylammonium bromide (CTAB) from Aldrich.
2. Ammonium hydroxide (NH_4OH), 28–32 wt.% aqueous solution.
3. Pure grade ethanol.
4. Silver nitrate ($AgNO_3$), purity higher than 99 %.
5. For the synthesis of Au nanostars, poly(vinylpyrrolidone) (PVP, Mw 8000) from Alfa Aesar is recommended. PVP from Sigma-Aldrich has been found to lead to particles with poor quality.
6. "Active silica" or 0.54 wt.% sodium silicate solution with a pH of 10–11: 1 mL of sodium silicate 27 wt.% is diluted 50 times with water and then the pH is adjusted to 10–11 by addition of a cation exchange resin, Dowex Macroporous Resin, for some minutes. Check the pH evolution with pH test strips.

3 Methods

3.1 Silica Coating of 15 nm Citrate-Stabilized Au Particles

This method, developed by Liz-Marzán et al., was probably the first successful coating of gold nanoparticles with uniform silica shells. Published in *Langmuir* in 1996 [22], this article has been widely used, as indicated by over 1000 citations, according to ISI Web of Knowledge©. The essence of this method (*see* scheme in Fig. 1a) is the modification of the metal cores with an amino-terminated silane coupling agent, such as 3-(aminopropyl) triethoxysilane (APS, *see* **Note 2**), which acts as a primer to facilitate uniform silica polymerization around the Au particles. Subsequently, the silica shell can be further grown in a controlled way through the well-known Stobër method [23]. This protocol allows a tight control on the silica shell thickness from 3 nm up to hundreds of nanometers.

We describe here a protocol that has been optimized for the silica coating of spherical Au nanoparticles with an average diameter of 17 nm but which could be adapted for the coating of other citrate-stabilized nanoparticles.

Gold nanospheres synthesis (*see* **Note 3**) [24]

1. Bring to a full rolling boil 500 mL of 0.5 mM $HAuCl_4$ in a 1 L Erlenmeyer flask under homogeneous and vigorous stirring (*see* **Note 4**).
2. Add rapidly 25 mL of an aqueous 1 wt.% aqueous solution of trisodium citrate dihydrate (0.25 g) (*see* **Note 5**). Maintain the

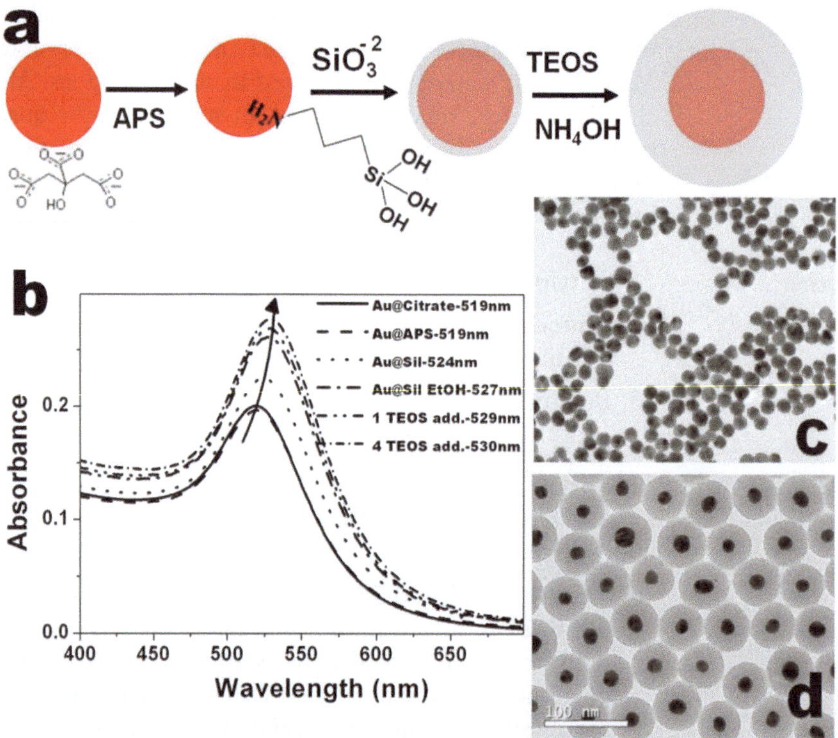

Fig. 1 (**a**) Scheme of the strategy for silica coating citrate-stabilized Au nanoparticles. (**b**) UV–Vis spectra of 15 nm Au-citrate nanoparticles during the different coating steps. The position of the LSPR band maximum is indicated in the graph. (**c**, **d**) TEM images of citrate-stabilized Au particles before and after silica coating, respectively. Magnification is the same for both images

boiling for 15 min and then slowly cool down to room temperature (*see* **Note 6**).

3. Measure the absorption spectrum by UV–Visible spectroscopy (*see* **Notes** 7 and **8**) (Fig. 1).

Silica coating

1. Add 2.5 mL of a freshly prepared 1 mM aqueous APS solution (*see* **Note 9**) to 500 mL of the gold colloid under vigorous magnetic stirring. Allow the reaction to proceed for 15 min (*see* **Note 10**) (Fig. 1b).
2. Add 20 mL of "active silica" under vigorous magnetic stirring, the resulting pH should be around 8.5. Allow active silica to polymerize onto the Au nanoparticles for 7 days (*see* **Notes 11** and **12**) (Fig. 1b).
3. Centrifuge the resulting silica-coated Au nanoparticles at 1,287 × *g* for 90 min to remove excess silicate. Collect the precipitate and centrifuge the supernatant twice under the same

conditions. Collect all precipitates and redisperse in 50 mL of Milli-Q water.

4. Add 200 mL of ethanol to 50 mL of Au@SiO_2 particles in water (*see* **Note 13**).
5. Add 3.0 mL of TEOS solution (10 % in ethanol, *see* **Note 14**) and 2 mL of ammonia (NH_4OH, 32 % in water) all under gentle stirring. Keep under gentle stirring for 12 h before adding again 3.0 mL of TEOS solution (*see* **Notes 15** and **16**).
6. Centrifuge the resulting silica-coated Au nanoparticles at 1,287 × *g*. Collect the precipitate and centrifuge the supernatant twice under the same conditions. Collect all precipitates and redisperse in Milli-Q water or ethanol to store (Fig. 1d).

3.2 Mesoporous Silica Coating of CTAB-Stabilized Au Nanoparticles

This method was developed by Gorelikov and Matsuura and published in *Nano Letters* in 2008 [25]. Since its publication this article has been cited over 100 times so far, according to Web of Knowledge©. Probably the main advantages of this method are related to its simplicity, since it is a single-step synthesis procedure (*see* scheme in Fig. 2a), with no need for any intermediate coating or removal of the surfactant bilayer, and to the high porosity of the obtained shell, which facilitates the interaction between the metal core and the surrounding medium. Although the protocol works very well for the deposition of thick silica shells, it fails to control very thin shells. It is also worth mentioning that the mesoporous silica shell may dissolve in basic aqueous solutions, but this can be avoided through a mild thermal treatment [26]. We describe here the protocol for coating Au nanorods (14 × 70 nm), but the same procedure can be adapted for other CTAB-stabilized Au nanoparticles.

Au nanorod synthesis (in acidic medium) (*see* **Note 17**) [27]

1. Add 0.3 mL of freshly prepared 10 mM sodium borohydride solution (*see* **Note 18**) (fast addition under vigorous stirring) to 5 mL of an aqueous solution containing 0.1 M CTAB and 0.25 mM $HAuCl_4$. Let the reaction proceed for 10 min (*see* **Note 19**).
2. Add 2.5 mL of 0.05 M $HAuCl_4$ to 250 mL of 0.1 M CTAB. Homogenize the mixture by shaking (*see* **Note 20**).
3. Add 4.75 mL of 1 M HCl, 2.0 mL of 0.1 M ascorbic acid (*see* **Note 21**), 3 mL of freshly prepared 0.01 M silver nitrate and 0.6 mL of seed solution (after each addition the solution should be gently shaked).
4. Introduce the resulting solution in a thermostatic bath at 27 °C for 2 h.
5. Measure the absorption spectrum by UV–Visible spectroscopy (*see* **Note 22**) (Fig. 2b).

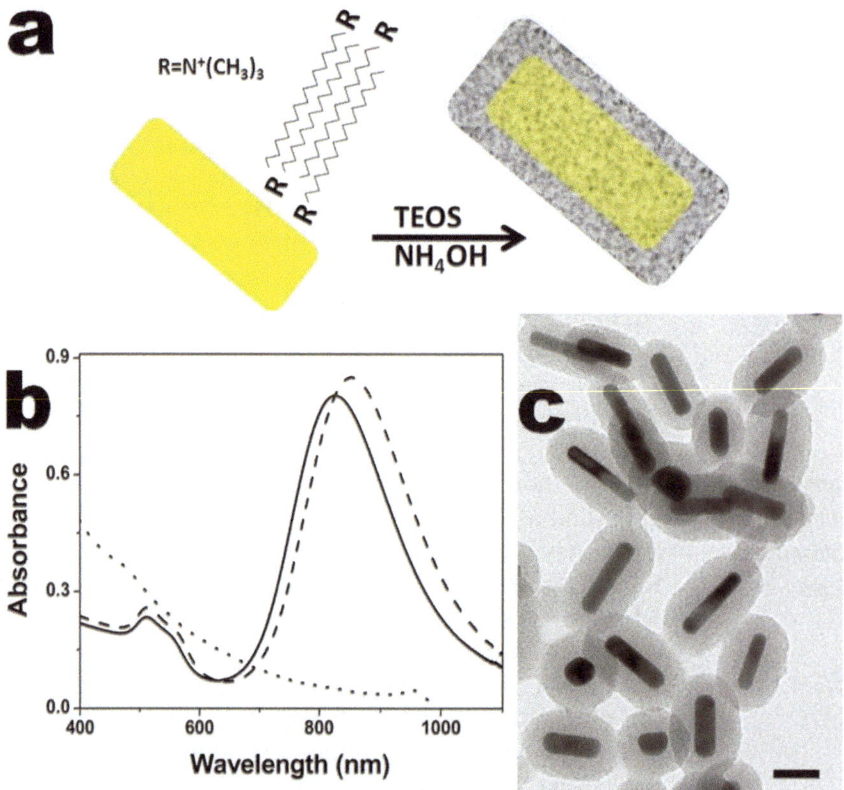

Fig. 2 (**a**) Scheme of the coating of CTAB-stabilized Au nanoparticles with mesoporous silica. (**b**) Vis–NIR spectra of 2–3 nm Au seeds (*dotted line*) and Au nanorods before (*solid line*) and after (*dashed line*) silica-coating. (**c**) TEM image of silica-coated Au nanorods. The scale bar represents 50 nm

Silica coating

1. Centrifuge 250 mL of 0.5 mM Au nanorods at 2,627 × *g* for 40 min in 10 mL tubes. Extract the supernatant, leaving ca. 1.2–1.4 mL in each tube (*see* **Note 23**).
2. Collect all particles and redisperse in water (final volume 100 mL).
3. Adjust the pH to 10 by adding 1 mL of 0.1 M NaOH.
4. Perform 3 additions of 300 μL TEOS (20 % in methanol, *see* **Note 14**). Allow 30 min after each addition.
5. After 2 h of the third addition check the UV–Vis–NIR spectrum of the dispersion (*see* **Notes 12** and **24**).
6. Wash Au@SiO_2 samples by centrifugation and redispersion in water, first at 1,681 × *g* for 30 min and then at 1,287 × *g* for 45 min.
7. Redisperse the colloids in ethanol and store in the fridge (*see* **Note 25**) (Fig 2c).

3.3 PVP-Mediated Silica Coating

Poly(vinylpyrrolidone) (PVP) is a widely used particle stabilizer and can also be used as a primer to promote silica coating of nanoparticles. Back in 2003, Graf et al. reported a general procedure [28] that has received wide attention (over 400 citations according to Web of Knowledge©). This method is based on the ability of PVP to adsorb onto a variety of surfaces (metals, oxides, hydroxides, polymers, etc., regardless of their surface charge). Since PVP is commonly used as capping agent during the synthesis of metal nanoparticles, not only to prevent aggregation but also as a shape-directing agent [29, 30] Stöber silica growth can be directly carried out on such particles, with no need for any intermediate step (*see* scheme in Fig. 3) [31, 32]. However, two main considerations should be taken into account to achieve successful coating, namely, PVP concentration (excess or defect of PVP onto the particles surface can lead to inhomogeneous coating, meaning that optimization of washing is often required) and PVP molecular mass (the choice of an appropriate molecular mass compared to particle size is crucial to avoid inhomogeneous coatings or aggregation, *see* Fig. 3).

Although plenty of examples have been described in the literature, we focus here on the silica coating of Au nanostars, which have become very popular because of their extraordinary morphology and associated properties [33].

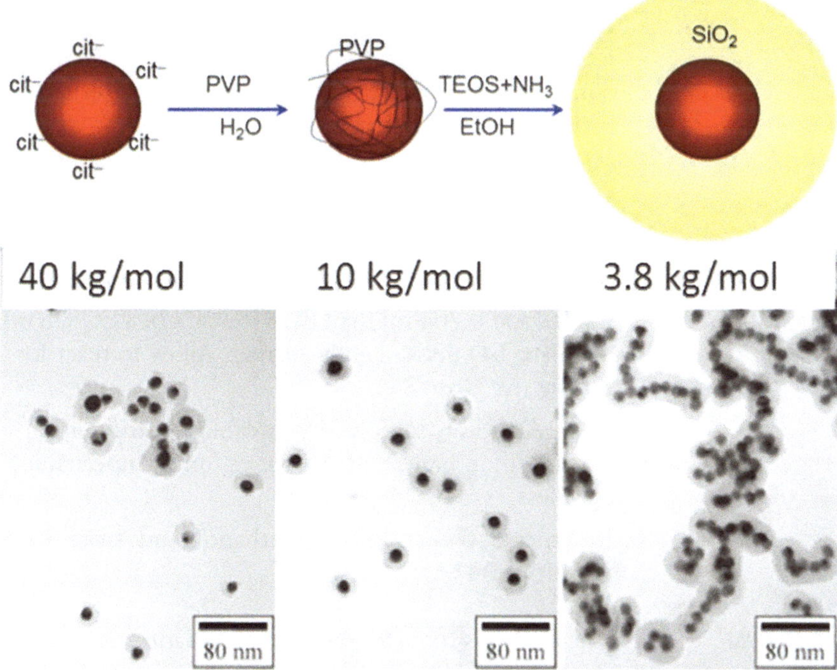

Fig. 3 *Top*: Scheme of the silica coating of citrate-stabilized Au nanoparticles using PVP. *Bottom*: TEM images of silica-coated Au nanospheres obtained using PVP with different molecular weights, as indicated. Adapted with permission from ref. 28. Copyright (2003) American Chemical Society

Synthesis of Au nanostars (*see* **Note 26**) [34, 35]

1. Prepare citrate-stabilized 15 nm Au nanospheres using the protocol described above (*see* Subheading 3.1).
2. Add 5 mL of an aqueous solution containing 0.1 g of PVP, previously sonicated for 15 min, to 50 mL of the as-prepared Au nanoparticles (C_{Au} = 0.5 mM). Allow to react overnight.
3. Centrifuge the colloid at 5,150 × *g* for 90 min. Collect the supernatant and centrifuge again at 5,150 × *g* until it becomes colorless.
4. Redisperse the resulting PVP protected particles in 5 mL of ethanol.
5. Calculate the Au concentration (*see* **Notes 27** and **28**)
6. Add 75 μL of 0.1 M $HAuCl_4$ aqueous solution to 15 mL of 12.5 mM PVP solution in DMF.
7. After ca. 2 min (*see* **Note 29**), add 171 μL of 2.1 mM Au seed solution (in ethanol, *see* **Note 30**) under continuous stirring. Allow to react for 30 min.
8. Check the LSPR band by UV–Visible–NIR absorption spectroscopy (*see* **Note 31**).

Silica coating

1. Centrifuge threefold the Au nanostars dispersion in 10 mL tubes for 60 min at 2,128 × *g* (the first washing should be performed at 3,179 × *g*) to remove excess PVP. Extract the supernatant and redisperse in 2 mL isopropanol (ethanol, however, is preferred during the washings).
2. Calculate the concentration of Au nanostars (*see* **Notes 28** and **32**).
3. Prepare 8.25 mL of 0.85 mM Au nanostars (*see* **Note 33**).
4. Add 1.44 mL of Milli-Q water, 0.106 mL of NH_4OH (30 % in water), and 0.204 mL of TEOS (5 vol.% of TEOS in isopropanol, *see* **Note 14**) under gentle stirring. Allow to react for 2 h under stirring (*see* **Note 34**).
5. Wash Au@SiO_2 samples by twofold centrifugation and redispersion in ethanol, at 1,681 × *g* until supernatant becomes colorless.
6. Redisperse the colloids in ethanol and store in the fridge (*see* **Note 34**).

3.4 Polyelectrolyte-Mediated Silica Coating

This method was reported by Pastoriza-Santos et al. and published in *Chemistry of Materials* in 2006 [36], currently counting 150 citations, according to Web of Knowledge©. The method was devised for the deposition of dense silica layers on CTAB-capped gold nanorods and comprises an initial surface modification step based on the polyelectrolyte layer-by-layer (LBL) assembly

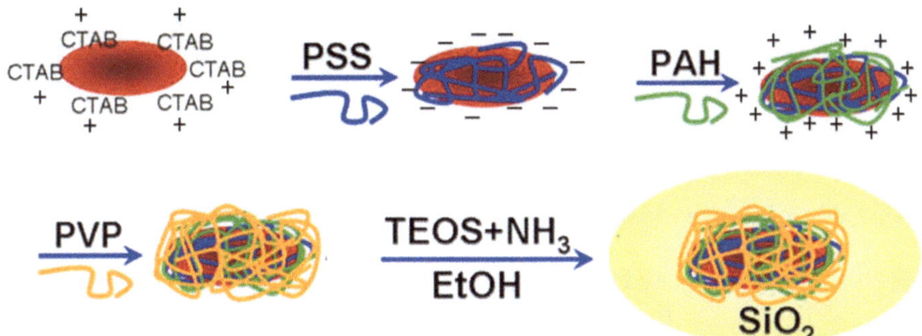

Fig. 4 Scheme of the silica coating of Au nanoparticles through the layer-by-layer approach [36]. The particles are first wrapped with polyelectrolytes of different charge and finally with PVP, which then allows Stöber silica coating

technique [37, 38], followed by Stobër silica coating (*see* scheme in Fig. 4) [23]. Although this protocol may become rather laborious and time consuming because of the different coating steps required, the fact that it is based on the LBL protocol makes it also rather versatile. The LBL technique is a powerful tool that allows us to modify the surface chemistry of almost any kind of hydrophilic nanoparticles. Therefore, the coating strategy may be readily extended to other types of nanoparticles. Other advantages are the tight control on the silica shell thickness and the stability of the shells in various dispersing media.

Next, we describe the protocol for application on CTAB-stabilized Au nanorods (ca. 13 × 50 nm).

Au nanorod synthesis (*see* **Note 35**) [39]

1. Prepare 2–3 nm Au seeds using the above detailed protocol (*see* Subheading 3.2, Au nanorod synthesis (in acidic medium), **step 1**).
2. Add 2.5 mL of 0.05 M $HAuCl_4$ to 250 mL of 0.1 M CTAB. Homogenize the mixture by shaking (*see* **Note 20**).
3. Add 1.87 mL of 0.1 M ascorbic acid (*see* **Note 21**), 2 mL of freshly prepared 5.0 mM silver nitrate and 3.0 mL of seed solution (after each addition the solution should be gently shaken).
4. Introduce the resulting solution in a thermostatic bath at 27 °C for 30 min (*see* **Note 36**).
5. Centrifuge at 6,726 × *g* for 30 min for partial removal of excess CTAB. Discard the supernatant and redisperse the precipitate in 125 mL of Milli-Q water (*see* **Note 37**).

LBL surface modification

1. Add dropwise 5 mL of a 0.5 mM Au nanorod hydrosol under vigorous stirring to 5 mL of a 2 g/L aqueous solution of poly(styrene sulfonate) sodium salt (PSS, MW 14,900), which also contains 6 mM NaCl (*see* **Note 38**). Stir the mixture for 3 h.

2. Centrifuge the resulting PSS-coated nanorods twice at 2,128 × *g* for 30 min to remove excess polyelectrolyte and redisperse in 5 mL of Milli-Q water.
3. Measure the Zeta Potential by Photon Correlation Spectroscopy and the UV–Vis–NIR absorption spectrum (*see* **Note 39**).
4. Add dropwise the nanorod solution under mild stirring to 5 mL of a 2 g/L poly(allylamine hydrochloride) (PAH, MW 15,000) aqueous solution (6 mM NaCl) (*see* **Note 38**). Stir the mixture for 3 h.
5. Centrifuge the resulting PAH-coated nanorods twice at 2,128 × *g* for 30 min to remove excess polyelectrolyte and redisperse in 5 mL of Milli-Q water.
6. Measure the Zeta Potential by Photon Correlation Spectroscopy and the UV–Vis–NIR absorption spectrum (*see* **Note 40**).
7. Mix 5 mL of PAH-coated gold nanorods with 5 mL of an aqueous solution of PVP (MW 10,000, 4 g/L) and stir overnight (*see* **Note 38**).
8. Centrifuge the mixture at 2,128 × *g* for 30 min, discard the clear supernatant, and redisperse the precipitate in 0.2 mL of water (*see* **Note 41**).
9. Add 1 mL of isopropanol (*see* **Note 41**).
10. Calculate the concentration of Au nanorods (*see* **Notes 28** and **42**).

Stöber silica coating [23]

1. Add 1.2 mL of PVP-coated gold nanorod dispersion under vigorous stirring to 0.46 mL of water.
2. Add 1.43 mL of ammonia solution in isopropanol (3.84 vol.%) under vigorous stirring and 0.40 mL of a TEOS solution in isopropanol (0.97 vol.%, *see* **Note 14**) under gentle stirring. Allow to react for 2 h under stirring (*see* **Note 43**).
3. Wash Au@SiO_2 samples by twofold centrifugation and redispersion in ethanol, at 1,681 × *g* until the supernatant becomes colorless.
4. Redisperse the colloids in ethanol and store in the fridge.

3.5 PEG-Mediated Silica Coating

This method was developed by Fernández-López et al. in 2009 and published in *Langmuir* [40]. Since its publication this article has been cited over 35 times according to Web of Knowledge©. The method relies on the high colloidal stability of polyethylene glycol (PEG) capped gold nanoparticles, both in water and ethanol. Thus, a thiolated PEG (O-[2-(3-mercaptopropionylamino) ethyl] O′-methyl-poly(ethylene glycol) (m-PEG-SH)) is used to replace the original surface capping agents, such as citrate or CTAB, and to subsequently facilitate the growth of uniform silica

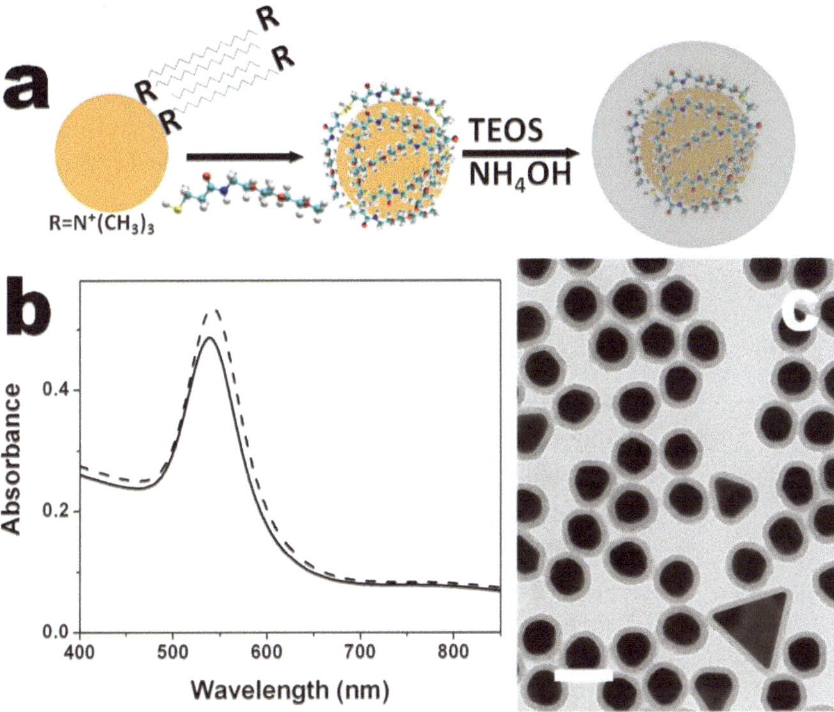

Fig. 5 (**a**) Scheme of the silica coating of CTAB-stabilized Au nanoparticles using thiolated PEG. (**b**) Vis–NIR absorption spectra of the Au nanospheres before (*solid line*) and after (*dashed line*) silica-coating. (**c**) TEM image of the silica-coated Au nanospheres. The scale bar represents 100 nm

shells around the particles via Stobër method [23] (*see* Fig. 5a). This method allows an accurate control of the silica shell thickness and the obtained coatings are highly uniform, even when very thin shells (few nanometers) are grown. It should be noted however that, although the silica shell is not mesoporous in this case, it can readily dissolve in basic aqueous solutions, which can be prevented through a mild thermal treatment [26].

As a representative example, we describe here the specific protocol for coating Au-CTAB spheres (ca. 60 nm), which can be easily adapted to other CTAB-stabilized Au nanoparticles.

Synthesis of 60 nm Au nanospheres [41]

1. Prepare 15 nm citrate-stabilized Au nanospheres (using the protocol described above (*see* Subheading 3.1).
2. Warm 500 mL of a solution containing $HAuCl_4$ (0.5 mM) and CTAB (0.015 M) up to 35 °C.
3. Add 5 mL of 0.1 M ascorbic (*see* **Note 21**) and finally 7.94 mL of Au seeds (ca. 0.5 mM) under mild stirring (*see* **Note 45**). Allow to react for 30 min.
4. Measure the absorption spectrum by UV–Visible spectroscopy (*see* **Note 46**) (Fig. 5b).

5. Centrifuge the as-synthesized particles at 1,681 × *g* for 20 min and after careful removal of the supernatant, redisperse the colloids in 10 mL of a 0.1 M CTAB aqueous solution.
6. Transfer into a glass test tube (10 mL) and introduce it in a beaker containing hot water (50 °C) for 15 min. Keep it in an upright position overnight.
7. Separate the supernatant, containing the Au spheres, from the precipitate containing nanorods and plates. Redisperse the Au spheres in 500 mL of 0.1 M CTAB.

mPEG-SH functionalization, ethanol transfer and silica coating

1. Centrifuge 20 mL of the Au colloid at 657 × *g* for 20 min and redisperse in 20 mL of water (*see* **Note 47**).
2. Centrifuge for 10 min at 657 × *g* and then centrifuge the supernatant for another 15 min at 420 × *g*. Collect all precipitates and redisperse in water to a final volume of 10 mL (*see* **Note 48**).
3. Add dropwise 0.3 mL of 0.25 mM m-PEG-SH (Mw 5000, *see* **Notes 38** and **49**) aqueous solution, under vigorous stirring. Allow to react for 30 min.
4. Centrifuge the resulting PEG-modified particles twice (657 × *g*, 20 min) to remove excess PEG and redisperse in 2 mL of ethanol.
5. Calculate the concentration of PEG-modified Au nanoparticles (*see* **Notes 28** and **50**).
6. Prepare 7.8 mL of 0.64 mM Au spheres (*see* **Note 51**).
7. Add 1.9 mL of Milli-Q water, 0.125 mL of NH_4OH (32 % in water), and 0.177 mL of TEOS (1 vol.% of TEOS in isopropanol, *see* **Note 14**) under gentle stirring. Allow to react for 2 h under stirring (*see* **Note 52**) (Fig. 6.5).
8. Wash Au@SiO_2 samples by twofold centrifugation and redispersion in ethanol, at 1,681 × *g* until the supernatant becomes colorless.
9. Redisperse the colloids in ethanol and store in the fridge (*see* Fig. 5c).

4 Notes

1. Glassware cleaning with aqua regia. This is an important step that should be carried out before starting the synthesis, since residues in the glassware may interfere with the synthesis. All glassware should be first washed with detergent and then with aqua regia, followed by extensive rinsing with water and Milli-Q water. Preparation of aqua regia: In a glass beaker, mix hydrochloric acid (HCl) and nitric acid (HNO_3) in a volume

ratio 3:1. The preparation should take place inside a fume hood, since the liquid and its vapors are extremely corrosive and highly oxidizing. Therefore, the use of appropriate gloves, lab coat, and chemical splash goggles is mandatory.

2. The use of mercaptopropyltrimethoxysilane (MPS) has also been reported by several authors.
3. Citrate stabilized Au spheres (ca. 0.5 mM, gold atomic concentration) were synthesized according to Turkevich's method [24].
4. It is important to select appropriate magnetic stirring conditions that allow uniform boiling.
5. This addition is probably the most critical step of this synthesis. It should be carried out as fast as possible and it is also convenient to warm up the citrate solution prior to addition, to avoid detention of the boiling. The citrate ions reduce Au(III) and Au nanoparticles are formed. Visually striking color changes are observed during the reaction: the pale yellow $HAuCl_4$ solution turns colorless upon citrate addition, suddenly changing to a very dark blue (even black) and slowly turning deep red.
6. It is advisable to wait 24 h before silica coating is started.
7. Alkali metals, Mg, Al, and noble metals such as Cu, Ag, and Au present localized surface plasmon resonance (LSPR) bands. The collective response of conduction band electrons to light produces a characteristic resonance band that lies in the UV–Visible–NIR range for Cu, Au, and Ag. (For more information *see* refs. [3, 5]).
8. The UV–Vis spectrum should present an LSPR band around 518–520 nm (*see* Fig. 6.1b). The particle size should be around 15–18 nm in diameter (see a typical transmission electron microscopy (TEM) image in Fig. 1c).
9. The first step comprises priming the gold surface with APS. Molecules with amine groups, as well as with mercapto groups complex strongly to Au metal.
10. No changes in the LSPR band should be observed (*see* Fig. 1b).
11. After 7 days the Au particles should be coated with a ca. 3–4 nm thick silica shell. A 5 nm LSPR red-shift is typically observed (*see* Fig. 6.1b and **Note 12**).
12. The growth of the silica shell produces the increase of the effective refractive index of the particles surrounding medium and this originates the observed plasmon band shift toward longer wavelengths [3].
13. Ethanol addition produces silicate condensation and a further 3 nm LSPR shift, (*see* Fig. 1b and **Note 12**).
14. Better results are obtained when TEOS is diluted in alcohol before the addition.

15. The total amount to be added can be easily calculated taking into account the Au nanoparticle size (ca. 17 nm for this case) and the final particle size. TEOS total amount (V_{TEOS}) is calculated using the formula $V_{TEOS} = \frac{\bar{V}_{TEOS}\bar{V}_{Au}}{\bar{V}_{SiO_2}} C_{Au} \left[\left(\frac{R_{tot}}{R_{Au}} \right)^3 - 1 \right]$ where R_{tot}, R_{Au}, C_{Au} and $\bar{V}$ are final radius, Au radius, and molar volume, respectively. Molar volume is calculated applying the relation $\bar{V} = \frac{M}{\rho}$, M being the molar mass and ρ the density.
16. A typical TEM image of Au@SiO_2 nanoparticles (after 4 TEOS additions) is shown in Fig. 1d. The silica coating process can be monitored by UV–Vis spectroscopy, since the growth of the shell leads to a gradual red-shift of the LSPR band (*see* Fig. 1b and **Note 12**).
17. This synthesis is based on a seeded growth method reported by Liu and Guyot-Sionnest [27], with some modifications. This method gives rise to Au nanorods with aspect ratio (length divided by the width) higher than 4. This process is scalable up to 250 mL but the gold nanorods synthesis is very sensitive to the quality of the preformed seeds. It is recommended to test the synthesis in 10 mL preparations prior to scale-up. The solvent in all the preparations is Milli-Q water and all CTAB based solutions were stored at 27–30 °C to avoid surfactant crystallization.
18. Sodium borohydride in water decomposes spontaneously into hydrogen gas and borates, therefore the solution should be prepared just before using and stored in an ice/water bath.
19. Since borohydride is a strong reducing agent, the reaction is completed in 10 min and the seeds are ready for further use within 24 h after preparation. The average particle size measured from TEM should be around 2–3 nm and the solution typically displays a yellow-brown color due to the small particle size (UV–Vis spectrum should not display a well-defined band or even a shoulder, which would both indicate the presence of larger particles, *see* Fig. 2b).
20. Magnetic stirring is not appropriate for the growth of Au nanorods, therefore the homogenization is obtained by hand shaking. The solution becomes deep yellow as a consequence of the formation of a complex between $HAuCl_4$ and CTAB.
21. The solution becomes colorless.
22. The final gold nanorod dimensions should be ca. 14×70 nm (aspect ratio 5) and the UV–Vis–NIR spectrum should display two LSPR bands, a very intense one located around 850 nm and a second one, much less intense, at ca. 525 nm (*see* Fig. 2b).

23. The CTAB concentration present in the Au nanorods solution is critical during silica coating, which requires careful washing of the nanoparticles to remove part of the CTAB in excess.
24. 40–50 nm red-shift of the main LSPR band should be observed. If the shift is not observed then the reaction should be allowed to proceed for 3 more hours.
25. A TEM image of the silica coated particles is shown in Fig. 2c, where the mesoporous nature of the coating is evidenced.
26. This synthesis is based on the N,N-dimethylformamide (DMF) reduction method, developed by Pastoriza-Santos, Liz-Marzán et al. [34, 35] Au nanostars are formed through the reduction of an Au salt in a solution of PVP (Mw 8000, from Alfa Aesar) in DMF, onto preexisting gold nanoparticles. Although DMF is the main reducing agent, PVP was also found to contribute to Au salt reduction. In our example, standard Au-citrate nanoparticles were used as seeds.
27. Au concentration should be ca. 4.2 mM (*see* **Note 28**).
28. Gold concentration is often measured from the value of absorbance at 400 nm, taking into account that absorbance is proportional to atomic Au concentration (Lambert-Beer Law). An absorbance of 1.2 corresponds to 0.5 mM concentration (cell path length of 1 cm).
29. It would be convenient to monitor the evolution (damping) of the absorption band characteristic of Au^{3+} at 325 nm by UV–Vis–NIR spectroscopy. DMF progressively reduces Au^{3+} to Au^{+}, leading to complete damping of the extinction band within 2 min (although the damping time could be dependent of the experimental conditions).
30. 2.1 mM Au seeds is obtained from a twofold dilution of the 4.2 mM Au colloids.
31. The UV–Vis–NIR spectrum should display two LSPR bands, one very intense around 800 nm and a second one, less intense, at ca. 550 nm, characteristic of tip and core plasmon modes, respectively. The overall particle diameter (including the tips) should be around 60–70 nm.
32. Au concentration should be ca. 3.75 mM (*see* **Note 28**).
33. Sample obtained by diluting the 3.75 mM Au nanostars colloids.
34. In 2 h particles should present a thin silica layer (ca. 5 nm). The shell can be further grown upon controlled TEOS addition. As in all other procedures, silica coating can be monitored by UV–Vis spectroscopy, following LSPR red-shift (*see* **Note 12**). *See* Fig. 1b as an example.
35. This synthesis is based on a seeded growth method developed by Nikoobakht and El-Sayed [39], with some modifications. The solvent was Milli-Q water in all preparations and all CTAB

based solutions were maintained at 30 °C to avoid surfactant crystallization. The synthesis of Au nanorods in non-acid medium gives rise to Au nanorods with aspect ratio below 4. This process is scalable up to 250 mL but gold nanorods synthesis is very sensitive to the quality of the preformed seeds. It is recommended to test the synthesis in 10 mL preparations prior to scale-up.

36. The gold nanorod dimensions should be ca. 13×52 nm (aspect ratio 4) and the corresponding UV–Vis–NIR spectrum should display two LSPR bands, one very intense centered around 750 nm and a second one, much less intense, at ca. 525 nm.
37. The zeta potential of the particles should be around +20 mV.
38. PSS, PAH, PVP, and PEG aqueous solutions should be sonicated for 30 min prior to use.
39. The zeta potential should be ca. −40 mV, and the Vis–NIR spectrum should not display significant changes.
40. PAH-coated particles display a zeta potential of ca. +45 mV, and again significant changes in the optical spectrum are not expected.
41. PVP-coated particles display a zeta potential of ca. +20 mV. Since PVP degrades in water, it is convenient to store the stock particle solution in alcohol (not in pure alcohol but in an alcohol–water mixture). Redispersion of those nanoparticles in pure alcohol (usually ethanol or isopropanol) could easily lead to particles aggregation (seen by color change).
42. The gold concentration at this stage should be around 2.47 mM (*see* **Note 28**).
43. The gold nanorods should be surrounded by a 12 nm thick silica shell. The shell can be further grown upon controlled TEOS addition. As in all other procedures, silica coating can be monitored by UV–Vis spectroscopy, following LSPR red-shift (*see* **Note 12**). *See* Fig. 1b as an example.
44. This synthesis is based on a seeded growth method reported by Rodríguez-Fernández et al. [41], with slight modifications.
45. Au nanospheres with sizes ranging from 15 to 300 nm can be readily prepared using this protocol just by varying the amount of seeds. The amount of seeds required to obtain nanoparticles with radius R_{Part} might be easily calculated using the following formula $R_{\mathrm{Part}} = R_{\mathrm{seed}} \left(\frac{C_{\mathrm{Ausalt}} + C_{\mathrm{Auseed}}}{C_{\mathrm{Auseed}}} \right)^{1/3}$ where C_{Ausalt} is the gold salt concentration and R_{seed} and C_{Auseed} are the radius and concentration of the seeds, respectively. To synthesize large particles ($\geq$90 nm) it is highly recommended to carry out the growth process in two steps (grow from 15 to 60 nm and from 60 to the final size).

46. Growth of the 15 nm seeds produces a shift of the plasmon band from 520 nm (LSPR of the seeds) to 536 nm (*see* Fig. 5b) and the absorbance at 400 nm should ca. 1.2 (1 cm optical path length). Apart from Au spheres, particles with other morphologies (rods and plates mainly) are obtained, but the nonspherical particles can be efficiently removed through CTAB-assisted shape selective separation [42].
47. The CTAB concentration (C_{CTAB}), is around 15 mM.
48. At this stage C_{CTAB} ~1 mM and C_{Au} ~1 mM.
49. The amount of m-PEG-SH is that required to provide 4 molecules per nm^2 of Au nanoparticle surface. This is the minimum amount to obtain homogeneous coatings; no significant effects on the coating were observed when higher concentrations were used. In the case of non-purified particles the minimum amount was calculated considering 15 molecules per nm^2.
50. Au concentration should be ca. 4.2 mM (*see* **Note 28**).
51. Sample obtained by diluting the 4.2 mM Au colloids.
52. The silica shell thickness should be around 7–10 nm but can be further grown upon controlled TEOS addition. Figure 5c shows a typical TEM image of Au@silica particles coated using PEG-mediated strategy. The progress of silica coating can be followed by UV–Vis spectroscopy, since the growth of the shell leads to a red-shift of the LSPR band (*see* **Note 12**) (Fig. 5b).

Acknowledgments

This work has been funded by the Spanish Xunta de Galicia/FEDER (grant 10 PXIB 314 218 PR) and MINECO/FEDER (grant MAT2010-15374). LML-M acknowledges the ERC for an Advanced grant (NANODIRECT, grant number CP-FP 213948-2). The authors would like to thank Sara Abalde-Cela, Cristina Fernández-López, Sergio Rodal and Jorge Pérez-Juste, Ana M. Sánchez-Iglesias for their contributions and comments.

References

1. See special issue devoted to gold published in Chem. Soc. Rev. (2008, volume 37)
2. Kreibig U, Vollmer M (1996) Optical properties of metal clusters. Springer, Berlin
3. Mulvaney P (1996) Surface plasmon spectroscopy of nanosized metal particles. Langmuir 12:788–800
4. Kelly KL, Coronado E, Zhao LL et al (2003) The optical properties of metal nanoparticles: the influence of size, shape, and dielectric environment. J Phys Chem B 107:668–677
5. Liz-Marzán LM (2006) Tailoring surface plasmons through the morphology and assembly of metal nanoparticles. Langmuir 22:32–41
6. Dreaden EC, Alkilany AM, Huang X et al (2012) The golden age: gold nanoparticles for biomedicine. Chem Soc Rev 41:2740–2779

7. Giljohann DA, Seferos DS, Daniel WL et al (2010) Gold nanoparticles for biology and medicine. Angew Chem Int Ed 49:3280–3294
8. Rosi NL, Mirkin CA (2005) Nanostructures in biodiagnostics. Chem Rev 105:1547–1562
9. Abalde-Cela S, Aldeanueva-Potel P, Mateo-Mateo C et al (2010) SERS biomedical applications of plasmonic colloidal particles. J Royal Soc Interface 7:S435–S450
10. Sepúlveda B, Angelomé PC, Lechuga LM et al (2009) LSPR-based nanobiosensors. Nano Today 4(3):244–251
11. Dykman L, Khlebtsov N (2012) Gold nanoparticles in biomedical applications: recent advances and perspectives. Chem Soc Rev 41:2256–2282
12. Ghosh P, Han G, De M et al (2008) Gold nanoparticles in delivery applications. Adv Drug Deliv Rev 60:1307–1315
13. Halas N (2008) Nanoscience under Glass: The versatile chemistry of silica nanostructures. ACS Nano 2:179–183
14. Liu SH, Han MY (2005) Synthesis, functionalization, and bioconjugation of monodisperse, silica-coated gold nanoparticles: robust bioprobes. Adv Funct Mater 15:961–967
15. Luo T, Huang P, Gao G et al (2011) Mesoporous silica-coated gold nanorods with embedded indocyanine green for dual mode X-ray CT and NIR fluorescence imaging. Opt Express 19:17030–17039
16. Rodríguez-Lorenzo L, Krpetic Z, Barbosa S et al (2011) Intracellular mapping with SERS-encoded gold nanostars. Integr Biol 3: 922–926
17. Zavaleta CL, Smith BR, Walton I et al (2009) Multiplexed imaging of surface enhanced Raman scattering nanotags in living mice using noninvasive Raman spectroscopy. Proc Natl Acad Sci USA 106:13511–13516
18. Yang X, Liu X, Liu Z et al (2012) Near-Infrared light-triggered, targeted drug delivery to cancer cells by aptamer gated nanovehicles. Adv Mater 24:2890–2895
19. Lal S, Clare SE, Halas NJ (2008) Nanoshell-enabled photothermal therapy: impending clinical impact. Acc Chem Res 41:1842–1851
20. Fales AM, Yuan H, Vo-Dinh T (2011) Silica-coated gold nanostars for combined surface-enhanced Raman scattering (SERS) detection and singlet-oxygen generation: A potential nanoplatform for theranostics. Langmuir 27: 12186–12190
21. Zhang Z, Wang L, Wang J et al (2012) Theranostics: mesoporous silica-coated gold nanorods as a light-mediated multifunctional theranostic platform for cancer treatment. Adv Mater 24:1418–1423
22. Liz-Marzán LM, Giersig M, Mulvaney P (1996) Synthesis of nanosized gold–silica core–shell particles. Langmuir 12:4329–4335
23. Stöber W, Fink A, Bohn E (1968) Controlled growth of monodisperse silica spheres in the micron size range. J Colloid Interface Sci 6:62–69
24. Enüstün BV, Turkevich J (1963) Coagulation of colloidal gold. J Am Chem Soc 85:3317–3328
25. Gorelikov I, Matsuura N (2008) Single-step coating of mesoporous silica on cetyltrimethyl ammonium bromide-capped nanoparticles. Nano Lett 8:369–373
26. Wong YJ, Zhu L, Teo WS et al (2011) Revisiting the Stöber method: Inhomogeneity in silica shells. J Am Chem Soc 133:11422–11425
27. Liu M, Guyot-Sionnest P (2005) Mechanism of silver(I)-assisted growth of gold nanorods and bipyramids. J Phys Chem B 109:22192–22200
28. Graf C, Vossen DLJ, Imhof A et al (2003) A general method to coat colloidal particles with silica. Langmuir 19:6693–6700
29. Wiley B, Sun Y, Mayers B et al (2005) Shape-controlled synthesis of metal nanostructures: the case of silver. Chem Eur J 11:454–463
30. Pastoriza-Santos I, Liz-Marzán LM (2009) N, N-dimethylformamide as a reaction medium for metal nanoparticle synthesis. Adv Funct Mater 19:679–688
31. Yin YD, Lu Y, Sun YG et al (2002) Silver Nanowires can be directly coated with amorphous silica to generate well-controlled coaxial nanocables of silver/silica. Nano Lett 2:427–430
32. Pastoriza-Santos I, Sánchez-Iglesias A, García de Abajo FJ et al (2007) Environmental optical sensitivity of gold nanodecahedra. Adv Funct Mater 17:1443–1450
33. Guerrero-Martínez A, Barbosa S, Pastoriza-Santos I et al (2011) Nanostars shine bright for you. Colloidal synthesis, properties and applications of branched metallic nanoparticles. Curr Op Colloid Interface Sci 16:118–127
34. Kumar PS, Pastoriza-Santos I, Rodríguez-González B et al (2008) High-yield synthesis and optical response of gold nanostars. Nanotechnology 19(1):015606
35. Barbosa S, Agrawal A, Pastoriza-Santos I et al (2010) Tuning size and sensing properties in colloidal gold nanostars. Langmuir 26: 14943–14950
36. Pastoriza-Santos I, Pérez-Juste J, Liz-Marzán LM (2006) Silica-coating and hydrophobation

of CTAB- stabilized gold nanorods. Chem Mater 18:2465–2467

37. Schneider G, Decher G (2004) From functional core/shell nanoparticles prepared via layer-by-layer deposition to empty nanospheres. Nano Lett 4:1833–1839
38. Gittins DI, Caruso F (2001) Tailoring the polyelectrolyte coating of metal nanoparticles. J Phys Chem B 105:6846–6852
39. Nikoobakht B, El-Sayed MA (2003) Preparation and growth mechanism of gold nanorods (NRs) using seed-mediated growth method. Chem Mater 15:1957–1962
40. Fernández-López C, Mateo-Mateo C, Alvarez-Puebla RA et al (2009) Highly controlled silica Coating of PEG-capped metal nanoparticles and preparation of SERS-encoded particles. Langmuir 25:13894–13899
41. Rodríguez-Fernández J, Pérez-Juste J, García de Abajo FJ et al (2006) Seeded growth of submicron Au colloids with quadrupole plasmon resonance modes. Langmuir 22: 7007–7010
42. Jana NR (2003) Nanorod shape separation using surfactant assisted self-assembly. Chem Commun (15) 1950–1951

Chapter 7

Surface Modifications by Polymers for Biomolecule Conjugation

Laura Sola, Marina Cretich, Francesco Damin, and Marcella Chiari

Abstract

Polymeric coatings, usually referred as tridimensional chemistries, provide homogenous surface derivatization methods presenting a high reactive group concentration and resulting in an increased binding capacity of targets. Furthermore, they act as linkers distributing the bound probe also in the axial position, thus causing a faster reaction with the target involved in biomolecular recognition and can be engineered to custom tailor their properties for specific applications.

Most approaches which aim at attaching polymers to a surface use a system where the polymer carries an "anchor" group either as an end group or in a side chain. This anchor group can reacts with appropriate sites at the substrate surface, thus yielding surface-attached monolayers of polymer molecules (termed "grafting to"). Another technique is to carry out a polymerization reaction in the presence of a substrate onto which monomers had been attached leading to the so called "grafting from" approach.

In this chapter, protocols to functionalize glass and silicon surfaces by "grafting to" as well as by "grafting-from" approach are shown using copolymers made of *N*,*N*-dimethylacrylamide (DMA) or Glycidyl methacrylate (GMA) as the polymer backbone, *N*-acryloyloxysuccinimide (NAS) as reactive group, and 3-(trimethoxysilyl)propyl methacrylate (MAPS) or 3-mercaptopropyl trimethoxy silane (MPS) as anchoring groups.

Key words Biomolecule conjugation, Polymeric coatings, Grafting to, Grafting from, Silicon, Glass

1 Introduction

Recent progresses in micro fabrication techniques and microarray technologies have led to the development of miniaturized and fully integrated solid phase analytical devices that allow to perform entire experiments, studies of complex cellular processes, high-throughput screening, and parallel diagnostic detection. They imply a scaling down of the entire analytical process (time, sample and reagent volumes, costs, etc.) while maintaining very high sensitivity. Considering their tiny dimensions, the aforementioned surfaces must be perfectly designed and controlled in order to maximize probe immobilization and target binding efficiency, to

Paolo Bergese and Kimberly Hamad-Schifferli (eds.), *Nanomaterial Interfaces in Biology: Methods and Protocols*, Methods in Molecular Biology, vol. 1025, DOI 10.1007/978-1-62703-462-3_7, © Springer Science+Business Media New York 2013

reduce background noise, and to prevent nonspecific molecular interaction.

When the modification of a suitable substrate must be performed to promote covalent attachment of oligonucleotides and proteins at specific locations, the control of surface chemical–physical characteristics is of remarkable importance in order to maximize probe density immobilization and to maintain their native and functional conformation. Besides, in order to obtain accurate analysis and an optimization of *signal-to-noise* ratio, the surface chemistry should minimize hydrophobic interactions, which are the main source of biomolecule nonspecific bonding.

Therefore, there are several requirements that must be taken into account to chemically change a surface:

- The chemically modified surface must be inherently inert and resists nonspecific adsorption
- The surface must contain functional groups for the facile immobilization of molecules of interest
- Bonding between a biomolecule and a solid surface must be strong enough to retain the molecule on the surface, but also sufficiently nonintrusive to have minimal effect on the delicate 3D structure
- The linking chemistry must allow the control of biomolecule orientation
- The local chemical environment must benefit the immobilized molecules to retain their native conformation

Polymeric coatings, usually referred as tridimensional chemistries, provide a homogenous surface derivatization presenting a high reactive group concentration and resulting, depending on the circumstances, in an increased binding capacity of targets or to its suppression. Furthermore, they act as linkers distributing the bound probe also in the axial position, thus causing a faster reaction with the target involved in biomolecular recognition. Additionally, 3D scaffolds can be engineered to custom tailor their properties for specific applications.

Most approaches which aim at attaching polymers to a surface use a system where the polymer carries an "anchor" group either as an end group or in a side chain. This anchor group can reacts with appropriate sites at the substrate surface, thus yielding surface-attached monolayers of polymer molecules (termed "grafting to") (Fig. 1) [1–3]. Another straightforward technique is to carry out a polymerization reaction in the presence of a substrate onto which monomers had been attached [1, 4, 5].

During the polymerization reaction, the surface-anchored monomers are incorporated into growing polymer chains (Fig. 1). Although "grafting to" reactions are easy to perform, it should be noted that several complication arise from the use of this strategy.

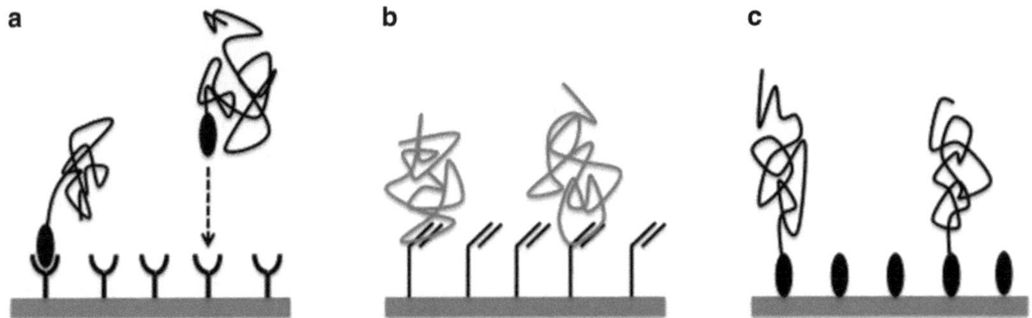

Fig. 1 Schematic illustration of different processes used for the attachment of polymers to surfaces: (**a**) "grafting to"; (**b**) grafting via incorporation of surface-bound monomeric units; (**c**) "grafting from/surface-initiated polymerization"

For example, in the case of functional polymers, the anchor groups present in the polymer chain and the sites exposed onto the surface must not compete and/or react with the functionalities. Another complication inherent to "grafting to" processes is an intrinsic limitation of the film thickness. With increasing coverage of the surface with attached chains, the polymer concentration at the interface quickly becomes larger than the concentration of polymers in solution. Additional chains, which are to become attached to the surface, must diffuse against this concentration gradient that ever increases with increasing grafting density of the attached polymer. Moreover, in order to accommodate further chains, these must change from coil conformation in solution to a stretched ("brush-like") conformation at the surface. Hence, the higher the graft density of the chains at the surface, the stronger will be the entropy penalty, and this rapidly precludes the attachment of further chains.

The term "polymer brush" refers to a system in which chains of polymer molecules are attached by one end to a surface. In spite of the fact that the "grafting to" method provides polymers attached in such fashion, due to the problems discussed above, it is not able to produce dense films in which the polymer chains are crowded and are stretched away from the surface. From the stretching of the polymer chains perpendicular to the surface, several new physical phenomena arise. Examples are ultralow friction surfaces [6] in which materials coated with surface-attached polymer chains do not become wetted by free polymer, even if the surface-attached and the free chains are chemically identical. Another characteristic of concentrated polymer brushes is noted to be size-exclusion effect [7], in which the graft chains are highly extended and highly oriented so that large molecules, sufficiently large compared with the distance between the nearest neighbor graft points, are physically excluded from the entire brush layer. In addition, in the case of functional brush polymers, high densities of functional groups can be obtained at the surface of the substrate through moving from the strictly two-dimensional arrangement of these groups present in typical surfaces to a more three-dimensional situation. An example which illustrates such a behavior is the attachment of DNA probe molecules to surface-attached polymer chains, which can significantly enhance the sensitivity of a DNA-chip [8].

An effective method to prepare high-density brushes is the "grafting-from" method, that is, the graft polymerization starting with initiating sites fixed on the surface. In this technique, the addition of monomer to growing chain ends or to primary radicals is not strongly hindered by the already grafted chains in a good solvent condition. Therefore, this technique is more promising to produce a polymer film with a larger thickness and a higher graft density than the "grafting to" technique [9–11]. The polymerization process obtained by "grafting-from" approach is also defined as Surface-Initiated Polymerization (SIP).

In this chapter we will show protocols to functionalize glass and silicon surfaces by "grafting to" as well as by "grafting-from" approach using copolymers made of *N,N*-dimethylacrylamide (DMA) or glycidyl methacrylate (GMA) as the polymer backbone, *N*-acryloyloxysuccinimide (NAS) as reactive group, and 3-(trimethoxysilyl)propyl methacrylate (MAPS) or 3-mercaptopropyl trimethoxy silane (MPS) as anchoring groups.

2 Materials

2.1 Synthesis of Poly(DMA-co-NAS-co-MAPS)

1. *N,N*-Dimethylacrylamide (DMA) (Scheme 1).
2. 3-(trimethoxysilyl)propyl methacrylate (MAPS).
3. *N*-Acryloyloxysuccinimide (NAS).
4. Aluminum oxide.
5. Tetrahydrofuran (THF).
6. α, α′-Azoisobutyronitrile (AIBN).
7. Petroleum ether.

2.2 Derivatization of Substrates with Poly(DMA-co-NAS-co-MAPS)

1. HARRICK Plasma Cleaner, PDC-002 (Ithaca, NY, USA).
2. Ammonium sulfate.
3. poly(DMA-*co*-NAS-*co*-MAPS) synthesized as described in Subheading 3.1.
4. Coating solution: 1 % w/v of poly(DMA-*co*-NAS-*co*-MAPS) in an aqueous solution of 20 % ammonium sulfate.

2.3 Synthesis of Poly(glycidyl methacrylate) (PGMA)

1. Glycidyl methacrylate.
2. α, α′-Azoisobutyronitrile (AIBN).
3. Tetrahydrofuran (THF).

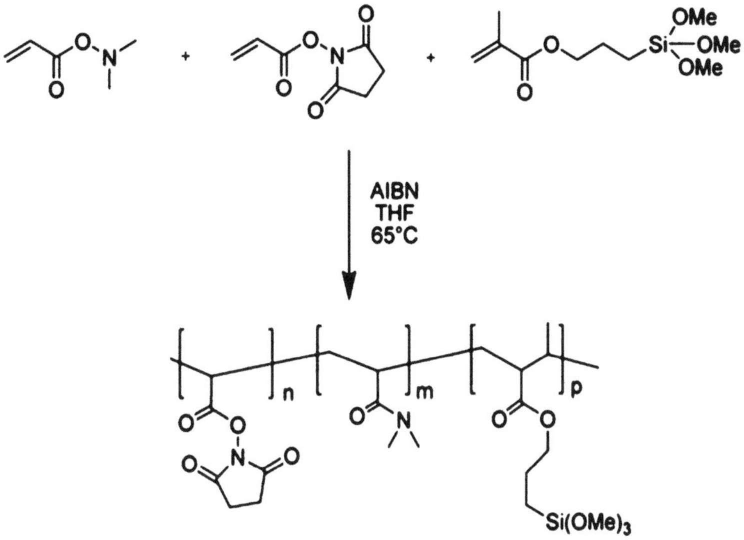

Scheme 1 Synthesis of poly(DMA-*co*-NAS-*co*-MAPS)

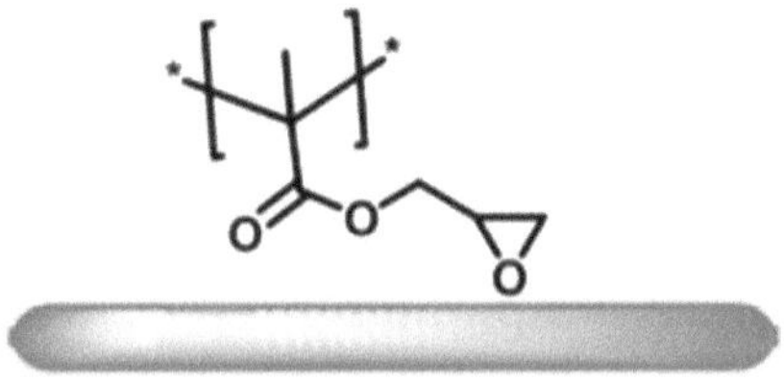

Scheme 2 Substrate derivatized by poly(glycidyl methacrylate) (PGMA)

2.4 Derivatization of Substrates with Poly(glycidyl methacrylate) (PGMA)

1. HARRICK Plasma Cleaner, PDC-002 (Ithaca, NY, USA) (Scheme 2).
2. PGMA synthesized as described in Subheading 3.3.
3. Dimethylformamide (DMF).

2.5 Derivatization of Substrates with Grafted Poly(DMA-co-NAS-co-MAPS) andGraftedPoly(DMA-co-NAS-co-MPS)

1. HARRICK Plasma Cleaner, PDC-002 (Ithaca, NY, USA).
2. *N*,*N*-Dimethylacrylamide (DMA).
3. *N*-Acryloyloxysuccinimide (NAS).
4. 3-(trimethoxysilyl)propyl methacrylate (MAPS).
5. 3-Mercaptopropyl trimethoxy silane (MPS).
6. Aluminum oxide.
7. α, α′-Azoisobutyronitrile (AIBN).
8. Toluene.
9. Tetrahydrofuran (THF).
10. *N*,*N*-Dimethylformamide (DMF).
11. IRGACURE 184 (Ciba Specialty Chemicals, Basel, Switzerland).

2.6 Derivatization of Substrates with Grafted Poly[DMA-b-(DMA-co-NAS)] via Free Radical Polymerization [12]

1. *N*,*N*-Dimethylacrylamide (DMA).
2. *N*-Acryloyloxysuccinimide (NAS).
3. (3-aminopropyl)-trimethoxysilane (APS).
4. 2-Bromopropionylbromide.
5. Phenylmagnesium bromide.
6. Triethylamine.
7. Carbon disulfide (CS_2).
8. Toluene.
9. Dichloromethane.
10. Tetrahydrofuran anhydrous (THF, 99 % inhibitor free).
11. Sodium sulfate anhydrous.
12. Octyl-trimethoxysilane (96 %, nOS).
13. α, α′-Azoisobutyronitrile (AIBN).
14. Aluminum oxide.
15. HARRICK Plasma Cleaner, PDC-002 (Ithaca, NY, USA).

2.7 Substrates

Glass, silicon and plastic substrates can be modified, including soda-lime or borosilicate glass, silicon slides, COC (Cyclic Olefin Copolymer), PET (Polyethylene terephthalate), PMMA (Poly(methyl methacrylate)), or PDMS (polydimethylsiloxane) chips. Substrates are typically 15×15 mm^2 square slides with a thickness from 1 to 5 mm. Microscope slide format is also frequently used.

3 Methods

The methods described in this section outline (a) synthesis of poly(DMA-*co*-NAS-*co*-MAPS), (b) coating of substrates with the poly(DMA-*co*-NAS-*co*-MAPS), (c) synthesis of PGMA, (d) coating of substrates with PGMA, (d) synthesis of grafted poly(DMA-*co*-NAS-*co*-MAPS), (e) synthesis of grafted poly(DMA-*co*-NAS-*co*-MPS), and (f) synthesis of grafted poly [DMA-*b*-(DMA-*co*-NAS)]

3.1 Synthesis of Poly(DMA-co-NAS-co-MAPS)

The copolymer made of DMA (with 97 % molar percentage), NAS (2 % molar percentage). and 3-(trimethoxysilyl)propyl methacrylate (MAPS, 1 % molar percentage) is synthesized by free radical copolymerization and used as a functional coating for biomolecule conjugation to the substrate.

1. In a 250 mL three neck round bottom flask, equipped with condenser, magnetic stirring, and nitrogen connection, dissolve *N*,*N*-dimethylacrylamide (DMA, 4.00 g, 4.00×10^{-2} mol; *see* **Note 1**), NAS (0.14 g, 8.30×10^{-4} mol), and the initiator

α,α′-azoisobutyronitrile (AIBN, 1.30×10^{-2} g, 7.90×10^{-5} mol), under nitrogen flow, in 20.00 mL of anhydrous tetrahydrofuran (THF) The concentration of the monomer feed in the solvent is 0.2 g/ml (20 % w/v).

2. Degas the solution by applying vacuum for 10 min.
3. Restore the nitrogen flow and add 3-(trimethoxysilyl)propyl methacrylate (MAPS, 1.03×10^{-1} g, 4.15×10^{-4} mol).
4. Stir magnetically and heat the solution to 65 °C for 2 h under nitrogen atmosphere
5. After the polymerization is completed, dilute 1:1 with anhydrous THF.
6. Precipitate the polymer by dripping the reaction mixture into a large excess of petroleum ether (about 1:10 by volume), while stirring.
7. Filter the obtained white solid on a Buchner funnel.
8. Dry the copolymer under a vacuum for 1–2 h at room temperature.
9. Store it at −20 °C in a dry environment.

3.2 Derivatization of Substrates with Poly(DMA-co-NAS-co-MAPS)

The coating of substrates requires two steps: (a) substrate surface pretreatment and (b) adsorption of the copolymer.

1. Surface pretreatment: The substrates are cleaned and activated by pretreatment with oxygen plasma for 10 min (high radio frequency).
2. Dissolve poly(DMA-*co*-NAS-*o*-MAPS) to a final concentration of 1 % w/v in an ammonium sulfate aqueous solution 20 % w/v (*see* **Note 2**).
3. Adsorption of poly(DMA-*co*-NAS-*co*-MAPS): Immerse the substrates into the coating solution for 30 min at room temperature in a PS (Polystyrene) chamber (*see* **Note 3**).
4. Wash the substrates vigorously with ddH_2O to remove the excess of the copolymer on the surface.
5. Dry the surfaces with nitrogen stream to avoid stain formation.
6. Cure the slides in a vacuum oven at 80 °C for 15 min (*see* **Note 4**).
7. Store the slides at room temperature in a desiccator and use the slides within 4 weeks after production (*see* **Note 5**).

3.3 Synthesis of Poly(glycidyl methacrylate) (PGMA)

1. In a round bottom flask, equipped with condenser, magnetic stirring, and nitrogen connection, dissolve glycidyl methacrylate 1 M in tetrahydrofuran (THF), under nitrogen flow.
2. Degas the solution by applying vacuum for 10 min.

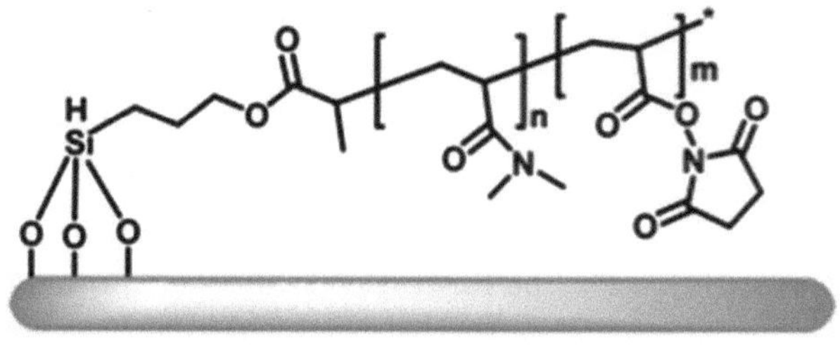

Scheme 3 Substrate derivatized by grafted poly(DMA-*co*-NAS-*co*-MPS)

3. Restore nitrogen flow and add 1 mM of α,α′-azoisobutyronitrile (AIBN).
4. Heat the solution to 65 °C overnight, under nitrogen atmosphere.
5. Cool the solution at room temperature and purify the polymer by precipitation in diethyl ether.
6. Filter the resulting white solid product on a Buchner funnel.
7. Dry the polymer under vacuum for 1 h at room temperature.

3.4 Derivatization of Substrates with Poly(glycidyl methacrylate) (PGMA)

The coating of the chips requires two steps: (a) surface pretreatment and (b) adsorption of the copolymer.

1. Surface pretreatment: The substrates are cleaned and activated by pretreatment with oxygen plasma for 10 min (high radio frequency).
2. Dissolve poly(glycidyl methacrylate) (PGMA) in anhydrous DMF at a final concentration of 5 % w/v.
3. Adsorption of the copolymer PGMA onto the substrates: Immerse the slides into the solution overnight at 70 °C. Rinse the substrate with DMF and then with anhydrous THF (*see* **Note 6**).
4. Dry each slide with a nitrogen stream to avoid stain formation.
5. Place the substrates in a vacuum oven at room temperature for 30 min.

3.5 Derivatization of Substrates with Grafted Poly(DMA-co-NAS-co-MAPS)

The coating of the substrates requires two steps: (a) silanization of the substrates with 3-(trimethoxysilyl)propyl methacrylate (MAPS) and (b) polymerization of poly(DMA-*co*-NAS) either by thermo-activation or by photo-activation (Scheme 3).

1. Pretreat the slides with oxygen plasma for 10 min (high radio frequency).
2. Prepare a solution of 3-(trimethoxysilyl)propyl methacrylate (MAPS) 1 % v/v in toluene.
3. Silanization with MAPS: Immerse the slides into the solution for 4 h, at room temperature.

4. Rinse each slide with fresh toluene and with THF(*see* **Note 7**).
5. Dry each slide with nitrogen stream to avoid stain formation.
6. Cure in a vacuum oven at 80 °C for 30 min (*see* **Note 4**).
7. For thermo-activated polymerization: Prepare the polymerization solution by dissolving DMA (8.90 g, 9.00×10^{-2} mol; *see* **Note 1**) and NAS (1.70 g, 1.00×10^{-2} mol) in 100 mL of *N*,*N*-dimethylformamide (DMF).
8. Immerse the previously silanized slides into the solution.
9. Degas the solution by purging Argon for 30 min.
10. Add AIBN (3.6×10^{-2} g, 2.20×10^{-4} mol).
11. Heat the solution to 65 °C overnight under nitrogen atmosphere.
12. Rinse each slide with fresh DMF and finally with THF (*see* **Note 6**).
13. Dry each slide with a nitrogen stream to avoid stain formation.
14. Dry the slides under vacuum at room temperature.
15. For photo-activated polymerization: Prepare the polymerization solution by dissolving DMA (4.42×10^{-1} g, 4.40×10^{-3} mol; *see* **Note 1**) and NAS (8.50×10^{-2} g, 5.02×10^{-4} mol) in 5 mL of DMF
16. Add IRGACURE 184 (3.50×10^{-2} g, 1.70×10^{-4} mol).
17. Immerse the previously silanized slides into the transparent chamber for photo-activated polymerization (*see* **Note 8**).
18. Expose the chamber, containing the slides, to UV light (365 nm) for 10 min.
19. Rinse with fresh DMF and finally with THF (*see* **Note 6**).
20. Dry each slide with a nitrogen stream to avoid stain formation.
21. Dry under vacuum at room temperature.

3.6 Derivatization of Substrates with Grafted Poly(DMA-co-NAS-co-MPS)

The coating of the substrates requires two steps: (a) silanization of the substrates with 3-mercaptopropyl trimethoxy silane (MPS) and (b) polymerization of poly(DMA-*co*-NAS) either by thermo-activation or by photo-activation.

1. Pretreat the slides with oxygen plasma for 10 min (high radio frequency).
2. Prepare a solution of 3-mercaptopropyl trimethoxy silane (MPS) 1 % v/v in toluene.
3. Silanization with 3-mercaptopropyl trimethoxy silane (MPS): Immerse the slides into the solution for 4 h at room temperature.
4. Rinse each slide with fresh toluene and with THF (*see* **Note 7**).

5. Dry each slide with a nitrogen stream to avoid stain formation.
6. Cure in a vacuum oven at 80 °C for 30 min (*see* **Note 4**).
7. Follow procedures described in Subheading 3.5 for thermo-activated polymerization (**steps 7–14**) or photo-activated polymerization (**steps 15–21**).

3.7 Derivatization of Substrates with Grafted Poly[DMA-b-(DMA-co-NAS)]

The procedure consists of three steps: (a) Synthesis of 1-oxo-1-(3-(trimethoxysilyl)propylamino)propan-2-yl-benzodithioate (RAFT Silane), (b) Surfaces functionalization with RAFT Silane, (c) polymerization.

1. Synthesis of 1-oxo-1-(3-(trimethoxysilyl)propylamino)propan-2-yl-benzodithioate (RAFT Silane): In a 100 mL two neck round bottom flask, equipped with a dropping funnel and nitrogen connection, dissolve, under nitrogen flow, (3-aminopropyl)-trimethoxysilane (APS, 4.86 mL, 0.027 mol) and triethylamine (TEA, 3.77 mL, 0.027 mol) in dichloromethane (30 mL).
2. Cool the round bottom flask to 0 °C with an ice bath.
3. Prepare a solution of 2-bromo propionyl bromide (2.83 mL, 0.027 mol) in dichloromethane (5 mL) and drip it into the round bottom flask, while stirring.
4. Stir the solution at room temperature for 2 h.
5. Filter the white precipitate of triethylamine bromohydrate.
6. Evaporate the solvent under vacuum.
7. Purify the residue by distillation using a Kugelrohr apparatus (175 °C, 1 mBar) to obtain the bromine derivative (**1**) as a colorless oil.
8. In a 100 mL two neck round bottom flask, equipped with dropping funnel and nitrogen connection, dissolve phenylmagnesium bromide (25.4 mL, 0.025 mol) in anhydrous THF (30 mL).
9. Cool the stirring solution to 0 °C with an ice bath.
10. Slowly add carbon disulfide (CS_2, 2.9 mL, 0.048 mol) and stir for 1.5 h at room temperature
11. Cool the solution again to 0 °C with an ice bath.
12. Dissolve the bromine derivative (**1**) obtained previously (7.6 g; 0.024 mol) in anhydrous THF (10 mL) and drip it into the round bottom flask.
13. Stir overnight at room temperature.
14. Evaporate the solvent under vacuum.
15. Dissolve the obtained residue in dichloromethane (15 mL) and wash twice with brine.
16. Dry the organic phases over anhydrous sodium sulfate (Na_2SO_4).

17. Evaporate under vacuum to obtain the RAFT silane as a red oil.
18. Surfaces functionalization with RAFT Silane : Pretreat surfaces with Oxygen plasma (high radio frequency).
19. Dissolve RAFT silane (0.5 mM) and octyl-trimethoxysilane (nOS, 2 mM) in toluene.
20. Immerse the pretreated slides into the solution, under nitrogen atmosphere, for 4 h.
21. Wash each slide with fresh toluene and with THF (*see* **Note 7**).
22. Dry each slide with a nitrogen stream to avoid stain formation.
23. Dry under vacuum at room temperature for 30 min.
24. Synthesis of grafted poly(DMA): Dissolve DMA (45.95 mL, 0.446 mol; *see* **Note 1**), *tert*-butyl dithiobenzoate (t-BDB, 0.262 g, 0.125×10^{-2} mol), and AIBN (0.041 g, 0.025×10^{-2} mol) in 100 mL of toluene.
25. Immerse the slides previously silanized with RAFT Silane.
26. Degas the solution by purging argon for 1 h.
27. Heat the solution to 80 °C overnight, under nitrogen atmosphere.
28. Rinse the slides by Soxhlet extraction for 12 h using THF.
29. Dry each slides using a nitrogen stream.
30. Synthesis of graft-poly[DMA-*b*-(DMA-*co*-NAS)]: Dissolve DMA (9.27 mL, 0.09 mol; *see* **Note 1**), NAS (1.69 g, 0.01 mol) and AIBN (9.2 mg, 0.056×10^{-3} mol) in 100 mL of DMF.
31. Immerse the slides previously coated with the first block of grafted poly(DMA) into the solution.
32. Degas the solution by purging Argon for 1 h.
33. Heat overnight at 80 °C, under nitrogen atmosphere.
34. Wash each slide with fresh DMF and with THF (*see* **Note 6**).
35. Dry each slide with a nitrogen stream to avoid stain formation.
36. Dry under vacuum at room temperature for 30 min.

4 Notes

1. DMA must be filtered on Aluminum oxide to remove inhibitor before use. Filtered DMA can be stored for no more than a week at 2–4 °C.
2. First prepare a stock solution of ammonium sulfate at a 40 % saturation level by adding 242 g of ammonium sulfate to 1 L of ddH_2O. Depending on the volume of the chamber and the number of the slides to be coated, weigh the exact amount of copolymer to obtain a water solution at a final concentration of

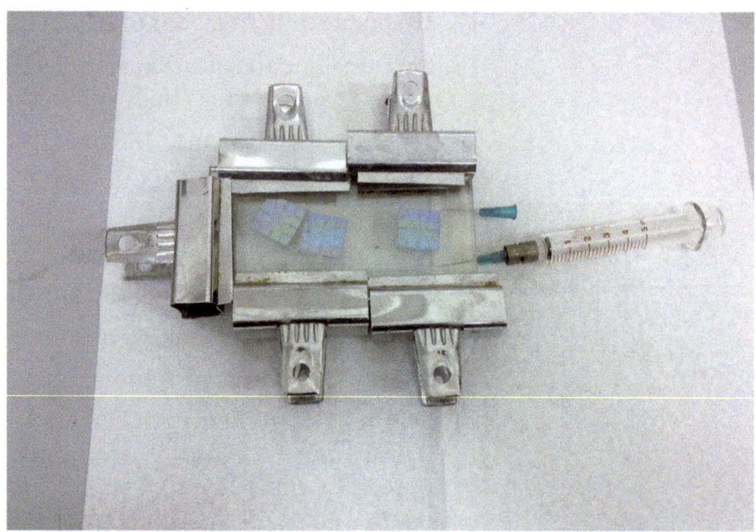

Fig. 2 Picture of the glass chamber used for UV initiated polymerization and derivatization of flat substrates

2 % w/v. When the copolymer is completely dissolved, dilute it 1:1 with the stock solution of ammonium sulfate at a 40 % saturation level.

3. Use preferably plastic (for example polystyrene) container as glass ones would be coated as well, competing with the surface of the slides that are intended to be coated
4. Curing step under vacuum is fundamental to assure the formation of covalent bonds between surface silanols and silane reagents.
5. Slides coated with functional groups are best stored in dry condition in a desiccator. This is fundamental when using the poly(DMA-*co*-NAS-*co*-MAPS) because its reactive monomer is very sensitive to humidity. Poly(DMA-*co*-NAS-*co*-MAPS) coated slides, can be stored in any plastic container in a desiccator at room temperature and used up to 4 weeks.
6. Rinse quickly each slide in both solvents. First rinse each slide in DMF to remove the excess of reagent, then immediately in THF, as this evaporates faster avoiding formation of stains.
7. Rinse quickly each slide in both solvents. First rinse each slide in toluene to remove the excess of silanizing agent, then immediately in THF, as this evaporates faster avoiding formation of stains.
8. To immerse the substrates into polymerization solution and be sure their surface is completely exposed to UV light prepare a chamber using two large glass slides (or any transparent material

which does not dissolve into DMF) and a silicon gasket to separate the bottom from the upper glass slide. Place the substrates that have to be coated within the area delimited by the gasket, then clamp the two glass slides. The silicon gasket will allow also the use a syringe to introduce the monomer solution into the obtained chamber. Use a second needle to allow removal of air bubbles so to be sure the chamber is completely filled with monomer solution (Fig. 2).

Acknowledgments

Financial support from the Italian Ministry of University and Research PRIN 2008 (2008JWKYXB_005), Italian Institute of Technology (project SEED "IPG-CHIP"), and Fondazione Cariplo (SpinBioMed, grant # 2008–2330) are gratefully acknowledged.

References

1. Krenkler KP, Laible R, Hamann K (1976) Polyreaktionen an Pigmentoberflächen. VII. Mitteilung: Reaktionen von Polymeren mit endständigen Chlorsilangruppen an Siliciumdioxidoberflächen. Angew Makromol Chem 53:101–123
2. Tsubokawa N et al (1990) Grafting onto carbon black: reaction of functional groups on carbon black with ACYL chloride-capped polymers. J Macromol Sci 27:445–457
3. Tsubokawa N, Kuroda A, Sone Y (1989) Grafting onto carbon black by the reaction of reactive carbon black having epoxide groups with several polymers. J Polym Sci A27:1701–1712
4. Dimitrenko AV et al (1990) Polymer—inorganic selective adsorbents for gas chromatography produced by graft polymerization. J Chromatogr A 520:21–31
5. Hashimoto K et al (1982) Graft copolymerization of glass fiber and its application. J Macromol Sci 18:173–190
6. Klein J et al (1993) Lubrication forces between surfaces bearing polymer brushes. Macromolecules 26:5552–5560
7. Yoshikawa C et al (2006) Protein repellency of well-defined, concentrated poly(2-hydroxyethyl methacrylate) brushes by the size-exclusion effect. Macromolecules 39:2284–2290
8. Pirri G et al (2006) Microarray glass slides coated with block copolymer brushes obtained by reversible addition chain-transfer polymerization. Anal Chem 78:3118–3124
9. Tsubokawa N et al (1989) Graft polymerization of acrylamide from ultra silica particles by use of a redox system consisting of ceric ion and reducing groups on the surface. Polym J 21:475–481
10. Prucker O, Rühe J (1998) Synthesis of poly(styrene) monolayers attached to high surface area silica gels through self-assembled monolayers of azo initiators. Macromolecules 31:592–601
11. Prucker O, Rühe J (1998) Mechanism of radical chain polymerizations initiated by azo compounds covalently bound to the surface of spherical particles. Macromolecules 31:602–613
12. Di Carlo G et al (2012) Synthesis and conformational characterization of functional diblock copolymer brusche for microarray technology. Appl Surf Sci 258:3750–3756

Chapter 8

Functionalization Protocols of Silicon Micro/Nano-mechanical Biosensors

Francesca Frascella and Carlo Ricciardi

Abstract

Functionalization is a key element in biodetection technologies such as micro/nano-mechanical sensors. Since assay sensitivity and stability drastically depends on a proper bioreceptor immobilization, the sensing surface must be first chemically modified with uniform, well-packed, and robust layers. Here, we describe three functionalization protocols that we developed for the surface modification with amino, aldehyde, and carboxyl groups of micro/nano-mechanical biosensors.

Key words Self-assembled monolayer, Organosilane, Functionalization, Molecular recognition, Biosensor

1 Introduction

Micro/Nano-mechanical sensors such as cantilevers have recently gained a significantly growing attention from the scientific community, due to their capability of label-free detection of bio/chemical molecules at extremely low concentrations [1]. Such applications require a chemical functionalization of the sensor surface to attach recognition molecules that have both high affinity and selectivity for the intended bio/chemical targets [2].

A widely used method to create well-organized and compact functionalizing monolayers is to exploit the properties of self-assembling of alkyl chains with thiol groups (R-SH) on gold substrates, or of alkyl silanes (R-Si-OH) on silicon-based substrates. Such covalently anchored layers are uniform, densely packed and robust, thus guaranteeing a high stability and sensitivity of the final sensor platform. Furthermore, Self-assembled monolayers (SAMs) represent an extremely versatile tool, being able to be synthesized with chains of variable length and groups with specific chemical properties, representing an ideal way of anchorage of biomolecules to the substrate.

Paolo Bergese and Kimberly Hamad-Schifferli (eds.), *Nanomaterial Interfaces in Biology: Methods and Protocols*, Methods in Molecular Biology, vol. 1025, DOI 10.1007/978-1-62703-462-3_8, © Springer Science+Business Media New York 2013

We here report on the functionalization protocols we developed for silicon micro/nano-mechanical biosensors, making use of the following:

- 3-Aminopropyltriethoxysilane (APTES) to have a sensing surface rich of amino (–NH_2) groups
- 3-Aminopropyltriethoxysilane (APTES) followed by Glutaraldehyde (GA) to have a sensing surface rich of aldehyde (–CHO) groups
- 3-Cyanopropyltriethoxysilane (CPTES) to have a sensing surface rich of carboxyl (–COOH) groups

2 Materials

Prepare all solutions using high quality pure water of high resistivity and low TOC (>10 MΩ cm and <50 ppb) and ACS reagents (essay ≥99.5 %), unless otherwise stated. Prepared solutions and reagents should be stored at room temperature (unless otherwise indicated).

Organosilane reaction involves handling of air-sensitive compound [3], so it is necessary to carry out the reaction in an inert atmosphere. A special equipment with a source of inert gas (preferably argon), a septum inlet, a bubbler, and syringes fitted only with small-gauge needles (no larger than 18-gauge) is required, while glass apparatus is of common type. Laboratory glassware, syringe and needle should be dried in an oven (140 °C for 4 h) prior to use.

2.1 Cleaning and Oxidation of the Surfaces

1. Piranha solution: mixture 3:1 of 98 % w/w H_2SO_4 and 30 % w/w H_2O_2 solution (hydrogen peroxide must be stored at 4 °C). Add 3 parts in volume of sulfuric acid to a graduated cylinder or a glass beaker (*see* **Note 1**). Then add slowly 1 part of hydrogen peroxide to the acid and mix. Once prepared, the solution has to be used as soon as possible.

2.2 Silanization Components

1. 3-Aminopropyltriethoxysilane 99 % (APTES) stored under argon atmosphere: 1 % v/v solution in anhydrous toluene 99.8 %.
2. Glutaraldehyde solution 25 % v/v (GA): 0.5 % v/v in dd-H_2O. Store at 4 °C.
3. Borate Buffer Saline (BBS) 0.1 M, pH 8.5. Dissolve 0.011 g of boric acid (H_3BO_3) in 5 mL dd-H_2O. Weigh 0.026 g of sodium chloride (NaCl) and dissolve it in 50 mL dd-H_2O. Mix the two solutions and adjust pH with small amount of sodium hydroxide (NaOH) solution (*see* **Note 2**). Store at 4 °C.
4. Basic pH adjusting solution: 0.1 M NaOH. Dissolve 0.1 g NaOH in 2.5 mL dd-H_2O.

5. Sodium cyanoborohydride $NaBH_3(CN)$ 95 %: 5 M solution in NaOH. Dissolve 0.31 g of $NaBH_3(CN)$ in 1 mL of NaOH 1 M (*see* **Note 3**).
6. (3-Cyanopropyl)triethoxysilane 98 % (CPTES) stored under Argon atmosphere: 1 % v/v solution in anhydrous toluene 99.8 %.
7. Hydrolysis solution: H_2SO_4 48 % solution in deionized water. Measure 49 mL of H_2SO_4 98 % w/w and make up to 100 mL with water.

3 Methods

3.1 Functionalization with Amino Groups: APTES (Fig. 1)

1. Grow a silicon oxide thin film (190 nm) on silicon substrates by thermal oxidation at 1,100 °C in O_2 atmosphere for 3 h.
2. Prior to monolayer preparation, treat the silicon substrates for 15 min in a freshly prepared piranha solution. Since the mixture is a strong oxidizer, it will remove most organic contaminants and hydroxylate the surface (adding of OH groups), creating the proper condition for the attachment and displacement of the alkoxy group from the silane, thus forming a covalent –Si–O–Si– bond (*see* **Note 1**).
3. Rinse the substrates with copious deionized water (3 times) and dry them carefully in a stream of nitrogen.
4. After that, incubate the freshly cleaned silicon substrates (*see* **Note 4**) in a round bottom flask with an oven dried magnetic stir bar inside, filled with 1 % v/v APTES solution in anhydrous toluene at 70 °C or until solvent reflux for 10 min. Add APTES via an oven-dried syringe and maintain the flask under an atmosphere of argon, following an anhydrous protocol (*see* **Note 5**).
5. Silane-coated substrates were rinsed deeply with toluene and carefully dried under N_2 flux.

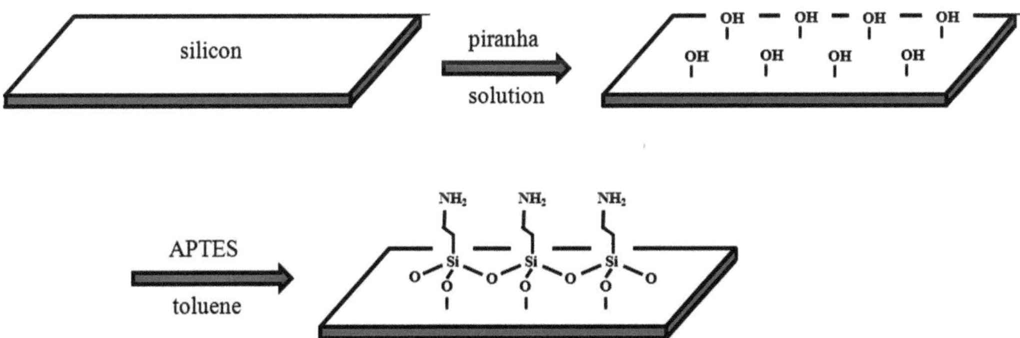

Fig. 1 Scheme of functionalization with amino groups—APTES

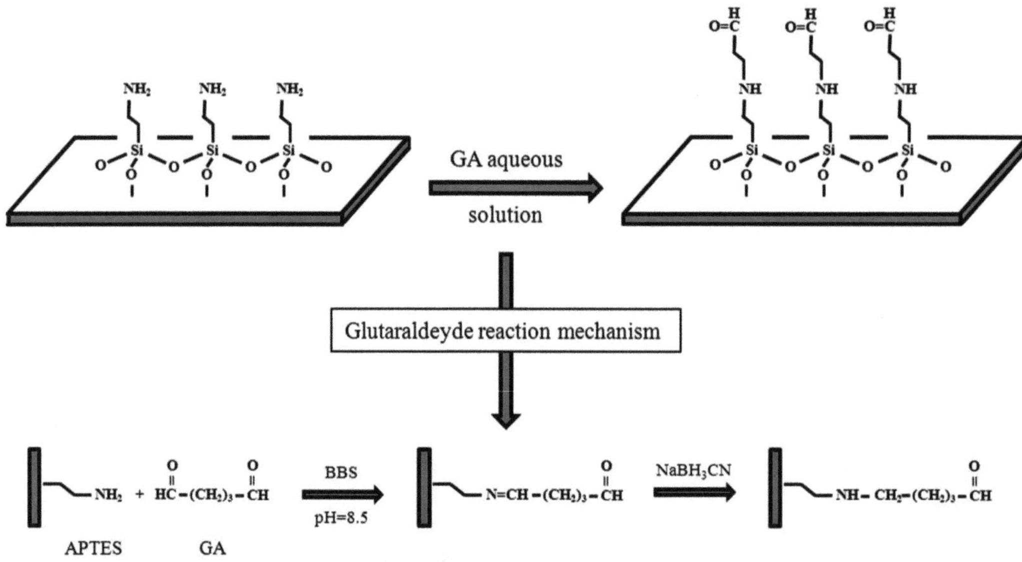

Fig. 2 Scheme of functionalization with aldehyde groups—APTES and GA

3.2 Functionalization with Aldehyde Groups: APTES and GA (Fig. 2)

1. Just after the reaction with APTES (**steps 1–5** in Subheading 3.1), incubate the samples in a 0.5 % v/v GA solution in BBS 0.1 M (pH 8.5) for 1 h using an orbital shaker at 40 rpm (*see* **Note 2**).
2. After 15 min, add 300 μL of sodium cyanoborohydride solution (5 M) in NaOH, in order to create a more stable bond, reducing the imine formed by the reaction between $-NH_2$ and -CHO groups (*see* **Note 6**).
3. After incubation, rinse the treated substrates several times with deionized water, dry them under nitrogen stream, and store them in a sealed desiccator until needed.

3.3 Functionalization with Carboxylic Groups: CPTES (Fig. 3)

1. Grow a silicon oxide thin film (190 nm) on silicon substrates by thermal oxidation 1,100 °C in O_2 atmosphere for 3 h.
2. Soak oxidized substrates into a freshly prepared piranha solution for 15 min (*see* **Note 1**), to remove organic contaminants.
3. Rinse several times the treated substrates with dd-H_2O and dry them in a nitrogen stream.
4. Incubate the freshly cleaned substrates with 1 % v/v CPTES toluene solution in reflux solvent condition (around 70 °C) for 10 min in experimental anhydrous conditions (*see* **Note 4**).
5. Rinse several times (3 times) the substrates with toluene and then dry them under nitrogen stream.
6. Treat the freshly silanized samples in H_2SO_4 48 % solution at 100 °C under reflux for 3 h in Argon flux, so to hydrolyze -CN group into -COOH (*see* **Note 7**).

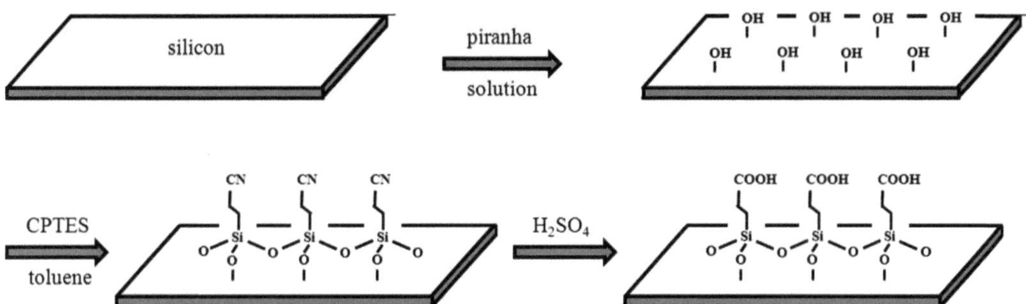

Fig. 3 Scheme of functionalization with carboxylic groups—CPTES

7. After incubation, rinse the treated substrates several times with deionized water, dry them under nitrogen stream, and store them in a sealed desiccator until needed.

4 Notes

1. Piranha solution is an extremely strong oxidant and should be handled very carefully. It is highly corrosive and oxidizing. Make sure that surfaces are reasonably clean, and completely free of organic solvents from previous wash steps, before coming into contact with piranha solution.

   Mixing the solution is exothermic, so prepare and store piranha solutions only in approved glass containers that can easily withstand temperatures of at least 200 °C.

   Piranha solution should be prepared by slowly adding small aliquots of hydrogen peroxide to sulfuric acid, never the reverse.

   Immersing the substrate into the solution (remember to use Teflon-coated tweezers) should be done slowly to prevent thermal shock that may crack the substrate material and one by one, giving time after each substrate addition for solution to quit foaming.

   Due to the self-decomposition of hydrogen peroxide, piranha solution should not be stored, and used as freshly prepared. The solution should be allowed to cool prior to disposal, so that oxygen gas could dissipate. To speed up the dissolution of piranha hydrogen peroxide, few grains of manganese dioxide can be added.

2. Remember to check and adjust pH when the solution is at room temperature. The gap between BBS starting pH (around 6) and the required pH (8.5) is not large, consequently a 1 M of NaOH solution presents an adequate ionic strength to allow a good pH adjust, by adding around 50 μL of basic solution every 100 mL BBS.

3. Sodium cyanoborohydride contact with air should be kept to a minimum because the compound is very hygroscopic. Remember to weigh into a good ventilated area and to use a fume cupboard, because the salt has a very unpleasant smell. $NaBH_3(CN)$ is used for the reduction of imines to amines. Selectivity is achieved in basic condition (BBS pH 8.5).
4. The substrate has to be freshly cleaned before the deposition of any silane, in order to obtain a good reaction yield.
5. To prepare a good anhydrous reaction you have to carry out the reaction in Ar flux and to use laboratory glassware dried in oven at 140 °C for at least 4 h, in order to remove the thin film of adsorbed water that, acting as a catalyst, causes APTES hydrolysis in ethanol and trisilanols. Pay attention not to assemble the syringe body and the plunger before being placed in oven. The glassware should be assembled while hot and fluxed with a stream of argon, as soon as fixed. A dry syringe should be separated from the environmental humidity by inserting the tip of the needle into a rubber septum. The reaction vessel must be secured through a mineral oil bubbler. At all times during the reaction, the system should be under a slow and continuous argon flow (slight positive pressure). Liquid reagents should be transferred with syringe by first pressurizing the reagent bottle with inert gas. This slight pressure is used to slowly fill the syringe with the reagent, without making air enters the bottle. The measured volume of reagent in the syringe has to be quickly transferred to the reaction flask by a rubber septum put on one neck [4].
6. Borate buffer pH 8.5 ensures that amino groups exposed at sensor surface are not in a protonated form ($-NH_2$), thus being available to react with GA aldehyde groups (-CHO).
7. Nitriles (-CN group) can be hydrolyzed, by heating under reflux with dilute acid, in two steps: first the protonated nitrile gives a primary amide and then the hydrolysis of this amide gives a free -COOH group. The reaction should be done under N_2 or Ar flow, in order to remove the formation of ethanol that would cause esterification of carboxylic groups [5].

Acknowledgments

The authors wish to thank the financial support of Regione Piemonte (Converging Technologies 2007—NAMATECH grant, Poli di Innovazione POR-FESR 2007–2013—CANESTRO grant, and Piattaforme Innovative POR-FESR 2007–2013—ITACA and SAFE-FOOD CONTROL grants) and Italian Ministry of Research and Education (FIRB 2010—NEWTON grant, Progetti Bandiera 2011—NANOMAX grant).

References

1. Eom K et al (2011) Nanomechanical resonators and their applications in biological/chemical detection: Nanomechanics principles. Phys Rep 503:115–163
2. Johnson BN, Mutharasan R (2012) Biosensing using dynamic-mode cantilever sensors: a review. Biosens Bioelectron 32:1–18
3. Aldrich Chemical Company (2010) Handling air-sensitive reagents. In: Technical Information Bulletin AL-134:1–9
4. Furniss BS et al (1989) Reaction involving air-sensitive compounds. In: Vogel's textbook of pratical organic chemistry, 5th edn. Wiley, New York 2:120–131
5. Yang C, Zibrowis B, Schüth F (2003) A novel-synthetic route for negatively charged ordered mesoporous silica SBA-15. Chem Comm 14: 1772–1773

Part II

Characterizing Bio–Nano Interfaces

Chapter 9

Stability and Aggregation Assays of Nanoparticles in Biological Media

James Chen Yong Kah

Abstract

Colloidal stability of nanoparticles in biological media is crucial to preserve their utility as aggregation often leads to undesirable biological response. A quantitative measurement of nanoparticles aggregation in solution would provide a valuable assessment of colloidal stability of bio–nano interfaces after surface functionalization. Here, we develop a quantitative technique based on optical absorption to assay the colloidal stability of plasmonic nanoparticles functionalized with different amphiphilic surface ligands in biologically relevant media.

Key words Gold nanorods, Colloidal stability, Aggregation, Amphiphilic ligand, Surface charge, Cell culture media

1 Introduction

The biological utility of nanoparticles often stems from their unique physicochemical properties that are often a function of their surface interactions with other nanoparticles or biological entities. Therefore, apart from key attributes required for their successful use in biology such as low toxicity, biocompatibility and effective biological clearance, the colloidal or dispersion stability of these nanoparticles is also crucial to preserve their intended physicochemical behavior and hence their utility in the physiological environment. Aggregation of nanoparticles in biological media often undermines their ability to perform their intended functions [1].

Colloidal stability of nanoparticles is maintained by two main mechanisms, namely, steric stabilization and charge stabilization, which can be probed by a wide range of techniques depending on the types of nanoparticles. Inorganic nanoparticles with unique optical properties are often probed by their optical emission such as photoluminescence intensity in the case of quantum dots [2–4] or their optical absorption in the case of plasmonic nanoparticles [5–9], since their optical properties are often a function of

Paolo Bergese and Kimberly Hamad-Schifferli (eds.), *Nanomaterial Interfaces in Biology: Methods and Protocols*, Methods in Molecular Biology, vol. 1025, DOI 10.1007/978-1-62703-462-3_9, © Springer Science+Business Media New York 2013

aggregation state. The colloidal stability of non-plasmonic nanoparticles made of materials such as carbon, chitosan, magnetic, and silica can be probed by their UV–Vis spectrum [10–12], zeta potential measurements [10, 13], hydrodynamic diameter based on dynamic light scattering [10, 13–16], or simply by observing their precipitation times for larger and heavier particles [13, 17].

Here, we develop a quantitative Aggregation Index (AI) as a descriptor to characterize the colloidal stability of plasmonic nanoparticles based on their optical absorption, specifically using gold nanorods coated with different amphiphilic ligands as an example [18]. The localized surface plasmon resonance (LSPR) of gold nanorods in the UV–Vis spectrum is highly sensitive to their aggregation state due to interaction of the plasmons between nanorods when they are in proximity. Aggregation of gold nanorods result in a red-shift and spectral broadening of the LSPR and the AI is a measure of the LSPR spectral broadening. Physically, it describes the equivalent bandwidth of the LSPR (with units of nm) for a spectral area normalized to the LSPR peak absorbance. A higher degree of aggregation corresponds to a higher AI value. The AI is concentration independent since it is normalized to the LSPR peak absorbance which varies with gold nanorods concentration.

While covalently attached ligands are predominantly used for nanoparticle passivation, amphiphiles are attractive due to the fact that they can be chemically tailored, and are versatiles, enabling a range of applications in delivery and therapy. Amphiphilic ligands have been used to passivate gold nanorods to mediate cytotoxicity and enable transfection of cells. The choice of surface ligands affects the colloidal stability of gold nanorods, which in turn affect a broad range of biological processes. Therefore, a quantitative AI would provide valuable indication of their effectiveness in passivating gold nanorods. We functionalized the surface of gold nanorods with four types of amphiphilic ligands having different headgroup charge (cetyltrimethylammonium bromide (CTAB): positive, polyoxyethylene [10] cetyl ether (Brij56): neutral, Oligofectamine™ (OF): positive and phosphatidylserine (PS): negative), to create gold nanorods with different surface charge. We then disperse these nanorods in biologically relevant media and demonstrate how we can develop an aggregation assay to probe their colloidal stability in these media.

2 Materials

Prepare all solutions of analytical grade reagents using MilliQ water with a resistivity of 18.2 MΩ cm at 25 °C. Otherwise stated, all reagents used in the preparation were used as received.

All waste disposal regulations should be strictly adhered to when disposing waste materials.

2.1 Synthesis of Gold Nanorods

1. 200 mM Cetyltrimethylammonium bromide (CTAB): Dissolve 1.458 g CTAB in 20 mL water (*see* **Note 1**).
2. 100 mM Sodium chloride (NaCl): Dissolve 58.44 mg NaCl in 10 mL water.
3. 50 mM Gold (III) chloride trihydrate ($HAuCl_4$): Dissolve 196.9 mg $HAuCl_4$ in 10 mL water (*see* **Notes 2** and **3**).
4. 10 mM Silver nitrate ($AgNO_3$): Dissolve 16.99 mg $AgNO_3$ in 10 mL water (*see* **Note 3**).
5. 100 mM l-Ascorbic acid (AA): Dissolve 176.1 mg AA in 10 mL water (*see* **Note 4**).
6. 312.5 μM Sodium borohydride ($NaBH_4$): Dissolve 1.182 mg $NaBH_4$ in 10 mL water. Dilute 10× by adding 100 μL of 3.125 mM $NaBH_4$ in 900 μL water (*see* **Note 5**).
7. Temperature controlled water bath.

2.2 Ligand Exchange with Amphiphilic Ligands

1. Amicon ultra–0.5 mL centrifugal filters unit using regenerated cellulose (ultracel-100) membrane with MWCO = 100,000.
2. 0.1 mM CTAB: Prepare 10 mM CTAB by dissolving 36.45 mg CTAB in 10 mL water and dilute the resulting solution 100× by adding 100 μL of 10 mM CTAB to 9.9 mL water.
3. 10 mM polyoxyethylene [10] cetyl ether (Brij56): Dissolve 68.3 mg Brij56 in 10 mL water (*see* **Note 6**).
4. Oligofectamine™ (OF) (catalog number 12252–011) (Invitrogen, Inc.) (*see* **Note 7**).
5. 20 mM phosphatidylserine (PS) (catalog number 840032P–25 mg) (Brain PS l-α-phosphatidylserine (Brain, Porcine, sodium salt): Dissolve 25 mg PS in 1.54 mLwater (*see* **Note 8**).

2.3 Stability Analysis in Biological Media

1. Phosphate buffered saline (PBS; 1×): 137 mM NaCl, 2.7 mM KCl, 10 mM $Na_2HPO_4 \bullet 2H_2O$, 2 mM KH_2PO_4, pH 7.4.
2. RPMI-1640 medium (catalog number 30–2001).
3. Cary 100 UV–Visible spectrophotometer.

3 Methods

Carry out all procedures at room temperature, unless otherwise specified.

3.1 Synthesis of Gold Nanorods

1. Prepare the 200 mM CTAB, 100 mM AA, and 312.5 μM $NaBH_4$ fresh before each gold nanorods synthesis. Cool the $NaBH_4$ to 4 °C.

2. Mix the reagents in the following order: 16.67 mL of 200 mM CTAB, 2.3 mL water, 300 μL of 100 mM NaCl, 240 μL of 50 mM $HAuCl_4$. Solution turns turbid orange upon addition of $HAuCl_4$. Swirl solution gently for 3 min and the solution will turn clear orange (*see* **Note 9**).
3. Add 240 μL of 10 mM $AgNO_3$, followed by 200 μL of 100 mM AA. Swirl the mixture briefly as the solution turns from orange to colorless to indicate the reduction of Au^{3+} to Au^{+} by AA.
4. Add 128 μL of 312.5 μM $NaBH_4$ to the mixture and swirl gently to ensure that the $NaBH_4$ is well mixed. Put the mixture immediately in water bath (≈26 °C) and leave it overnight for the gold nanorods to form (*see* **Note 10**).
5. Aliquot 1 mL of the gold nanorods colloid into 1.5 mL centrifuge tubes and centrifuge them at 12,000 rpm (13,400 × *g*) for 30 min to remove the excess reactants.
6. Resuspend the pellet in 1 mL of water and store the resulting gold nanorods passivated with CTAB (NR-CTAB) at room temperature (*see* **Note 11**).

3.2 Ligand Exchange with Amphiphilic Ligands

1. Centrifuge 1 mL of gold nanorods colloid at 12,000 rpm (13,400 × *g*) for 30 min. Remove the supernatant and resuspend pellet in 1 mL of 0.1 mM CTAB.
2. Repeat the centrifugation at 12,000 rpm (13,400 × *g*) for 30 min. Remove the supernatant and resuspend pellet in 500 μL of 10 mM Brij56 to replace the CTAB ligands on gold nanorods with Brij56.
3. Vortex the mixture briefly and sonicate for 1 min before incubating it for 1 h at 37 °C to allow the ligand exchange to take place.
4. Remove excess Brij56 by centrifuging the gold nanorods using Amicon ultra-0.5 mL centrifugal filters at 5,000 rpm (2,300 × *g*) for 15 min (*see* **Note 12**). Discard the filtrate containing excess Brij56 and recover the pellet (≈30 μL) into a new centrifuge tube.
5. Resuspend the pellet in 470 μL of water and repeat the centrifugation at 5,000 rpm (2,300 × *g*) for 15 min. Discard the filtrate and recover the pellet (≈30 μL) into a new centrifuge tube. Add 470 μL of water to the pellet to give 500 μL of gold nanorods passivated with Brij56 (NR-Brij56) and store the solution at room temperature.
6. To prepare gold nanorods passivated with OF (NR-OF), add 50 μL of OF reagent to 30 μL of the NR-Brij56 pellet obtained in **step 5**. Vortex the mixture briefly and sonicate for 1 min before incubating the mixture overnight at 37 °C to allow the ligand exchange from Brij56 to OF to take place (*see* **Note 13**).

7. To prepare gold nanorods passivated with PS (NR-PS), add 100 μL of 20 mM PS to 30 μL of the NR-Brij56 pellet obtained in **step 5**. Vortex the mixture briefly and sonicate for 1 min before incubating the mixture overnight at 37 °C to allow the ligand exchange from Brij56 to PS to take place (*see* **Note 13**).
8. Remove the excess OF and/or PS by centrifuging the gold nanorods using Amicon ultra-0.5 mL centrifugal filters at 5,000 rpm (2,300 × *g*) for 15 min. Discard the filtrate containing excess OF or PS and recover the pellet (≈30 μL) into a new centrifuge tube.
9. Resuspend the pellet in 470 μL of water and repeat the centrifugation at 5,000 rpm (2,300 × *g*) for 15 min. Discard the filtrate and recover the pellet (≈30 μL) into a new centrifuge tube. Add 470 μL of water to the pellet to give 500 μL of gold nanorods passivated with OF (NR-OF) or PS (NR-PS) and store the solution at room temperature.

3.3 Stability Analysis in Biological Media

1. Warm the 1× PBS and RPMI-1640 medium to room temperature.
2. Add 50 μL of the gold nanorods passivated in different amphiphilic ligands (NR-CTAB, Brij56, OF and PS) as prepared in Subheading 3.2 to 450 μL of 1× PBS and 450 μL of RPMI-1640 medium in a centrifuge tube (*see* **Note 14**).
3. Vortex the mixture briefly and sonicate for 1 min before incubating it for 2 h at 37 °C.
4. Acquire the UV–Visible absorption spectrum of the amphiphilic ligand coated gold nanorods in 1× PBS and RPMI-1640 medium from 400 nm to 900 nm.

3.4 Measuring the Aggregation Index (AI)

1. Calculate the total area under the absorption spectrum of the gold nanorods LSPR from $\lambda_{min} = 600$ to $\lambda_{max} = 900$ nm (*see* **Note 15**). This can be done by applying the trapezium rule to sum the area of trapezoidal strips from the raw spectral data as given by equation below:

$$\text{Area} \approx \frac{h}{2}\left[I_{\lambda,\min} + I_{\lambda,\max} + 2\left(I_{\lambda,\min+h} + I_{\lambda,\min+2h} + \cdots + I_{\lambda,\max-h}\right)\right]$$

where h = sampling interval of the scan in nm
$I_{\lambda,min}$ = Absorbance at wavelength of λ_{min}
$I_{\lambda,min+h}$ = Absorbance at wavelength of (λ_{min+h})
$I_{\lambda,max}$ = Absorbance at wavelength of λ_{max}

2. Divide the total area determined in **step 1** by the LSPR peak absorbance, $I_{\lambda,max}$ (Fig. 1) to obtain the AI of the various amphiphilic ligand coated gold nanorods of interest.

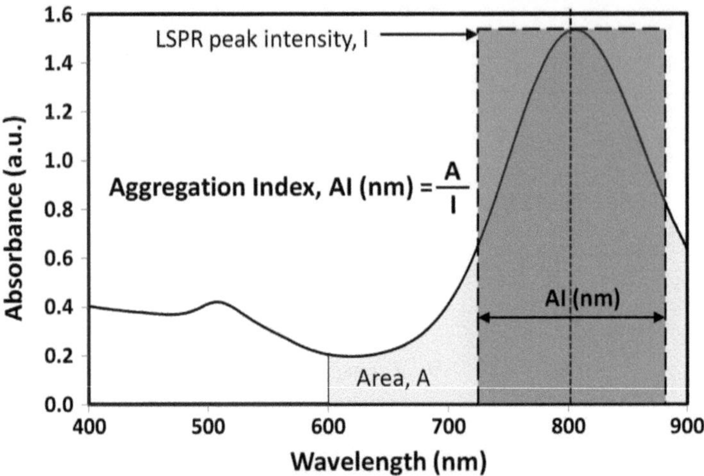

Fig. 1 Derivation of the Aggregation Index (AI) from the UV–vis absorption spectrum of the gold nanorods

4 Notes

1. We found that by warming the solution to 40 °C would aid the dissolution of CTAB.
2. $HAuCl_4$ is highly acidic and would corrode metal spatula. Use either a plastic spatula or a metal spatula wrapped with a layer of parafilm when weighing out the $HAuCl_4$.
3. The stock solutions of both $HAuCl_4$ and $AgNO_3$ should be stored in the dark at room temperature.
4. Use the 100 mM l-Ascorbic acid within 15 min of preparation to ensure that the reduction of Au^{3+} to Au^{+} proceeds well to yield good synthesis results.
5. $NaBH_4$ oxidizes easily at room temperature and loses it reducing potential with time. Use the 312.5 μM $NaBH_4$ within 15 min of preparation to ensure that $NaBH_4$ is sufficiently active to reduce the $HAuCl_4$. The $NaBH_4$ should be cooled to 4 °C just before its use for synthesis to slow the reduction reaction and yield good synthesis results.
6. Dissolve Brij56 in a water bath under sonication at 40 °C for about 40 min. Swirl the mixture occasionally to aid the dissolution of the transparent gel-like Brij56. Prepare the Brij56 fresh for each ligand exchange.
7. Do not freeze the OF, but store it at 4 °C. Let the OF warm to room temperature before proceeding with the ligand exchange.
8. Lipids composed of fatty acids containing one or more double bonds such as PS are not stable as powders. These lipids are extremely hygroscopic as powders and will quickly absorb moisture and become gummy upon opening the container.

This could result in hydrolysis or oxidation of the materials. Therefore, the PS should be prepared fresh and used just before each ligand exchange.

9. The prepared CTAB must be cool to ~30 °C before using it for synthesis. If the temperature is too high, it will increase the rate of reduction of Au^{3+} and hence the likelihood of forming spherical gold nanoparticles instead of gold nanorods. If the temperature is too low, the CTAB will crystallize out of the solution leaving a lower concentration of CTAB in the solution for the synthesis process. This will affect the quality of gold nanorods formed.
10. We found that by leaving the gold nanorods solution at a slightly elevated temperature of >25 °C would prevent the CTAB from crystallizing out of the solution the following day. This would result in a more consistent and reproducible quality of the gold nanorods in terms of their CTAB passivation and optical properties. It is also crucial that movement or vibration of the mixture is minimized to ensure proper growth of the gold nanorods.
11. The size of the gold nanorods can be determined from Transmission Electron Microscopy (TEM) and their concentration can be determined from optical absorption. Using the protocol as described, the gold nanorods are synthesized with a size of ~42 × 11 nm, with a typical concentration of ~1.5 nM. The washed gold nanorods can be stored at room temperature and will remain stable for several months.
12. The gold nanorods require a high centrifugation speed to separate them from the supernatant. However, we found that a high centrifugation speed >5,000 rpm after ligand exchange with Brij56 causes the nanorods to aggregate. The use of Amicon ultra-0.5 mL centrifugal filters allow us to separate the gold nanorods from supernatant by centrifuging at lower speed in a shorter time to minimize aggregation of the nanorods.
13. The use of Brij56 as an intermediate ligand is necessary for ligand exchange to OF and PS. Direct addition of OF and PS to NR-CTAB resulted in irreversible aggregation for PS and poor ligand exchange for OF.
14. To assess the colloidal stability of gold nanorods in other types of media, simply disperse 50 μL of the gold nanorods colloid in 450 μL of the media of interest.
15. This wavelength range is determined based on a gold nanorods LSPR peak of 800 nm. In general, for gold nanorods having other LSPR peak (λ_{peak}), the lower limit of the wavelength range (λ_{min}) is determined from the minimum point in the spectrum between the transverse peak and longitudinal peak. The upper limit of the wavelength range (λ_{max}) is determined from the relationship $\lambda_{max} = \lambda_{peak} + (\lambda_{peak} - \lambda_{min})$ or the upper wavelength scan limit of the spectrophotometer, whichever is lower.

Acknowledgments

This work was supported by the NSF grant DMR #0906838.

References

1. Alkilany AM, Murphy CJ (2010) Toxicity and cellular uptake of gold nanoparticles: what we have learned so far? J Nanoparticle Res 12:2313–2333
2. Boldt K, Bruns OT, Gaponik N, Eychmuller A (2006) Comparative examination of the stability of semiconductor quantum dots in various biochemical buffers. J Phys Chem B 110: 1959–1963
3. Noh M, Kim T, Lee H, Kim CK, Joo SW, Lee K (2010) Fluorescence quenching caused by aggregation of water-soluble CdSe quantum dots. Colloids Surf a-Physicochem Eng Aspects 359:39–44
4. Yang P, Murase N, Yu JH (2011) SiO(2) beads with quantum dots: preparation and stability investigation for bioapplications. Colloids Surf a-Physicochem Eng Aspects 385:159–165
5. de Puig H, Federici S, Baxamusa SH, Bergese P, Hamad-Schifferli K (2011) Quantifying the nanomachinery of the nanoparticle-biomolecule interface. Small 7:2477–2484
6. Ferhan AR, Guo LH, Kim DH (2010) Influence of ionic strength and surfactant concentration on electrostatic surfacial assembly of cetyltrimethylammonium bromide-capped gold nanorods on fully immersed glass. Langmuir 26:12433–12442
7. MacCuspie RI (2011) Colloidal stability of silver nanoparticles in biologically relevant conditions. J Nanoparticle Res 13:2893–2908
8. Sethi M, Joung G, Knecht MR (2009) Stability and electrostatic assembly of Au nanorods for Use in biological assays. Langmuir 25:317–325
9. Stebounova LV, Guio E, Grassian VH (2011) Silver nanoparticles in simulated biological media: a study of aggregation, sedimentation, and dissolution. J Nanoparticle Res 13:233–244
10. Chen ZP, Zhang Y, Xu K, Xu RZ, Liu JW, Gu N (2008) Stability of hydrophilic magnetic nanoparticles under biologically relevant conditions. J Nanosci Nanotechnol 8:6260–6265
11. Heister E, Neves V, Lamprecht C, Silva SRP, Coley HM, McFadden J (2012) Drug loading, dispersion stability, and therapeutic efficacy in targeted drug delivery with carbon nanotubes. Carbon 50:622–632
12. Santander-Ortega MJ, Peula-Garcia JM, Goycoolea FM, Ortega-Vinuesa JL (2011) Chitosan nanocapsules: Effect of chitosan molecular weight and acetylation degree on electrokinetic behaviour and colloidal stability. Colloids Surf B Biointerfaces 82:571–580
13. Park JY, Choi ES, Baek MJ, Lee GH (2009) Colloidal stability of amino acid coated magnetite nanoparticles in physiological fluid. Mater Lett 63:379–381
14. Fang C, Bhattarai N, Sun C, Zhang MQ (2009) Functionalized nanoparticles with long-term stability in biological media. Small 5:1637–1641
15. Lin YS, Abadeer N, Haynes CL (2011) Stability of small mesoporous silica nanoparticles in biological media. Chem Commun 47: 532–534
16. Wiogo HTR, Lim M, Bulmus V, Yun J, Amal R (2011) Stabilization of magnetic iron oxide nanoparticles in biological media by fetal bovine serum (FBS). Langmuir 27:843–850
17. Cirtiu CM, Raychoudhury T, Ghoshal S, Moores A (2011) Systematic comparison of the size, surface characteristics and colloidal stability of zero valent iron nanoparticles pre- and post-grafted with common polymers. Colloids Surf a-Physicochem Eng Aspects 390:95–104
18. Kah JC-Y, Zubieta A, Saavedra RA, Hamad-Schifferli K (2012) Stability of gold nanorods passivated with amphiphilic ligands. Langmuir 28:8834–8844

Chapter 10

Electrochemical Measurements of DNA Melting on Surfaces

Irina Belozerova, Dongbiao Ge, and Rastislav Levicky

Abstract

Thermal denaturation, or melting, measurements are a classic technique for analysis of thermodynamics of nucleic base driven associations in solution, as well as of interactions between nucleic acids and small molecule ligands such as drugs or carcinogens. Performed on surface-immobilized DNA films, this well-established technique can help understand how energetics of surface hybridization relate to those in solution, as well as provide high-throughput platforms for screening of small molecule ligands. Here we describe methods for measuring DNA melting transitions at solid/liquid interfaces with focus on the role of immobilization chemistry, including a common "immobilization-through-self-assembly" approach that is effective at moderate temperatures, and a thermo-stable approach based on polymer-supported DNA monolayers that can be used at elevated temperatures. We also discuss conditions necessary for reversible measurements, as signified by superimposition of the association (cooling) and dissociation (heating) transitions of immobilized DNA strands.

Key words DNA melting, Surfaces, Electrochemistry, Monolayers, Stability, Thermodynamics

1 Introduction

Temperature-induced denaturation of double-stranded nucleic acids, or "melting" [1], has been widely used for analysis of binding between small molecule ligands and DNA [2–7] as well as for investigations of factors affecting the stability of nucleic acid secondary structures such as nucleotide mismatches, ionic strength, and presence of chaotropic agents [8–16]. Adaptation of DNA melting to solid supports [17–33] enables affordable, time-effective high-throughput analysis as well as provides a valuable tool for fundamental study of how interactions of nucleic acids in interfacial environments compare to their better understood solution behaviors.

Paolo Bergese and Kimberly Hamad-Schifferli (eds.), *Nanomaterial Interfaces in Biology: Methods and Protocols*, Methods in Molecular Biology, vol. 1025, DOI 10.1007/978-1-62703-462-3_10, © Springer Science+Business Media New York 2013

Surface melting transitions can be performed by heating a double-stranded DNA layer in which one strand of each duplex, the "probe" strand, is immobilized by one end to the solid support while the other, the "target" strand, is free to diffuse into solution once melted off. Nonequilibrium melting results when there is an imbalance between the reverse (i.e., dehybridization) and forward (i.e., hybridization) reaction rates; this condition can be exacerbated by heating too fast or by maintaining a low solution concentration of targets. Data from such experiments are not suited to quantitative extraction of thermodynamic parameters because of the nonequilibrium conditions of measurement. Equilibrium melting curves result when the surface population of bound targets remains always in equilibrium with the free target species in solution. Equilibrium melting transitions produce superposition of heating (dissociation) and cooling (hybridization) curves, so that the same sequence of surface states is reproduced as a function of temperature whether the states are reached by heating or cooling. Data obtained under equilibrium conditions can be used to derive thermodynamic parameters of probe-target hybridization, as well as those of small ligand binding.

DNA melting curves on solid supports have been measured by a variety of techniques, including surface plasmon resonance (SPR) [17–21], fluorescence [22–29], dynamic light scattering [30], and electrochemical methods [31–33]. In some cases reversibility of the transitions was verified by superposition of heating and cooling traces [19], though often such verification of equilibrium was omitted possibly due to difficulties with maintaining the probe layer structure at elevated temperatures.

Here we describe electrochemical methods for monitoring surface DNA helix–coil transitions under near equilibrium conditions. Two surface chemistries, both applicable to gold electrode supports, are described: (1) self-assembled (SA) DNA monolayers, in which probes attach to the electrode directly via thiolate bonds, and which are suitable for service at up to 50 °C, and (2) polymer-anchored (PA) DNA monolayers, in which probes are attached through thioether bonds to preassembled nanometer-thick films of poly(mercaptopropyl)methylsiloxane (PMPMS) on the electrode (Fig. 1). In method (2), the polymer mercaptopropyl side chains form multivalent anchoring that is further reinforced through cross-linking of the PMPMS film before covalent immobilization of the DNA probes through thiol-maleimide conjugation. We have used the PMPMS chemistry at temperatures of up to 90 °C.

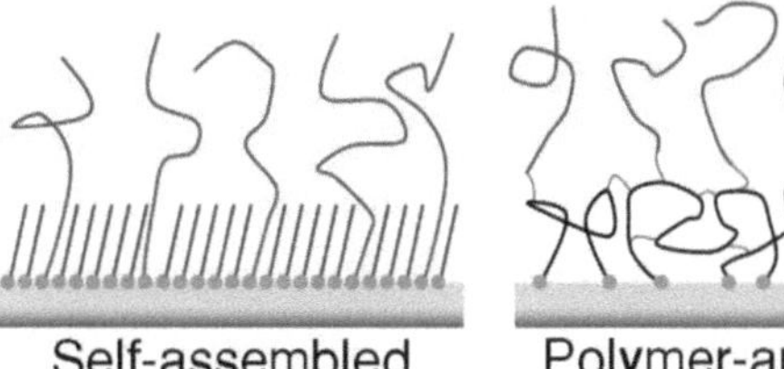

Fig. 1 DNA immobilization. *Left*: self-assembled DNA monolayers; *right*: polymer-anchored DNA monolayers

2 Materials

2.1 DNA Oligonucleotides and Electroactive Label

1. Disulfide-modified DNA oligonucleotide probe (Eurofins MWG Operon or Integrated DNA Technologies), purified by reverse phase high-performance liquid chromatography (HPLC) (*see* **Note 1**). Store probe DNA aqueous solutions at −20 °C.
2. Single-stranded, amine end-modified DNA target oligonucleotides (Eurofins MWG Operon or Integrated DNA Technologies, with HPLC purification).
3. Amine-reactive electroactive ferrocene label: *N*-succinimidyl ferrocenecarboxylate (FcCA, TCI America).
4. Ferrocene-labeled target oligonucleotides. Store dry at −20 °C (*see* **Note 2**).

2.2 Buffers and Solutions

All solutions are prepared by dissolving dry reagents in deionized water with a purity corresponding to a resistivity of 18.2 MΩ cm or higher at 25 °C.

1. Electrode polishing solution: (1) 0.5 M sulfuric acid, 10 mM potassium chloride; (2) 0.5 M sulfuric acid.
2. Probe deposition solution: typically 1–2 μM of disulfide-terminated (SA layers) or thiol-terminated (PA layers) probe in 0.5 M sodium phosphate buffer (SPB), at pH 7.0–7.4.
3. Electrode passivation solution for SA layers: typically 0.5–1.0 mM 6-mercapto-1-hexanol (MCH, 97 % purity, Sigma-Aldrich) in 0.1 M SPB, pH 7.0–7.4. Prepare fresh passivation solutions daily.

4. PMPMS deposition solution for PA layers: 10 mM (by monomer residue) of poly(mercaptopropyl)methylsiloxane (PMPMS, degree of polymerization 40, purity 95 %, Gelest, Inc.) in toluene.
5. Disulfide reducing solution: typically 10 mM dithiothreitol (DTT) in 10 mM tris(hydroxymethyl)aminomethane, 1 mM ethylenediaminetetraacetic acid, pH 8.0 (TRIS–EDTA buffer). The reducing solution is typically used for 100 μM concentrations of disulfide probe DNA.
6. HPLC solvent: 12–60 % methanol linear gradient spread over 22 min in 100 mM hexafluoroisopropanol, 4.5 mM triethylamine aqueous buffer, pH 8.0.
7. PMPMS cross-linking solution: 5 mM 1,11-bis-maleimidotriethyleneglycol (BM(PEG)$_3$, Pierce Biotechnology) in dichloromethane (DCM).
8. Measurement (for melting curves) buffer: typically 0.1–1.0 M SPB (pH 7.0–7.4), 25–50 nM FcCA-labeled target, 0.1 mM MCH (SA layers only) (*see* **Note 3**).

2.3 Electrochemical Supplies and Equipment

1. Working electrode: 3 mm diameter polycrystalline gold rotating disk electrode (RDE) or 1.6 mm diameter stationary gold electrode; counter electrode: coiled platinum wire; reference electrode: silver/silver chloride reference electrode (Ag/AgCl/3 M NaCl).
2. Electrode polishing accessories: 1 μm diamond slurry, 0.05 μm polishing alumina, nylon and velvet polishing pads (Bioanalytical Systems, Inc.).
3. Equipment: CH Instruments 660C workstation or equivalent; Bioanalytical Systems RDE-2 rotating electrode station or equivalent; jacketed RDE or stationary electrode glass cell (e.g., from Bioanalytical Systems, Inc.).
4. Temperature control: refrigerated water bath (RTE-740 Digital Plus, Thermo Scientific NESLAB), with software temperature control.

3 Methods

3.1 Electrode Cleaning

Mechanically polish gold disk electrodes with 1 μm diamond slurry on a nylon polishing pad for 2 min, rinse the electrodes with copious amounts of methanol followed by deionized water (18.2 MΩ cm), and then sonicate them in methanol and water for 3 min each (e.g., Ultrasonic Cleaner FS20D, Fisher Scientific). Repeat polishing on a velvet pad using 0.3 μm alumina powder and the same sonication procedure (*see* **Note 4**). Assemble the electrochemical cell with the platinum auxiliary/counter electrode, the silver/silver chloride reference electrode, and the mechanically polished working electrode, and immediately perform electrochemical polishing in electrode polishing solution #1 for 10 cyclic voltammetry cycles between −0.2 V and 1.75 V (potentials are quoted vs. Ag/AgCl/3 M NaCl reference) at a scan rate 0.1 V/s, followed by polishing in electrode polishing solution #2 for 50 cycles from −0.2 V to 1.75 V at 0.1 V/s. Assess good polishing during potentiodynamic cycling in polishing solution #2 by observing superimposition of the cyclic voltammograms. Rinse the electrodes with deionized water. If desired, determine the roughness factor r (r = actual area/geometric area) from the double-layer capacitance in 0.1 M sodium fluoride at −0.84 V as reported previously [34]. Keep the electrode surface wet after electrochemical polishing to avoid surface contamination.

3.2 SA DNA Layer Formation

Form the DNA layer by immersing freshly polished and rinsed with deionized water electrodes into the probe deposition solution at 45 °C for 90 min. Prevent solution evaporation by sealing the solution container housing the electrode (e.g., microcentrifuge tube). Passivate the remaining electrode surface with MCH via immersing into the SA electrode passivation solution at 45 °C overnight (*see* **Note 5**).

3.3 PA DNA Layer Formation

3.3.1 Preparation of PMPMS-Coated Electrodes

Wash the electrodes with deionized water, completely dry them under a nitrogen stream, rinse them with anhydrous toluene solvent, and transfer them, while still wet, into PMPMS deposition solution for 1 h (*see* **Note 6**). After chemisorption of PMPMS, rinse the modified electrodes extensively with toluene and dry them under a compressed nitrogen stream.

3.3.2 Preparation of PMPMS-DNA Layers

1. Reduce disulfide probe to its thiol terminus by treating the probe DNA with the disulfide reducing solution for 2.5 h. After reduction, pass the solution mixture once through a NAP-10 (GE Healthcare) separation column, followed by final purification using reverse phase HPLC (e.g., Clarity 3 μm Oligo RP column from Phenomenex) with the HPLC solvent gradient at a flow rate of 0.5 mL min^{-1}.
2. Prepare PMPMS-anchored DNA probe monolayers by first immersing PMPMS-coated disk electrodes in PMPMS cross-linking solution under stirring for 5 h (*see* **Note 7**), followed by thorough rinsing with DCM and drying under compressed nitrogen. Immediately place the dried, maleimide-activated electrodes into probe deposition solution at room temperature overnight. After probe immobilization, rinse electrodes with deionized water and SPB prior to use.

3.4 DNA Hybridization-Melting

Following layer preparation and electrode rinsing, assemble the electrochemical jacketed cell with the platinum auxiliary electrode and the Ag/AgCl reference electrode. If desired, first set up the reference electrode in a double junction configuration by inserting it into a glass tube with a porous Vycor tip that has been filled with the same measurement buffer as the cell. This serves to limit Cl^- leakage from the reference electrode's internal reservoir into the cell. Fill the cell with the measurement buffer, after first deoxygenating the buffer by bubbling nitrogen through it for 20 min. Maintain a nitrogen blanket over the measurement solution throughout the experiment (*see* **Note 8**). Connect the cell to the thermostatted water bath for temperature control and preheat the cell to high limit temperature at which no hybridization is observed (*see* **Note 9**). Immerse the working electrode into the measurement solution. Set the RDE rotation speed to 1,500 rpm (*see* **Note 10**). Start the cooling cycle at a sufficiently slow temperature scan rate to ensure reversible operation (*see* **Note 11**). Following each temperature step, perform a cyclic voltammetry (CV) scan. Typical CV settings are a scan rate of between 1 and 20 V/s and over a potential range sufficient to fully measure the ferrocene electroactivity, e.g., 0 V to 0.75 V for FcCA. It is advisable to control the surface potential between CV scans to ensure that ferrocene labels remain unoxidized during these wait times, for example by keeping the electrode at 0 V vs. the Ag/AgCl/3 M NaCl reference. After the cooling scan, perform a heating scan following analogous procedures but increasing rather than decreasing the temperature between CV scans. From each CV trace, obtain an unnormalized coverage of hybridized targets (i.e., probe-target duplexes) by integrating the area of the ferrocene peaks. Plot the integrated areas vs. temperature to obtain the melting curve (Fig. 2). If desired, the peak areas can be converted to quantitative coverages (targets/area) by dividing the peak areas by the CV scan rate, the elementary charge of one electron, and the electrode area [35] (Fig. 3). Examine superimposition of cooling and heating transitions to confirm that equilibrium was maintained during measurement.

4 Notes

1. In our experience, for self-assembled monolayers, shorter oligonucleotide lengths that melt below 50 °C, e.g., 12-mers, are compatible with this immobilization chemistry. Polymer-anchored monolayers employ longer oligonucleotides, e.g., 22-mers, for which melting process is completed below 90 °C. Our probes are typically modified with a disulfide modification at the 5′ end, which can be connected to the rest of the molecule by an alkyl, oligothymine, or other linkers to provide elevation above the substrate and to improve strand accessibility for hybridization with targets.
2. Procedures for labeling oligonucleotides with electroactive tags are described elsewhere [35, 36]. It is advantageous to

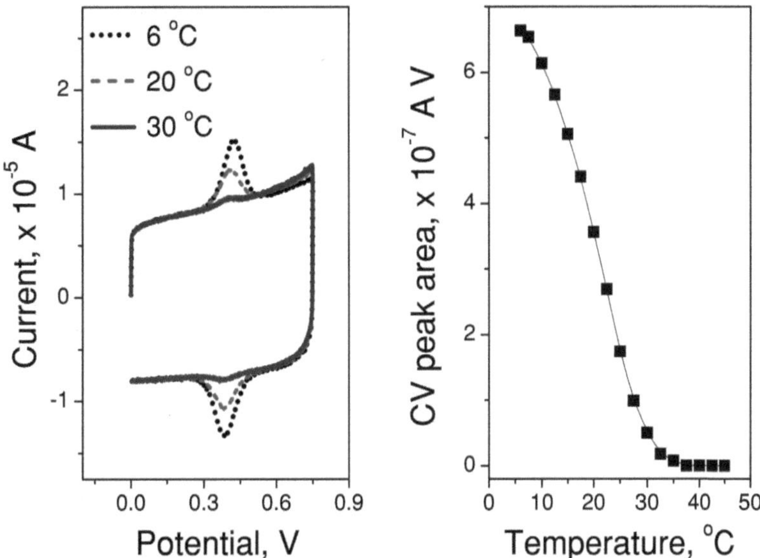

Fig. 2 *Left*: cyclic voltammograms of self-assembled DNA monolayers hybridized to ferrocene-labeled target at various temperatures. *Right*: CV peak area as a function of temperature, tracing out the melting transition of the immobilized DNA

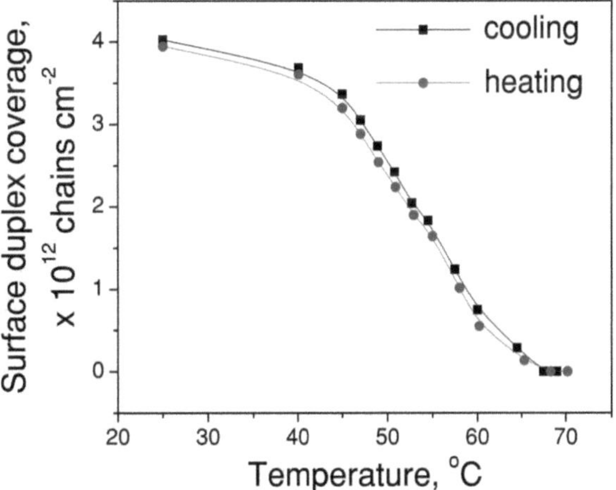

Fig. 3 Hybridization (cooling) and melting (heating) transitions of polymer-anchored DNA monolayers in the presence of ferrocene-labeled target, illustrating good superposition of the cooling and heating scans

place the label such that, when hybridized, the label is in proximity to the surface to facilitate electron transfer and thus to improve signal.

3. Aqueous solutions other than SPB, may be used, although chloride-containing buffers should be avoided to prevent ferrocene degradation, via its oxidized ferricenium state, by

nucleophilic Cl^-. For SA layers, presence of 0.1 mM MCH in the experimental solution provides regeneration of the MCH passivation to counter any MCH desorption that may occur during thermal cycling.

4. Polishing pads should be thoroughly rinsed before use to remove any remaining polishing grit and to avoid scratching of gold by grit clumps. Only one size (and type) of polishing slurry should be used on each polishing pad (i.e., 1 μm, 0.5 μm). Sufficient amount of the polishing slurry should be applied upfront, adding additional slurry once polishing has been started is not recommended. The electrode has to be positioned perpendicularly to the pad surface and polishing motions performed in circular fashion (with alternating clockwise/counterclockwise direction), or using figure-eight motions, in order to avoid uneven polishing and erosion of the metal below the surrounding plastic coating. The surface of the electrode is not to be touched and should be kept wet with a drop of deionized water during transfers between all cleaning solutions so as to minimize adsorption of ambient contaminants.
5. It is advisable to perform SA layer formation at the high temperature limit in order to expose the forming probe layer to elevated temperatures so as to decrease formation of less stably bound probes.
6. We find that a 60 min immobilization time is sufficient to prepare continuous PMPMS anchor films of good quality for subsequent attachment of DNA probes.
7. $BM(PEG)_3$ serves to cross-link unbound PMPMS free thiol groups for formation of a stable two-dimensional anchor polymer network on the surface. Any unreacted maleimide groups will remain available for attachment of thiol-terminated DNA probes. By varying the cross-linking reaction time, prior to DNA deposition, the probe DNA coverage can be adjusted, e.g., in our hands, cross-linking time of 1 h yields final DNA coverage of about 2×10^{12} cm^{-2}, while a 5 h cross-linking reaction yields about 4.0×10^{12} cm^{-2} probe coverage.
8. Presence of dissolved O_2 facilitates degradation of ferrocene labels; therefore, it is desirable to exclude O_2 from the measurement buffer as much as possible. If proper deoxygenation steps are followed, the O_2 concentration in deoxygenated SPB should be less than about 10 μM at room temperature.
9. Measurements should be initiated at temperatures above the expected range of melting. Starting at the higher temperatures promotes structural relaxation of the DNA layer into the hybridized state as it cools, improving reversibility of the observed melting transitions.

10. Use of a rotating disk electrode geometry helps overcome mass transfer limitations associated with transport of the target DNA to the electrode, as well as improves temperature uniformity. Similarly, for stationary electrodes, it is advisable to continuously stir the measurement solution via a magnetic stir bar.
11. Cooling and heating can be performed in temperature increments of about 1–2 °C, followed by 5–15 min equilibration times at each new temperature. Averaged over time, this corresponds to effective scan rates of about 0.1–0.4 °C min^{-1}. It was observed that the optimal scan rate depends on the solution ionic strength, i.e., in higher salinity medium faster scan rates can be applied without causing hysteresis between cooling and heating scans, thus retaining reversibility of measurement (e.g., in 0.1 M SPB a 0.16 °C min^{-1} scan rate is employed, while 1.0 M SPB could use a scan rate of 0.30 °C min^{-1}).

Acknowledgments

This work was supported by NIH grant R33HG003089 and NSF grant DMR 07-06170.

References

1. Doty P (1957) The physical chemistry of deoxyribonucleic acids. J Cell Comp Physiol 49:27–57
2. Demeunynck M, Bailly C, Wilson WD (eds) (2003) Small molecule DNA and RNA binders: from synthesis to nucleic acid complexes. Wiley-VCH, Weinheim
3. Crothers DM (1971) Statistical thermodynamics of nucleic acid melting transitions with coupled binding equilibria. Biopolymers 10:2147–2160
4. Lazurkin YS, Frank-Kamenetskii MD, Trifonov EN (1970) Melting of DNA: its study and application as a research method. Biopolymers 9:1253–1306
5. McGhee JD (1976) Theoretical calculations of the helix-coil transition of DNA in the presence of large, cooperatively binding ligands. Biopolymers 15:1345–1375
6. Spink CH, Wellman SE (2001) Thermal denaturation as tool to study DNA-ligand interactions. In: Chaires JB, Waring MJ (eds) Methods in enzymology, vol 340. Academic Press, New York, NY, pp 193–211
7. Wilson WD et al (1997) Evaluation of drug-nucleic acid interactions by thermal melting curves. In: Fox KR (ed) Drug-DNA interaction protocols. Humana Press, Totowa, NJ, pp 219–240
8. Kwok P-Y (2001) Methods for genotyping single nucleotide polymorphisms. Annu Rev Genomics Hum Genet 2:235–258
9. Howell WM et al (1999) Dynamic allele-specific hybridization: a new method for scoring single nucleotide polymorphisms. Nat Biotechnol 17:87–88
10. Nelson JW, Martin FH, Tinoco I Jr (1981) DNA and RNA oligomer thermodynamics: the effect of mismatched bases on double-helix stability. Biopolymers 20:2509–2531
11. Sadhu C, Dutta S, Gopinathan KP (1984) Influence of formamide on the thermal stability of DNA. J Biosci 6:817–821
12. Blake RD, Delcourt SG (1996) Thermodynamic effects of formamide on DNA stability. Nucleic Acids Res 24:2095–2103
13. Spink CH, Chaires JB (1999) Effects of hydration, ion release, and excluded volume on the melting of triplex and duplex DNA. Biochemistry 38:496–508
14. Nordstrom LJ et al (2006) Effect of ethylene glycol, urea, and N-methylated glycines on DNA thermal stability: the role of DNA base pair composition and hydration. Biochemistry 45:9604–9614
15. Spink CH, Garbett N, Chaires JB (2007) Enthalpies of DNA melting in the presence of osmolytes. Biophys Chem 126:176–185

16. Del Vecchio P et al (1999) The effects of polyols on the thermal stability of calf thymus DNA. Int J Biol Macromol 24:361–369
17. Peterlinz KA et al (1997) Observation of hybridization and dehybridization of thiol-tethered DNA using two-color surface plasmon resonance spectroscopy. J Am Chem Soc 119:3401–3402
18. Fuchs J et al (2010) Effects of formamide on the thermal stability of DNA duplexes on biochips. Anal Biochem 397:132–134
19. Fuchs J et al (2010) Salt concentration effects on equilibrium melting curves from DNA microarrays. Biophys J 99:1886–1895
20. Fiche JB et al (2007) Temperature effects on DNA chip experiments from surface plasmon resonance imaging: isotherms and melting curves. Biophys J 92:935–946
21. Fiche JB et al (2008) Point mutation detection by surface plasmon resonance imaging coupled with a temperature scan method in a model system. Anal Chem 80:1049–1057
22. Nasef H et al (2010) Melting temperature of surface-tethered DNA. Anal Biochem 406: 34–40
23. Russom A et al (2006) Rapid melting curve analysis on monolayered beads for high-throughput genotyping of single-nucleotide polymorphisms. Anal Chem 78:2220–2225
24. Watterson JH et al (2000) Effects of oligonucleotide immobilization density on selectivity of quantitative transduction of hybridization of immobilized DNA. Langmuir 16:4984–4992
25. Wang X, Krull UJ (2005) Synthesis and fluorescence studies of thiazole orange tethered onto oligonucleotide: development of a self-contained DNA biosensor on a fiber optic surface. Bioorg Med Chem Lett 15:1725–1729
26. Ozel AB et al (2012) Target concentration dependence of DNA melting temperature on oligonucleotide microarrays. Biotechnol Prog 28:556–566
27. Vainrub A et al (2007) Predicting DNA duplex stability on oligonucleotide arrays. Methods Mol Biol 382:393–403
28. Forman JE et al (1998) Thermodynamics of duplex formation and mismatch discrimination on photolithographically synthesized oligonucleotide arrays. ACS Sym Ser 682:206–228
29. Khomyakova E et al (2008) On-chip hybridization kinetics for optimization of gene expression experiments. Biotechniques 44: 109–117
30. Xu J, Craig SL (2005) Thermodynamics of DNA hybridization on gold nanoparticles. J Am Chem Soc 127:13227–13231
31. Nasef H, Beni V, O'Sullivan CK (2010) Electrochemical melting-curve analysis. Electrochem Commun 12:1030–1033
32. Nasef H, Beni V, O'Sullivan CK (2010) Labelless electrochemical melting curve analysis for rapid mutation detection. Anal Methods 2:1461–1466
33. Meunier-Prest R et al (2003) Direct measurement of the melting temperature of supported DNA by electrochemical method. Nucleic Acids Res 31:e150
34. Oesch U, Janata J (1983) Electrochemical study of gold electrodes with anodic oxide films- I. Formation and reduction behavior of anodic oxides on gold. Electrochim Acta 28:1237–1246
35. Tercero N et al (2009) Morpholino monolayers: preparation and label-free DNA analysis by surface hybridization. J Am Chem Soc 131:4953–4961
36. Ge D, Levicky R (2011) A comparison of five bioconjugatable ferrocenes for labeling of biomolecules. Chem Commun 46: 7190–7192

Chapter 11

Formation and Characterization of the Nanoparticle–Protein Corona

Marco P. Monopoli, Andrzej S. Pitek, Iseult Lynch, and Kenneth A. Dawson

Abstract

Over the last decade the existence of "the corona," a natural interface between nanomaterials and living matter in biological milieu, evolved from a vague concept into broadly recognized fact. This robust shell arises (to some extent) on the surface of all nanoparticles (NPs), even the ones designed to avoid its formation upon contact with biological fluids and confers a biological identity to the nanomaterials such that they can engage with cellular machinery. The NP corona consists of those proteins (and other biomolecules such as lipids and sugars) residing on the NP surface for a sufficient timescale to influence the NP's properties and interactions with living systems. This chapter aims to provide simple protocols, as well as notes on potential pitfalls, to help researchers to perform basic experiments in this field as the basis for a more mechanistic approach to study and understand NP–protein corona complexes. This work has been supported by INSPIRE (Integrated NanoScience Platform for Ireland) funded by the Irish Government's Programme for Research in Third Level Institutions, Cycle 4, National Development Plan 2007–2013, and 3MICRON (NMP-2009-LA-245572), NAMDIATREAM (NMP4-LA-2010-246479) and QualityNano (INFRA-2010-262163) funded by the European Commission 7th Framework Programme.

Key words Nanoparticle, Biological fluids, Protein corona, Dynamic light scattering, Differential centrifugal sedimentation, Z-potential, TEM

1 Introduction

It is now well recognized that nanomaterials exposed to a biological milieu will strongly interact with biological molecules, including proteins, sugars, and lipids, forming "a corona" [1–10]. While the exposed layer of biomolecules rapidly exchanges, an inner layer, the hard corona, remains associated with NP surface for so long that it is believed to give a new biological identity to the particle in the biological environment. Thus the inner hard corona is of high scientific relevance and is the most studied (Fig. 1) [4, 11].

Nanomaterial surface properties, surface curvature (thus size), material, morphologies, surface modification, etc., will

Paolo Bergese and Kimberly Hamad-Schifferli (eds.), *Nanomaterial Interfaces in Biology: Methods and Protocols*, Methods in Molecular Biology, vol. 1025, DOI 10.1007/978-1-62703-462-3_11, © Springer Science+Business Media New York 2013

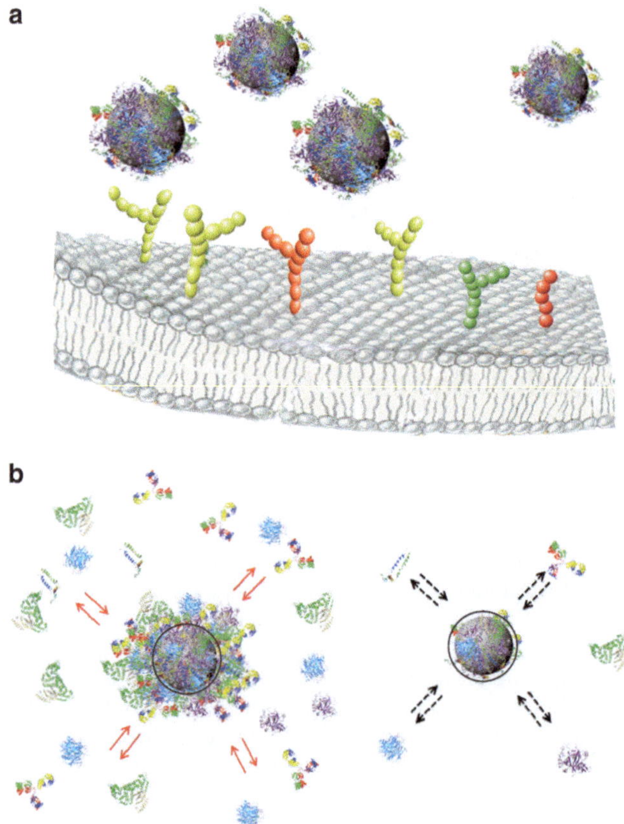

Fig. 1 (**a**) Schematic drawing of the NP–protein corona complexes and their interaction scenarios at the bio–nano interface at cellular level, where the long-lived protein corona modulates the interaction with cellular receptors while the pristine NPs surface is not available for binding. (**b**) Schematic drawing of the dynamic protein corona in rapid exchange with its environment and of NPs hard (slowly exchanging) corona. Samples of proteins were re-drawn from the RCSB PDB (www.pdb.org). Adapted with permission from ref. 4. Copyright (2010) American Chemical Society

strongly influence the protein corona composition. For example, studies performed in human plasma have shown that from the ~4,000 proteins in the solution, only a few tens of proteins form the hard corona as their affinity to the NP surface is extremely high, often despite their low abundance in the biological fluid [3, 8, 9, 12–19].

Recent studies have shown that it is possible to understand and characterize the protein corona in more detail using emerging approaches or adapting older techniques of physical chemistry, microscopy, and proteomics for use at the nanoscale [6, 20, 21].

The aim of this chapter is to provide simple protocols and guides and to highlight potential pitfalls in order to help researchers to perform basic experiments in this field in a manner that ensures that all steps are correctly performed. We provide a guide for all steps, starting from how blood proteins (typically used in most corona studies, although the approaches are equally applicable to other biofluids such as lung lavage fluid or cerebrospinal fluid) should be collected and processed, the experimental conditions, how to successfully isolate monodisperse NPs and their strongly bound hard corona from rich biological complexes, and all possible precautions that have to be taken to ensure reproducibility.

Understanding the nature of the NP–protein corona complex, and its impacts on NP dispersion stability, surface presentation, and cellular uptake is central to interpreting data from all studies involving NP interactions with living systems, as it provides key information on dose–response (Fig. 1) [22–27].

The following sections outline a complete and coherent methodology, which allows researchers to prepare, isolate, and characterize the NP–protein corona complexes and perform physicochemical characterization of the NP-corona complexes (with and without excess of proteins).

1.1 NP–Protein Corona Preparation and Characterization

The standard procedure for preparation of *NP–protein corona* complexes consists of three main stages:

- Biological fluid (plasma or serum) preparation.
- NP incubation with biological fluid.
- Separation of NP–protein corona complexes from excess plasma (i.e., proteins not bound to NPs surface) and washes to remove loosely bound proteins.

Physico-chemical characterization of the NPs in their pristine state and after exposure to biological media is of the highest importance for understanding impacts of NPs. The presence of agglomerates and multiple populations of dimers, trimers, and larger agglomerates, and poorly stable samples will lead to different biological consequences. Analysis of the protein corona without a well-thought-through approach will lack relevance, as it will inevitably study the average protein corona of the multiple populations rather than being cognizant of the different subpopulations.

The basic NP dispersion characteristics such as hydrodynamic diameter, zeta-potential, and polydispersity index in the relevant medium should be determined both before and after the “corona prep” procedure, and in a time resolved manner. The comparison of size and zeta-potential before and after protein corona formation provides information such as the thickness of the corona formed and its effect on the NP surface charge due to protein adsorption, since most NPs are negatively charged in the presence of a protein corona at physiological pH. Such changes in NP surface charge as a result of protein adsorption might lead to a drop of the electrostatic stability of the system but may also increase the steric stability of the system [28].

Dispersion proprieties of the NPs and NP-corona complexes are generally determined by Dynamic light scattering (DLS) and by differential centrifugal sedimentation (DCS).

DLS is the technique most commonly used for hydrodynamic diameter determination, allowing NP size measurement based on the light scattered by them due to their Brownian motion, which is strictly dependent on the size. After applying specific correlation functions, DLS leads to estimation of the NPs' diffusion constant, and after application of the Einstein–Stokes equation (below) calculates the NPs' hydrodynamic radius, R_h.

$$D = \frac{k_B T}{6\pi\eta R_h},$$

where:

D—diffusion constant, k_B—Boltzmann's constant, *T*—absolute temperature, η—viscosity, R_h—hydrodynamic radius of the particle.

Zeta potential (ZP), which characterizes the net charge of the particle surface in the given dispersant, provides information regarding the surface charge of the NPs and the change of surface charge after incubation in different media.

The DCS Disk Centrifuge is a particle size analyzer for measuring particles in the range of 0.01–40 μm. The analyzer measures particle size distributions using centrifugal sedimentation, within an optically clear spinning disk that is filled with fluid. The primary information from the analytical disk centrifuge is the time taken by the particles to travel from the center of the disk through a defined viscous sucrose gradient under a strong centrifugal force, to a detector placed at the outer rim of the disk. For materials with homogenous density and simple shape (for example, spherical particles) one can directly relate this time to a particle size [4]. A significant advantage of DCS measurement is that the instrument can successfully resolve multiple populations over a wide size ranges from a few nanometers to microns within the same sample. Moreover, it allows measurement of NPs incubated with fluids of any kind (in situ) as the biomolecule background will not significantly impact the NP measurement due to their very different sedimentation times. An example of the degree of size resolution possible with DCS is given in Fig. 2, where 100 nm PSCOOH NP–protein complexes (a) in situ (full plasma) and (b) hard corona complexes were measured after 1 h (full line) and 6 h (dotted line), where NP corona complexes are shifted from pristine NPs (line with open circles) due to the increase of size and the change of NPs density.

1.2 Proteomic Characterization of Protein Corona Composition

Another relevant aspect of the protein corona is its composition. There are a number of techniques in the field of molecular biology, which can be adopted and utilized to determine the identities of the protein constituents of the corona. One of the most popular

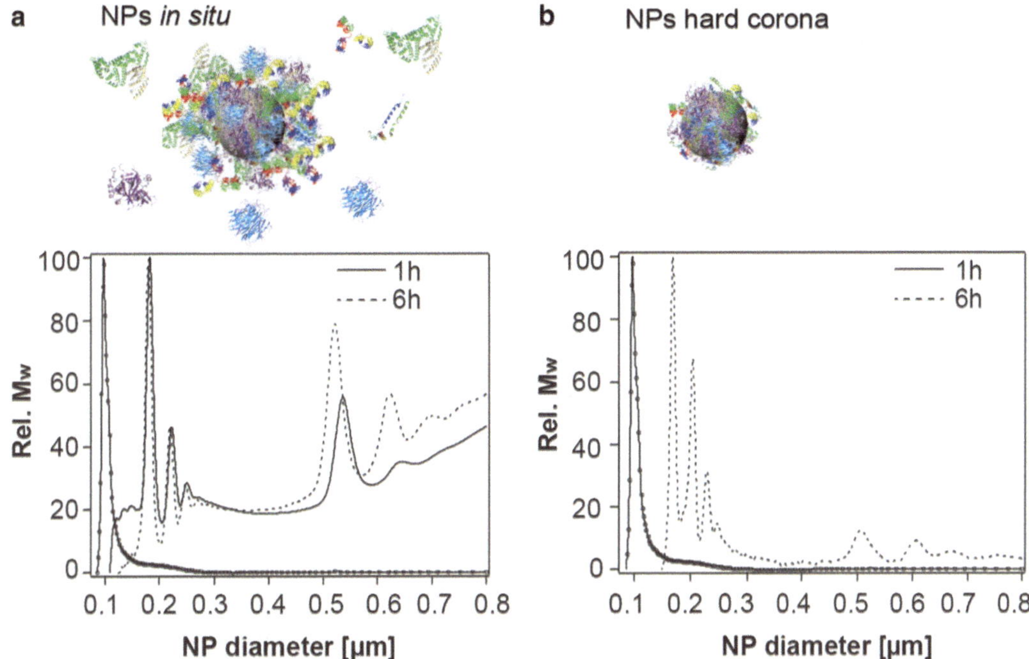

Fig. 2 DCS characterization of nominal "100 nm" carboxylated NPs size and their corona complexes size, after incubation in human plasma, both in situ (full corona) and after washes (hard corona complexes). Reprinted with permission from ref. 4. Copyright (2010) American Chemical Society

protein separation techniques is SDS-PAGE, where the proteins are separated via a polyacrylamide gel by size, which allows for the comparison of protein bands between different samples. After protein corona isolation, a typical proteomics shotgun approach requires the gel lane(s) to be sliced into several bands (typically ten for a 10 cm standard SDS-PAGE), an in gel trypsin digestion to be performed, followed by MS analysis following one of several well-established protocols available in the literature [3, 29, 30]. Since SDS-PAGE and gel based MS analysis procedures are commonly available in several laboratories, in this book chapter we describe potential pitfalls in **Notes 28–33** only, as detailed protocols are easily found in the literature [30].

2 Materials

1. Tubes for blood collection for plasma preparation: BD Vacutainer whole blood tubes with spray-coated K_2EDTA, lavender closures.
2. Tubes for blood collection for serum preparation: BD Increased Silica Act Clot Activator, red closure.
3. Protein LoBind Tube: Eppendorf.

4. Nanoparticles: various sizes of polystyrene sulfonate/carboxylate/amino modified fluorobeads, various sizes of silica yellow green NPs: Life Science, PolyScience, Kisker, Sigma, BBI.
5. Phosphate saline buffer: 0.01 M phosphate buffer, 0.0027 M potassium chloride and 0.137 M sodium chloride, pH 7.4.
6. Sucrose solutions for gradient preparation (DCS) 2–24 % w/w sucrose in H_2O, PBS (or any NP dispersant media).
7. Dodecane.
8. DLS sizing cuvette: Plastic disposable low volume cuvette.
9. Z-potential measurement cuvette: low volume disposable zeta cuvette.
10. DCS calibration standard, PVC 0.377 and PS 0.497: Analytics LTD.
11. Carbon coated copper TEM grids: Agar Scientific.
12. Protein concentration assay: Micro BCA assay: Pierce.
13. EDTA, Sodium Chloride, Acrylamide/Bis-acrylamide 40 % solution BioReagent, Trizma base, Glycine, Sodium Lauryl Sulfate (SDS).
14. 3× SDS-PAGE loading buffer: 60 mM Tris–HCl (pH 6.8 at 25 °C), 2 % (w/v) SDS, 0.1 M DTT, 10 % glycerol, 0.01 % (w/v) bromophenol blue: New England Biolabs.
15. Brilliant blue: Thermo Fisher.
16. Silver staining kit: 2D-Silver Stain-II Cosmo Bio.
17. Silver staining kit: Plos One GE Healthcare.

3 Methods

3.1 Biological Fluid Preparation: Plasma Preparation

1. Identify at least six seemingly healthy donors, ensuring an equal number of male and females.
2. Draw blood into EDTA coated Vacutainer tubes, taking 30–50 ml blood from each donor being sure to collect the full volume for a correct ratio between blood and EDTA (*see* **Notes 1** and **2**).
3. Gently invert the Vacutainer tubes ten times immediately after collection to ensure mixing of anticoagulant with the blood, and store the tubes at room temperature prior to the next step.
4. After all blood samples have been collected, centrifuge at 1,200 × *g* for 10 min at room temperature. Avoid using the brake to stop the centrifuge in order to avoid sample mixing.
5. Pool the supernatants (the plasma) into 50 ml tubes and discard the pellet that remains at the bottom of the centrifuge tubes (cells).

6. Centrifuge the pooled plasma sample at 2,500 × *g* for 15 min at room temperature.
7. Aliquot the supernatant into cryovials (typically 1 ml) and place them into −80 °C freezer until use (*see* **Note 3**).
8. The whole procedure should not take more than 3 h from the blood collection to freezing storage.

3.2 Biological Fluid Preparation: Serum Preparation

1. Identify at least six seemly healthy donors ensuring an equal number of male and females.
2. Draw blood into Vacutainer tubes for serum preparation taking 30–50 ml blood from each donor (*see* **Notes 1** and **2**).
3. Gently invert each tube five or six times, then incubate in an upright position at room temperature for 30–45 min (no longer than 60 min) to allow clotting.
4. Invert tubes to ensure that blood is no longer liquid, indicating that it is clotted.
5. Centrifuge the tubes at 1,300 × *g* for 10 min at room temperature. Avoid using the brakes to stop the centrifuge in order to avoid sample mixing.
6. Pool the supernatants (serum) into 50 ml tubes and discard the pellet at the bottom of the tubes.
7. Inspect serum for turbidity. Turbid samples should be centrifuged and aspirated again to remove remaining insoluble matter.
8. Aliquot the supernatant into cryovials (typically 1 ml) and place them into −80 °C freezer until use (*see* **Note 3**).

3.3 NPs Incubation with Biological Fluid

1. Ensure that biological fluid samples are fully defrosted and vortex briefly to ensure that mixture is homogeneous.
2. Bring the biological fluid of choice to the incubation temperature (typically 4, 20, or 37 °C) (*see* **Note 5**).
3. Spin for 3 min at 16,000 × *g* at 20 °C to pellet possible protein aggregates which might have been in the solution. Discard the pellet and keep the supernatant for the incubation step.
4. Add the biological fluid to the mixing vessel (*see* **Note 6**) and if necessary, add PBS to dilute it to the required protein content and vortex briefly to ensure homogeneity.
5. Add NPs (*see* **Notes 7–9**) and by repeated agitation with use of pipette, ensure that the sample is well mixed.
6. Put samples in orbital shaker at 250 rpm setting, at the required incubation temperature and incubate for time required by the experiment (*see* **Note 10**).

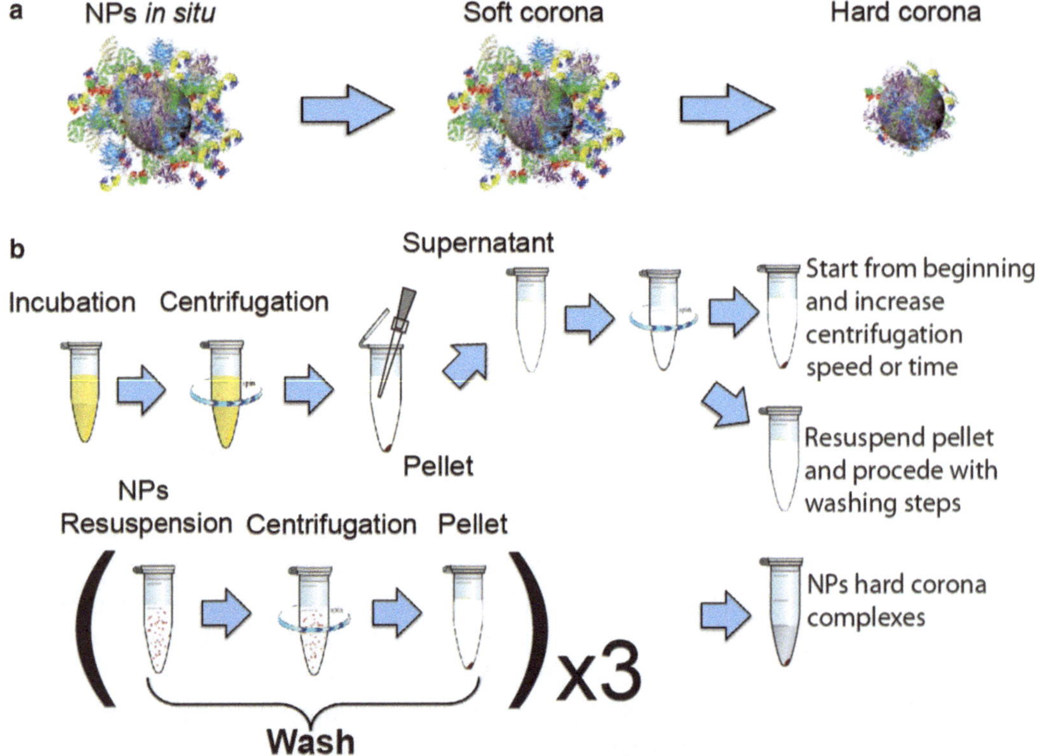

Fig. 3 Corona preparation steps. (**a**) Schematic representation of the different stages of corona formation and separation. (**b**) Diagram of optimization step and isolation of NP-hard corona complexes from unbound and loosely bound proteins. Proteins adapted from Protein Data Bank (http://www.pdb.org)

3.4 Hard Corona Preparation by Centrifugation (See Note 11)

1. Set temperature of the centrifuge to match the incubation temperature (Fig. 3).
2. Immediately after incubation time has been completed, place the NP-containing tubes in a bench centrifuge (*see* **Note 12**).
3. Spin the NP–protein complexes down to ensure that all NPs have been pelleted. The spin velocity and time needs to be optimized according to the NPs size and density (*see* **Note 11**).
4. NP-hard Corona complexes should be visible as a pellet at the bottom of the tube. Gently remove the supernatant without disturbing the pellet (immediate removal of supernatant prevents NPs from going back into dispersion). Transfer supernatant into a new vessel and centrifuge for 2 h at full speed (centrifugation control). If further NPs pellet is detected after this step, this indicates that the first centrifugation step has to be increased (time or centrifugation speed), so discard samples and start from incubation step, increasing centrifugation speed or time. If no pellet is detected after the centrifugation control, proceed to **step 5**.

5. Resuspend the NP-corona complexes in 1,000 μl of PBS by mixing with a pipette until solution is homogenous.
6. Spin NP-corona complexes in centrifuge as in previous step at the centrifugation speed optimized in **step 4**.
7. Gently remove supernatant (wash 1) and store it at −20 °C and resuspend pellet as described in **step 5**.
8. Spin NP-corona complexes in centrifuge as in previous step at centrifugation speed optimized in **step 4**.
9. Gently remove supernatant (wash 2) and store it at −20 °C. Resuspend pellet as in **step 5** and transfer the samples to a new vessel (*see* **Note 13**).
10. Spin NP-corona complexes in centrifuge as in previous step at centrifugation speed optimized in **step 4**.
11. Gently remove supernatant (wash 3) and store it at −20 °C (*see* **Note 14**).
12. Pellet now contains NP-hard corona and it should be processed according to the analysis that will be performed. The hard corona samples should be prepared freshly before each of the analytical procedures listed in Subheadings 3.5–3.9.

3.5 NP Size Characterization by Dynamic Light Scattering (DLS) (See Notes 15 and 16)

1. Dilute NPs to a final concentration of 50–100 μg/ml in 500 μl of the medium of choice (water, PBS, cell culture medium, diluted plasma, etc.). This NPs concentration provides a high quality of measurement data for most NP kinds/materials (*see* **Note 17**).
2. Set up the SOP according to the instrument manual paying special attention to the following parameters: Measurement type: Size; Material: NP material from the list; Solvent: appropriate solvent from the list; Temperature: depending on the experimental conditions; Equilibration time: make sure to set equilibration time long enough for the sample to reach the measurement temperature and equilibrate; No of measurements: at least 3; Number of runs: set on auto (minimum 10); Single run duration: 10 s; Attenuator: in most cases set on Auto.
3. Run the SOP and read the hydrodynamic diameter (D_h) (*see* **Note 18**). Note standard deviation and average PDI (Polydispersity index).

3.6 Differential Centrifugal Sedimentation (DCS) (See Note 19)

1. Prepare two sucrose solutions (*see* **Note 20**) appropriate for the specific NPs' material density: 2 and 8 % (w/w) for particles with density between 1.0 g/cm^3 and 1.1 g/cm^3, or 8 and 24 % (w/w) for particles with density above 1.3 g/cm^3.
2. Use one of the existing or set up a new operating procedure in DCS software, following the manual and paying special attention

to values of: NP material density, standards' density and size, and size range of measurement.

3. Inject first portion of sucrose gradient into the disk. Set RPM value to manual and set the velocity required according to the NPs' nominal size and density (*see* **Note 21**).
4. Wait until the disk is spinning at full velocity and start injecting the step gradient following the manufacturer's procedure.
5. Finish gradient with 0.5 ml of dodecane to prevent water evaporation. Leave the instrument for approximately 0.5 h to allow gradient to equilibrate.
6. Each measurement consists of a calibration run (with use of standard calibration NPs) and the sample run.
7. To ensure gradient stability, run a test sample twice and compare the results with older measurements. If peaks overlap gradient can be considered as stable, otherwise the test measurement should be repeated after an additional 15 min of gradient equilibration. Failing in this measurement requires preparation of a new gradient (*see* **Note 22**).
8. Prepare 100 μl of NPs solution in the dispersant fluid of choice (water, PBS, or biological fluid) at a concentration of 0.05–1 mg/ml, depending on the NPs' refractive index and light adsorption (*see* **Note 23**).
9. Run samples (*see* **Notes 24** and **25**).

3.7 Z-Potential

1. Dilute NPs to a final concentration of 50–100 μg/ml in 1 ml of the medium of choice (water, PBS, cell culture medium, diluted plasma, etc.). This dilution provides high quality measurement data for most NP kinds/materials.
2. Degas the sample by pouring it into a syringe, covering its top with a cap and creating a vacuum inside (it usually helps to gently knock the side of syringe). Repeat the procedure until no visible air bubbles appear during the degassing procedure.
3. Use the syringe to transfer 1 ml of sample into a disposable zeta potential cuvette (avoid creating air bubbles) and place cuvette in the instrument.
4. Set up the SOP according to the instrument manual paying special attention to the following parameters: Measurement type: Zeta Potential; Material: NPs material from the list; Solvent: appropriate solvent from the list; Temperature: depending on the experimental conditions; Equilibration time: make sure to set equilibration time long enough for the sample to reach measurement temperature and equilibrate; No of measurements: at least 3; Number of runs: manual: from 10 to 100; Attenuator: in most cases set on Auto.
5. Run the SOP and read the samples' ZP value as an average of three measurements with standard deviation.

3.8 TEM Protocol

1. NP-hard corona complexes (NP / HC) are prepared as described in Subheadings 3.3 and 3.4. Resuspend the pellet into the desired concentration (typically 0.01 mg/ml) in 2.5 % glutaraldehyde in water and leave for 1 h.
2. Place the TEM grid in a plastic dish, with the shiny carbon side upwards.
3. Drop cast 10 μl of NP/HC solution onto the grid and remove the excess solution by soaking with filter/blotting paper. Leave to dry for 24 h.
4. Place the grid in TEM and carry out the standard operating procedure analysis (*see* **Note 26**).
5. Acquire images at different magnifications from 100 k× to 160 k×, ensuring that at least ten images per area are acquired for qualitative analysis.
6. For comparison, also acquire images of the pristine NPs.

3.9 Protein Assay (Micro BCA Protein Assay) Optimized for NP/Corona Complexes

1. Preparation of Albumin standards:

   Take 0.25 ml of 2 mg/ml BSA stock and dilute with 4.75 ml of PBS buffer to obtain 100 μg/ml standard solution. Dilute this standard solution with PBS in order to obtain standard solutions of: 75, 50, 37.5, 25, 12.5, 6.25, and 0 μg/ml (pure PBS). Transfer 250 μl of each standard into microcentrifuge tubes.

   Perform a progressive dilution of the NP corona sample in PBS to ensure that absorbance values (*see* **step 8**) are in the protein standard range. Transfer 250 μl of each sample into microcentrifuge tubes.
2. Working reagent (copper solution) preparation: mix reagents A, B, and C in the ratio 25:24:1.
3. Add working reagent (WR) to the previously prepared standards and samples in a ratio 1:1, i.e., add 250 μl of WR to each 250 μl sample and standard.
4. Incubate for 60 min at 60 °C.
5. Cool at RT for ~45 min.
6. Spin down the samples at 20,000 ×*g* for 1 h (or the time more appropriate to NPs from previous studies) to pellet the NPs.
7. Transfer 200 μl of each supernatant (in duplicate) into 96 well plate and measure absorbance at 562 nm, using a plate reader spectrophotometer.
8. Plot standard curve of Concentration versus Absorbance and fit linear function to obtain data. Use the obtained equation to calculate protein concentration in unknown samples.
9. Controls: NPs in PBS; NPs in BSA standard (*see* **Note 27**).

4 Notes

1. Subheading 3.1, **steps 1** and **2** follow the HUPO Plasma Proteome Project guidelines [31].
2. Ethical permission is required in order to collect and prepare plasma/serum. It collection must be carried out by a qualified medical doctor and the whole procedure has to be approved by local ethical approval committee.
3. Storage and monitoring of the aging of the biological fluid is crucial, as old or partially degraded biological fluids can compromise the outcome of the study. Typically, storage at −80 °C is preferable for long term sample storage (more than 3 weeks) and the cycle of freezing and thawing must be avoided to reduce protein degradation. As per the HUPO guidelines, samples should only be frozen and thawed once [31]. Protease inhibitor cocktail can be used, however these might result in protein structure change and artifacts [31].
4. Plasma is generally preferred to serum for studies to predict NP fate in vivo, as it is representative of most of the proteins and biomolecules of the blood, however, extreme care is needed to avoid blood coagulation and protein aggregation. Any plasma dilution has to be performed in PBS containing 1 mM EDTA.
5. The incubation temperature of NPs in biological fluid is an important factor that influences the kinetics of protein association and dissociation to and from the NP surface. Previous studies have shown that when NPs are introduced into a biomolecule rich medium, their surface will immediately attract entities from its closest environment—an "early corona" formed mainly by abundant proteins with low affinity. The protein corona composition evolves with time until it reaches equilibrium, where proteins with strong affinity towards the NP surface form the "mature" protein corona [3, 17, 28, 32].
6. To reduce plasma protein and NPs stickiness to the sides of the tubes during incubation, use of low protein binding Eppendorf tubes is highly recommended.
7. Before proceeding with NPs incubation with biological fluid, ensure that the NPs are stable and monodisperse by common physicochemical approaches. Bath sonication for three minutes before use is suggested to disperse any possible agglomerated NPs.
8. To mimic "in vivo or in vitro" conditions as much as possible, NPs should be exposed to different biological media with different protein concentrations. A typical approach would be to keep the NP concentration constant (and thus the surface area)

and incubate with biological fluid diluted with different amounts of PBS [3], where the ratio between NP surface area and proteins is varied [3]. A theoretical five layers of protein excess for each individual NP will ensure full surface coverage of each individual NP, while a lower protein concentration generally promotes NPs agglomeration.

9. The NP amount used for each experiment depends on the detection method that will be used to study the NP-corona complex. For DLS measurements typically 50–100 μg/ml in 500 μl, for DCS analysis each injection requires 100 μl of 0.1–0.01 mg/ml, depending on the NP's absorbance. For protein corona detection by MS a higher NP amount is generally required. 1 mg/ml of NPs in 1 ml of solution will allow detecting protein corona by blue Coomassie staining, silver staining will have to be performed when lower NP amounts have been used for the experiment.
10. Incubation time and temperature are strictly dependent on the aim of the study, however possible degradation of the biological fluid during the incubation step must be monitored. Also the biological fluid alteration due to the exposure/incubation conditions should be monitored.
11. Subheading 3.4 allows isolation of NP hard corona complexes from unbound biomolecules in the biological media and from the loosely bound proteins (dynamic corona) (Fig. 3). Optimization steps are crucial in order to ensure that NP loss during the centrifugation step is kept to a minimum and that loosely bound proteins are completely washed off during the procedure. The hard corona is typically of high biological interest, while the outer layer of the protein corona is composed of proteins in rapid exchange with the solution. Centrifugation speed optimization: the first centrifugation is generally longer than the following ones due to the higher density of the biological fluid compared to PBS. Centrifugation time and speed depend on NP size and density and prolonged centrifugation time (over 1 h) and at speed above 18,000 × *g* are generally avoided as this will promote biomolecule pelleting and thus increase of background.
12. The temperature of the incubation step and of the centrifugation steps, while preparing hard corona samples, should be similar.
13. To avoid carryover of biomolecules from the biological fluid used in the incubation step, a change of vessel/tube is strongly recommended in Subheading 3.4 at least in **step 9**. Also long centrifugation times (more than 60 min) and speeds higher than 18,000 × *g* should be avoided to prevent co-elution of the plasma protein background.

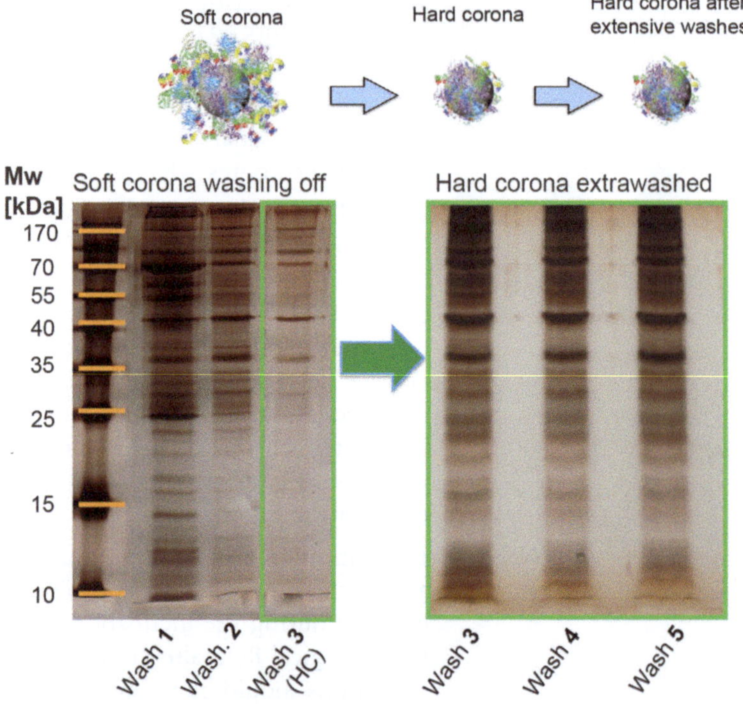

Fig. 4 SDS-PAGE of 30 nm gold NP hard corona complexes, after different numbers of washing steps. While the abundance of NP corona proteins decreases after first to third washes (the soft corona is gradually removed), it does not change after any additional washing steps (the remaining strongly bound hard protein corona does not elute from the NP surface). Proteins adapted from Protein Data Bank (http://www.pdb.org)

14. Typically three washes are needed to obtain NP-hard corona complexes (*see* Subheading 3.4), however to ensure protocol efficacy, NP-hard corona washes can be over washed without resulting in protein corona changes (Fig. 4).

15. DLS is a powerful and fast general-purpose method for characterization of NP dispersions, however its resolution is limited (due to the fact that the scattering signal generated by the whole dynamic range of NP sizes present in the sample is analyzed at the same time) and does not allow for observation of discrete peaks of similar sizes, or small shifts in size of a range of nanometer. Therefore, a complementary high precision size-analysis technique, such as DCS is needed (*see* Subheading 3.6).

16. Subheading 3.5 is applicable to pristine NPs in different media and to NP-hard corona complexes; however, if performing measurements in situ in the presence of complex biological media, the background signal of free biomolecules might

interfere with the NPs' scattering, and as such the background solution should also be run for comparison. Use of 0.2 μm filtered solutions is highly recommended to remove dust or contaminants from solutions and potentially to remove NP aggregates from solution.

17. The optimal sample concentration for DLS measurements differs for different materials with different optical properties (e.g., it is possible to measure polystyrene NP dispersions of lower concentrations than silica NPs, which have a lower refractive index and adsorption) therefore it's difficult to suggest a universal concentration a priori. However, the value of instrument attenuator when it's set on "auto" indicates the intensity of the light scattered by the sample. The higher the attenuator value (ranges from 1 to 11) the lower the light scattering generated by the sample. NP concentration should be adjusted to obtain an attenuator value ~7.
18. The reported hydrodynamic diameter (D_h) value should be an average of at least three subsequent measurements of min. 10 × 10 s runs. For monodisperse samples the average D_h corresponds to *z*-average value, while for complex/polydisperse samples D_h can correspond to the maximum of a particular size peak.
19. The advantage of DCS in respect to DLS is the fact, that in the centrifugal gradient NPs are separated and detected size-by-size, allowing for resolution of multiple peaks in polydisperse samples (where DLS would show only one broad peak). Another advantage of DCS is its capacity to perform measurements in essentially any possible dispersant media, even in the presence of a large excess of biomolecules, such as in full plasma, cMEM (and other biological fluids) in situ, which allows us to have an insight into the NP–protein complexes properties, as they are in vivo.
20. DCS gradient solutions should be prepared in the buffer in which the NP samples are prepared/stored. Typically water is used for characterization of pristine NPs and PBS for NP–protein corona complexes.
21. RPM of 14,000 should be used for gold NPs of size around 100 nm.

    RPM of 18,000 should be used for silica NPs of size around 100 nm.

    RPM of 22,000–24,000 should be used for polystyrene NPs of size around 100 nm.
22. Stability of the DCS gradient has to be monitored with time by running test samples and ensuring size measurement reproducibility (as in Subheading 3.6, **step** 7).

23. The detection limit of the DCS instrument is correlated to the refractive index and light absorption of the NPs. Typically, a NP concentration of 0.5–1 mg/ml is indicated for particles with low light absorption (e.g., Silicon dioxide NPs) and 0.05–0.1 mg/ml for particles with high absorptivity (such as gold or polystyrene NPs). However, a dilution series experiment has to be performed to find optimal NPs sample concentration to ensure a sufficient signal to noise ratio.
24. Change of DCS sucrose gradient is required after measurement of samples injected with an excess of biomolecular fluid (in situ) or after 6 h of measurements.
25. Following the manufacturer's procedure, a core shell model can be applied to analyze the obtained measurement results for NPs and their hard corona complexes. The core shell model allows correlation of NPs' apparent size shifts with the thickness of the biomolecule shell adsorbed on to the NP surface [4].
26. For high-resolution images, the use of field emission high resolution TEM is recommended, otherwise tungsten filament or lanthanum hexaboride filament TEM systems can be also used to achieve acceptable images.
27. The Micro BCA assay is designed for determination of protein concentration in an unknown sample. In order to evaluate the degree of NP interference with the accuracy of obtained results, it's necessary to perform two control samples: NPs in PBS (where ideally the assay should produce the result of 0 mg/ml protein concentration), and NPs in BSA standard of known concentration (where assay should ideally produce a result in agreement with the BSA standard concentration).
28. For SDS-PAGE analysis, to ensure that all proteins from the NP corona are stripped off the NPs' surface, the use of high denaturing condition buffer is strongly recommended. Typically NP protein corona complexes prepared as described in Subheading 3.4 are resuspended in loading buffer (*see* Subheading 2) and heated for 5 min at 100 °C, and samples are directly loaded onto the SDS-PAGE gel, where they migrate under an electric field and are separated by size in the pores of the polyacrylamide gel. In some rare cases more denaturing buffer might be applied in order to ensure complete protein corona desorption from the NPs' surface.
29. The sample where the biological fluid is prepared and processed in an identical way to the NP–protein corona samples (*see* Subheadings 3.3 and 3.4), in the absence of the NPs, is a control of highest importance. It is crucial to determine whether the centrifugation and washing steps promote the isolation of a

plasma protein background. This sample will also have to be loaded into the SDS-PAGE along with NP–protein corona.

30. Washes obtained as described in Subheading 3.4 should be run by SDS-PAGE (*see* Subheading 3.4) to monitor the presence of the dynamic corona. A successful protocol should show a strong presence of proteins in wash 1 that becomes attenuated in wash 2 and is not present in wash 3. It is recommended to load in the same gel a protein corona sample, to obtain relative protein amounts from washes to proteins in the protein corona.
31. To reduce gel to gel variability error, it is recommended to compare gel bands only within the same gel.
32. The choice of the gel staining technique is crucial: Blue Coomassie has a dynamic range linear for 20-fold changes of protein abundance, but the main limitation is the fact that it fails to detect low abundance proteins; the lowest detection limit is 10–50 ng, depending on the proteins (Fig. 4). However, the Coomassie stain protocol is quick to use and simple. A parallel experiment using a high sensitive staining approach, such as silver nitrate staining, or a fluorescent staining, is recommended to ensure that low abundant proteins are detected.
33. It is recommended to run the same sample in different pore size acrylamide gels (varying the acrylamide percentage) to ensure high resolution of proteins of different size ranges, as large proteins (~150–55 kDa) are better resolved in 8 % acrylamide gel and smaller proteins (55–10 kDa) are better resolved in 15 % acrylamide gel.

References

1. Cedervall T, Lynch I, Lindman S, Berggard T, Thulin E, Nilsson H, Dawson KA, Linse S (2007) Understanding the nanoparticle–protein corona using methods to quantify exchange rates and affinities of proteins for nanoparticles. Proc Natl Acad Sci USA 104:2050–2055
2. Lynch I, Dawson KA (2008) Protein–nanoparticle interactions. Nano Today 3:40–47
3. Monopoli MP, Walczyk D, Campbell A, Elia G, Lynch I, Baldelli Bombelli F, Dawson KA (2011) Physical–chemical aspects of protein corona: relevance to in vitro and in vivo biological impacts of nanoparticles. J Am Chem Soc 133:2525–2534
4. Walczyk D, Bombelli FB, Monopoli MP, Lynch I, Dawson KA (2010) What the cell "sees" in bionanoscience. J Am Chem Soc 132:5761–5768
5. Kapralov AA, Feng WH, Amoscato AA, Yanamala N, Balasubramanian K, Winnica DE, Kisin ER, Kotchey GP, Gou P, Sparvero LJ, Ray P, Mallampalli RK, Klein-Seetharaman J, Fadeel B, Star A, Shvedova AA, Kagan VE (2012) Adsorption of surfactant lipids by single-walled carbon nanotubes in mouse lung upon pharyngeal aspiration. ACS Nano 6:4147–4156
6. Walkey CD, Chan WC (2012) Understanding and controlling the interaction of nanomaterials with proteins in a physiological environment. Chem Soc Rev 41:2780–2799
7. Nel AE, Madler L, Velegol D, Xia T, Hoek EM, Somasundaran P, Klaessig F, Castranova V, Thompson M (2009) Understanding biophysicochemical interactions at the nano-bio interface. Nat Mater 8:543–557
8. Tenzer S, Docter D, Rosfa S, Wlodarski A, Kuharev J, Rekik A, Knauer SK, Bantz C, Nawroth T, Bier C, Sirirattanapan J, Mann W, Treuel L, Zellner R, Maskos M, Schild H, Stauber RH (2011) Nanoparticle size is a critical physicochemical determinant of the human blood plasma corona: a comprehensive

quantitative proteomic analysis. ACS Nano 5:7155–7167

9. Martel J, Young D, Young A, Wu CY, Chen CD, Yu JS, Young JD (2011) Comprehensive proteomic analysis of mineral nanoparticles derived from human body fluids and analyzed by liquid chromatography-tandem mass spectrometry. Anal Biochem 418:111–125
10. Zeng Z, Patel J, Lee SH, McCallum M, Tyagi A, Yan M, Shea KJ (2012) Synthetic polymer nanoparticle–polysaccharide interactions: a systematic study. J Am Chem Soc 134:2681–2690
11. Milani S, Baldelli Bombelli F, Pitek AS, Dawson KA, Radler J (2012) Reversible versus irreversible binding of transferrin to polystyrene nanoparticles: soft and hard corona. ACS Nano 6:2532–2541
12. Dobrovolskaia MA, Patri AK, Zheng J, Clogston JD, Ayub N, Aggarwal P, Neun BW, Hall JB, McNeil SE (2009) Interaction of colloidal gold nanoparticles with human blood: effects on particle size and analysis of plasma protein binding profiles. Nanomedicine (Lond) 5:106–117
13. Zheng M, Li ZG, Huang XY (2004) Ethylene glycol monolayer protected nanoparticles: synthesis, characterization, and interactions with biological molecules. Langmuir 20:4226–4235
14. Sund J, Alenius H, Vippola M, Savolainen K, Puustinen A (2011) Proteomic characterization of engineered nanomaterial–protein interactions in relation to surface reactivity. ACS Nano 5:4300–4309
15. Deng ZJ, Mortimer G, Schiller T, Musumeci A, Martin D, Minchin RF (2009) Differential plasma protein binding to metal oxide nanoparticles. Nanotechnology 20:455101
16. Maiorano G, Sabella S, Sorce B, Brunetti V, Malvindi MA, Cingolani R, Pompa PP (2010) Effects of cell culture media on the dynamic formation of protein–nanoparticle complexes and influence on the cellular response. ACS Nano 4:7481–7491
17. Zhang H, Burnum KE, Luna ML, Petritis BO, Kim JS, Qian WJ, Moore RJ, Heredia-Langner A, Webb-Robertson BJ, Thrall BD, Camp DG 2nd, Smith RD, Pounds JG, Liu T (2011) Quantitative proteomics analysis of adsorbed plasma proteins classifies nanoparticles with different surface properties and size. Proteomics 11:4569–4577
18. Lundqvist M, Stigler J, Elia G, Lynch I, Cedervall T, Dawson KA (2008) Nanoparticle size and surface properties determine the protein corona with possible implications for biological impacts. Proc Natl Acad Sci USA 105:14265–14270
19. Schaefer J, Schulze C, Marxer EE, Schaefer UF, Wohlleben W, Bakowsky U, Lehr CM (2012) Atomic force microscopy and analytical ultracentrifugation for probing nanomaterial protein interactions. ACS Nano 6:4603–4614
20. Mahmoudi M, Lynch I, Ejtehadi MR, Monopoli MP, Bombelli FB, Laurent S (2011) Protein–nanoparticle interactions: opportunities and challenges. Chem Rev 111:5610–5637
21. Akesson A, Cardenas M, Elia G, Monopoli MP, Dawson KA (2012) The protein corona of dendrimers: PAMAM binds and activates complement proteins in human plasma in a generation dependent manner. RSC Adv 2:11245–11248
22. Lesniak A, Fenaroli F, Monopoli MP, Aberg C, Dawson KA, Salvati A (2012) Effects of the presence or absence of a protein corona on silica nanoparticle uptake and impact on cells. ACS Nano 6:5845–5857
23. Ge C, Du J, Zhao L, Wang L, Liu Y, Li D, Yang Y, Zhou R, Zhao Y, Chai Z, Chen C (2011) Binding of blood proteins to carbon nanotubes reduces cytotoxicity. Proc Natl Acad Sci USA 108:16968–16973
24. Hu W, Peng C, Lv M, Li X, Zhang Y, Chen N, Fan C, Huang Q (2011) Protein corona-mediated mitigation of cytotoxicity of graphene oxide. ACS Nano 5:3693–3700
25. Deng ZJ, Liang M, Monteiro M, Toth I, Minchin RF (2011) Nanoparticle-induced unfolding of fibrinogen promotes Mac-1 receptor activation and inflammation. Nat Nanotechnol 6:39–44
26. Monopoli MP, Bombelli FB, Dawson KA (2011) Nanobiotechnology: nanoparticle coronas take shape. Nat Nanotechnol 6: 11–12
27. Pitek AS, O'Connell D, Mahon E, Monopoli MP, Baldelli Bombelli F, Dawson KA (2012) Transferrin coated nanoparticles: study of the bionano interface in human plasma. PLoS One 7:e40685
28. Casals E, Pfaller T, Duschl A, Oostingh GJ, Puntes V (2010) Time evolution of the nanoparticle protein corona. ACS Nano 4:3623–3632
29. Shevchenko A, Wilm M, Vorm O, Mann M (1996) Mass spectrometric sequencing of proteins silver-stained polyacrylamide gels. Anal Chem 68:850–858
30. Simpson RJ (2008) Proteins and proteomics: a laboratory manual. Cold Spring Harbor Laboratory. http://www.proteinsandproteomics.org/home_1.html
31. Rai AJ, Gelfand CA, Haywood BC, Warunek DJ, Yi J, Schuchard MD, Mehigh RJ, Cockrill SL, Scott GB, Tammen H, Schulz-Knappe P, Speicher DW, Vitzthum F, Haab BB, Siest G,

Chan DW (2005) HUPO Plasma Proteome Project specimen collection and handling: towards the standardization of parameters for plasma proteome samples. Proteomics 5:3262–3277

32. Casals E, Pfaller T, Duschl A, Oostingh GJ, Puntes VF (2011) Hardening of the nanoparticle–protein corona in metal (Au, Ag) and oxide (Fe_3O_4, CoO, and CeO_2) nanoparticles. Small 7:3479–3486

Chapter 12

Electrophoretic Implementation of the Solution-Depletion Method for Measuring Protein Adsorption, Adsorption Kinetics, and Adsorption Competition Among Multiple Proteins in Solution

Hyeran Noh, Naris Barnthip, Purnendu Parhi, and Erwin A. Vogler

Abstract

The venerable solution-depletion method is perhaps the most unambiguous method of measuring solute adsorption from solution to solid particles, requiring neither complex instrumentation nor associated interpretive theory. We describe herein an SDS-gel electrophoresis implementation of the solution-depletion method for measuring protein adsorption and protein-adsorption kinetics. Silanized-glass particles with different surface chemistry/energy and hydrophobic sepharose-based chromatographic media are used as example adsorbents. Electrophoretic separation enables quantification of adsorption competition among multiple proteins in solution for the same adsorbent surface, demonstrated herein by adsorption-competition kinetics from binary solution.

Key words Protein adsorption, Adsorption kinetics, Adsorption competition, Electrophoresis, Solution depletion

1 Introduction

Fouling is a natural phenomenon by which working surfaces of devices, instruments, or machines in contact with aqueous solutions become coated with a layer or layers of adventitious contamination that compromises intended performance. Fouling has broad technologic importance with considerable socioeconomic impact spanning environment (e.g., occlusion of pipes and filters used in civil engineering), medicine (e.g., reduction in biosensor sensitivity), and transportation (e.g., resistance to the flow of water across boat hulls). Adsorption of surface-active glycosylated proteins and peptidoglycans onto surfaces is among the first events of the fouling process. In particular, protein adsorption is of singular importance in the development and application of medical devices because adsorbed proteins can catalyze, mediate, or moderate the biological response

Paolo Bergese and Kimberly Hamad-Schifferli (eds.), *Nanomaterial Interfaces in Biology: Methods and Protocols*, Methods in Molecular Biology, vol. 1025, DOI 10.1007/978-1-62703-462-3_12, © Springer Science+Business Media New York 2013

to various biomaterials used in fabrication of medical devices [1]. Above-mentioned applications usually involve multiple proteins in solution rather than the single purified solutions typically used in studies of the mechanism of protein adsorption. For example, blood contains more than 1,000 proteins spanning 10 decades in concentration, with five proteins dominating the concentration profile (albumin, IgG, fibrinogen, transferrin, and IgA) [2]. As a consequence, understanding adsorption competition among multiple proteins in solution to the same adsorbent surface remains an important unsolved problem [1].

We describe herein an electrophoretic implementation of the solution-depletion method for measuring protein adsorption and protein-adsorption kinetics [3–7]. The solution-depletion method is one of the most unambiguous methods of measuring adsorption, requiring no complex instrumentation, protein labeling, or separation of protein from adsorbent particles. Interpretation is simply the arithmetic of mass balance. Electrophoresis is used a multiplexing, separation and quantification tool that enables study of protein-adsorption competition between two or more proteins in solution [8, 9]. Commercially prepared SDS gels and staining kits greatly simplifies the methodology.

1.1 Mass Balance Equations

Solution-depletion measures the concentration of the *i*th protein in solution before-and-after contact with particulate adsorbent as quantified by $D_i \equiv \left(W_{B_i}^{o} - W_{B_i}\right)$. Here, $W_{B_i}^{o}$ is the initial protein solution concentration (in mg/mL for example), and W_{B_i} is the final protein solution concentration after contact with adsorbent under the desired experimental conditions (time, temperature, stirring, static solution, etc.). The depletion D_i has the same units as $W_{B_i}^{o}$ and W_{B_i} and represents the change in bulk-solution concentration due to adsorption.

The absolute mass of protein adsorbed $m_i = D_i V_B$ if $W_{B_i}^{o}$ and W_{B_i} are measured in w/v units and V_B is the volume of the bulk protein solution. These relationships assume that V_B is a system constant (e.g., not adsorbed by adsorbent to a significant extent). In this case, it is clear that m_i and D_i are directly proportional and adsorption isotherms can be constructed by plotting either m_i or D_i as a function of solution concentration $W_{B_i}^{o}$ at constant adsorbent surface area [3]. Measurement of adsorption isotherms over the concentration range of interest is essential to a complete understanding of the adsorption process [1].

Adsorption isotherms for solution concentrations $15 > W_{B_i}^{o} > 0$ mg/mL show that blood proteins adsorb to various test adsorbents in proportion to $W_{B_i}^{o}$, approximating a Henry isotherm up to saturation of the adsorbent surface which occurs at a particular solution concentration $\left(W_{B_i}^{o}\right)^{\max}$ [1, 3]. Henry-like isotherms are obtained for any number of proteins adsorbing simultaneously from solution when individual protein solution concentrations are measured before and after contact with adsorbent. For example, adsorption of proteins from serum to hydrophobic adsorbents exhibit Henry-like isotherms when scaled as per-cent solution concentration [10] (herein we subscribe to a 65° advancing contact angle as the boundary between hydrophilic and hydrophobic materials [1]).

Further interpretation of adsorption and adsorption isotherms measured by solution depletion depends on the adsorption paradigm in effect. If this paradigm construes adsorption to occur at a planar surface [11], then the surface concentration $\Gamma = m_{\mathrm{i}} / A$ in mass-per-unit adsorbent surface area A. Because proteins can absorb in multilayers (depending on solution concentration and protein size), it can be convenient to construe adsorption as a partitioning of protein from bulk solution into a three-dimensional (3D) interphase region that separates the physical-adsorbent surface from bulk solution [1]. This interphase thus has a finite volume V_{I} with a specific protein concentration W_{I} and $D \equiv \left(W_{\mathrm{B}}^{\mathrm{o}} - W_{\mathrm{B}}\right) = W_{\mathrm{I}}\left(\frac{V_{\mathrm{I}}}{V_{\mathrm{B}}}\right)$. For a single-protein solution $W_{\mathrm{I}} = W_{\mathrm{I}_i}$ and $D_i \equiv \left(W_{\mathrm{B}_i}^{\mathrm{o}} - W_{\mathrm{B}_i}\right) = W_{\mathrm{I}_i}\left(\frac{V_{\mathrm{I}_i}}{V_{\mathrm{B}}}\right)$, but in the case of adsorption from multiple-protein, solution $W_{\mathrm{I}} = \sum_i W_{\mathrm{I}_i}$. Needless to say, perhaps, mass balance equations become more complicated with the number of proteins simultaneously adsorbing from solution [5]. At steady state, the partition coefficient for the *i*th protein $P_i \equiv \left(W_{\mathrm{I}_i} / W_{\mathrm{B}_i}\right)$, and it follows that $W_{\mathrm{B}_i} = W_{\mathrm{B}_i}^{\mathrm{o}} - W_{\mathrm{I}_i}\left(\frac{V_{\mathrm{I}_i}}{V_{\mathrm{B}}}\right) = W_{\mathrm{B}_i}^{\mathrm{o}}\left[\frac{V_{\mathrm{B}}}{V_{\mathrm{B}} + P_i V_{I_i}}\right]$ for $0 \le W_{\mathrm{B}_i} \le W_{\mathrm{B}_i}^{\max}$. A related isotherm equation is $D_i = W_{\mathrm{B}_i}^{\mathrm{o}}\left[\frac{P_i V_{\mathrm{I}_i}}{V_{\mathrm{B}} + P_i V_{\mathrm{I}_i}}\right]$ for $0 \le W_{\mathrm{B}_i} \le W_{\mathrm{B}_i}^{\max}$. As mentioned above, it has been found that solution-depletion isotherms approximate a linear Henry isotherm and thus have a slope $S_i \equiv \left[\frac{P_i V_{\mathrm{I}_i}}{V_{\mathrm{B}} + P_i V_{\mathrm{I}_i}}\right]$ or, by rearrangement, $P_i V_{\mathrm{I}_i} = V_{\mathrm{B}}\left(\frac{S_i}{1 - S_i}\right)$ [3]. Comparison of adsorption experiments with two different proteins *i* and *j* to the same adsorbent surface area (in separate experiments) allows solution of these equations in terms of interphase volume ratio $\left(\frac{V_{\mathrm{I}_i}}{V_{\mathrm{I}_j}}\right)$ and partition-coefficient ratio $\left(\frac{P_i}{P_j}\right)$ as further elaborated in ref. 3.

2 Materials

2.1 Reagents, Proteins, and Solutions

Solvents such as ethanol and chloroform were reagent grade and used as received from VWR, Inc. Other chemicals such as H_2O_2 and H_2SO_4 were also used as received from VWR unless otherwise specified. Water was 18 MΩ deionized obtained from a Millipore Simplicity water purification system. Phosphate buffer saline (PBS) used as diluent was prepared from powder (Sigma; 0.14 M NaCl, 3 mM KCL) in 18 MΩ deionized water with final pH = 7.4. Conical microtubes (0.5 mL, Safe-lock microcentrifuge tubes, Eppendorf; approximately 2 cm^2 internal surface area) were used in solution-handling steps outlined below.

Silanes applied in this work to derivatize glass-particle adsorbents used as received from Gelest Inc., (Morrisville, PA) were octadecyltrichlorosilane (OTS), 3-aminopropyltriethoxysilane (APTES), *n*-propyltriethoxysilane (PTES), and vinyltriethoxysilane (VTES). OTS-treated glass particles were optionally coated in a 0.2 % solution of 1,1-pentadecafluorooctylmethacrylate in tricholorotrifluoroethane ("Nyebar," Nye Lubricants, Fairhaven, MA) to render OTS-treated glass particles slightly more hydrophobic than OTS-treated counterparts.

Piranha solution was prepared by slowly mixing 1 part 30 % H_2O_2 in three parts concentrated H_2SO_4 while cooling in an ice bath *see* **Note 1**.

Human sourced proteins were used as received from various vendors without further purification. Test proteins spanned a molecular weight (MW) range from lysozyme (15 kDa, [3]) to IgM (1,000 kDa, [11, 12]). Protein solutions were prepared in PBS by 80:20 or 90:10 dilution of protein concentrate freshly prepared before adsorption measurements. Solution concentrations typically varied between 0 and 10 mg/mL depending on protein under study. Electrophoresis was performed using 15- or 26-lane NuPAGE Novex Tris–Acetate pre-cast gels (Invitrogen Corp.; 500 kDa capacity) to separate and quantify proteins. NuPAGE Novex Bis–Tris gels were used for low MW proteins. Electrophoresis was performed in Xcell Sure Lock Mini-Cell or Xcell 4 SureLock Midi-Cell (Invitrogen Corp.) for 70 min at 150 V (Tris–Acetate) or 35 min at 220 V (Bis–Tris). Gels were stained with SimplyBlue SafeStain (Invitrogen Corp.) for 1 h and destained with deionized (18 MΩ) water for several hours while mixing on a standard hematology rocker. Band intensity was quantified using a Gel-doc system (Bio-Rad Laboratories Inc.).

2.2 Test Adsorbents

Glass-particle adsorbents were 425–600 μm diameter glass particles (Sigma Aldrich) in either cleaned or silanized form. The nominal specific area used in this work was 5×10^{-3} m^2/g (based on 512.5 μm mean diameter and 168 μg/particle). The actual surface area measured by the Brunauer–Emmett–Teller (BET) method was 0.25 ± 0.09 m^2/g (Micromeritics ASAP 2000 using liquid nitrogen as the probe gas, *see* **Note 2**) was 50× larger than the nominal value, possibly reflecting dispersity in particle size and porosity. Glass coverslips (Fisher, 22 × 30 × 0.1 mm) were used as a witness samples to glass-surface treatments amenable to contact angle measurements. Glass used in this study was activated before silanization by immersion in piranha solution followed by rinsing in 18 MΩ deionized water. After air-drying, particles were subjected to 5–10 min air discharge in a commercial plasma generator (Herrick, Wippany NY).

Octyl Sepharose 4 Fast Flow adsorbent with 90 μm diameter was obtained from Amersham Biosciences (40 % or 75 % by volume of 90 μm nominal diameter particles dispersed in 20 % ethanol–water solution, *see* **Note 3**). Sepharose particles were washed in PBS to remove ethanol by 3× sequential centrifugation (40 RPM for 1 min in a Hettick microtube fixed-rotor centrifuge, VWR) and resuspension in PBS.

3 Methods

3.1 Preparation of Glass-Particle Adsorbents

1. Clean/activate glass particles and coverslip witness samples by 30 min immersion in 80 °C piranha solution followed by three sequential washes in each of 18 MΩ deionized water and ethanol.
2. Air dry the piranha-solution-oxidized glass and subsequently oxidize by air–plasma treatment of a single layer of particles (or glass coverslip witness samples) held in a 15 mm Pyrex glass Petri dish directly before silanization. Glass surfaces treated in this manner will be fully water-wettable (0° water contact angle).
3. Silanize activated glass particles (and coverslip witness samples) using silane reagents selected from below. Typical results compiled in Table 1.

Table 1
Typical outcome of glass silanization procedures

Surface treatment	Advancing contact angle range (°)
Nyebar	109–118
OTS	106–109
PTES	93–97
VTES	85–90
APTES	42–58
Hydrolyzed glass	0

3a. 1.5 h reaction with 5 % v/v OTS in chloroform (*see* **Note 4**).

- Rinse the silanized samples by three sequential washes in chloroform before curing in a vacuum oven at 110 °C for 12 h.

3b. Optional Dip coat with Nyebar

- Dip cured OTS-silanized glass adsorbents in Nyebar solution for 10 min and air dry to produce surface slightly more hydrophobic than rendered by OTS treatment alone.

3c. 20 min reaction with 5 % APTES dissolved in 95:5 v/v ethanol–water solutions.

- APTES solution in 95:5 v/v ethanol–water should be allowed to hydrolyze 12 h before use (*see* **Note 4**).
- Rinse APTES-treated glass with absolute ethanol 3× and cure 12 h in a vacuum oven at 110 °C.

3d. 20 min reaction with 5 % PTES dissolved in 90:10 v/v ethanol–water solution.

- PTES in 90:10 v/v ethanol–water solution containing 0.5 % glacial acetic acid should be allowed to hydrolyze 12 h before use (*see* **Note 4**).
- Rinse PTES-treated glass with absolute ethanol 3× and cure 12 h in a vacuum oven at 110 °C.

3e. 20 min reaction with 5 % VTES dissolved in 90:10 v/v ethanol–water solution.

- VTES in 90:10 v/v ethanol–water solution containing 0.5 % glacial acetic acid should be allowed to hydrolyze 12 h before use (*see* **Note 4**).
- Rinse VTES-treated glass with absolute ethanol 3× and cure 12 h in a vacuum oven at 110 °C.

3.2 Preparation of Octyl Sepharose Adsorbents

1. Prepare fresh octyl sepharose adsorbent (40 % or 75 % solids) before each depletion experiment by 3× washing in PBS (to remove ethanol) using a sequential centrifugation/resuspension protocol that processed 1 mL of as-received suspension.
2. Replace 500 μL of supernatant with 500 μL PBS, ending with a 60:40 (or 75:25) v/v stock suspension in PBS.
3. Pipette 50 μL (or 20 μL) stock corresponding to 30 μL fluid, 20 μL beads (or 5 μL fluid, 15 μL beads) into a 0.5 mL microtube.
4. Remove 25 μL fluid after centrifugation.

3.3 Measurement of Adsorption from Single-Protein Solutions

1. Prepare desired protein solutions (25 μL) in PBS at varying concentrations (from 0.1 to 10 mg/mL).
2. Resuspend octyl sepharose adsorbent (or 22 mg of glass-adsorbent particles) by gentle pipette aspiration in 25 μL protein

solution so that the final solution volume is 30 μL. Quantities of adsorbent suggested here are general guides *see* **Note 5**.

3. Leave the solution with adsorbent undisturbed in 0.5 mL conical microtubes for desired equilibration time at ambient temperature. Solution and adsorbent can be optionally continuously mixed using a rotating mixer (*see* **Note 6**).
4. Sample 20 μL of supernatant protein solution for gel electrophoresis, once for steady-state measurements or every 5 min for kinetics experiments. Kinetic experiments use a separate microtube for each time interval.

3.4 Measurement of Adsorption from Binary-Protein Solutions

1. Prepare protein solutions (25 μL) in PBS at varying concentrations of both proteins (*see* **Note 7**).
2. Resuspend octyl sepharose beads (or 22 mg of glass adsorbent particles) by gentle pipette aspiration in 25 μL protein solution so that the final solution volume is 30 μL.
3. Leave the solution with beads undisturbed in 0.5 mL conical microtubes. Solution and adsorbent can be optionally continuously mixed using a rotating mixer (*see* **Note 8**).
4. Sample 20 μL of supernatant protein solution for gel electrophoresis, once for steady-state measurements or every 5 min for kinetics experiments (*see* **Note 8**). Kinetic experiments use a separate microtube for each time interval.

3.5 Electrophoresis

1. Use 15- or 26-lane commercial gels and electrophoresis equipment to separate and quantify proteins.
2. Quantify band intensity using a commercial gel reader of band optical density (OD).
3. Prepare a standard curve for each gel using the first 6–7 lanes by applying proteins at known solution concentration (*see* **Note 9**). Plot OD vs. standard protein concentration to obtain working calibration curve.

4 Notes

1. Piranha solution is a strong oxidizer used to remove organic impurities on glass-particle adsorbent surfaces and to hydroxylate glass surfaces to enhance subsequent silanization reactions. Piranha solution must be prepared with great care with appropriate safety precautions including use of a chemical hood and personal protection equipment. In particular, admixture of H_2O_2 with H_2SO_4 is exothermic and should be performed slowly with cooling in water ice. Piranha solution should be freshly prepared before use and never stored due to self-decomposition of hydrogen peroxide. Glass particles and coverslips should be immersed slowly in piranha solution to prevent thermal shock.

2. Reliable estimate of adsorbent surface area by BET can be problematic because glass-particle size used herein is quite large and specific area correspondingly low. Measurement of hydrated octyl sepharose particle surface area by BET is not possible. Nevertheless, the "volumetric" method of interpreting protein adsorption can be applied which does not require accurate measurement of adsorbent surface area [1, 3].

3. Results described in refs. [3–7] with Amersham octyl sepharose are specific to this particular chromatographic packing material. Subsequent work with alternative sources of similar chromatographic media (GE Healthcare Bio-Sciences) did not yield similar results, presumably because alternative material was more porous and ABsorbed a significant amount of protein over-and-above the amount ADsorbed to the Amersham product. Adsorption isotherms obtained with alternative sepharose sources did not achieve a saturation $\left(W_{B_i}^{o}\right)^{max}$ at any solution concentration tested.

4. Trichlorosilanes are water sensitive and highly reactive [13–15]. All manipulations should be carried out with appropriate safety precautions including use of a chemical hood and personal protection equipment. Dry conditions obtained using a chemical glove bag or glove box can make silanization with trichlorosilanes more successful, but use of refluxing $CHCl_3$ obtained in open beakers on a hot plate housed in an ordinary chemical hood can be used to minimize reaction of with atmospheric water. The primary side reaction responsible for poor silanization outcomes appears to be polymerization of trichlorosilane, possibly facilitated by the presence of water, which can lead to opaque, multilayer films on glass coverslip witness samples. Typically, these films are rough and exhibit advancing water contact angles exceeding 120° due to the superhydrophobic effect [16, 17].

   Hydrolysis of ethoxy silanes in ethanol–water solution is critical to formation of complete monolayers. An overabundance of water will result in excessive polymerization in the solvent phase, while a deficiency of water will result in the formation of an incomplete monolayer.

5. Adsorbent capacity (maximum solution depletion at fixed surface area) is a function of protein size (MW) [3, 11]. As a consequence, it may be necessary to adjust surface-area-to-solution-volume to achieve a full adsorption isotherm up-to-and-including surface saturation. *See* ref. 11 for details. Thorough mixing of hydrophobic silanized glass-particle adsorbents with protein solution at high-surface-area-to-solution volume is difficult due to the non-wetting properties of adsorbent. Mixing can be enhanced by manual tapping of tubes with fingers or on a bench top. Mixing protein solutions with octyl sepharose is not problematic and similar adsorption isotherms are obtained with both types of hydrophobic adsorbents, suggesting that mixing obtained with silanized glass as above is sufficient.

6. Mass adsorption from single-protein solution near mg/mL concentrations is rapid, probably within milliseconds and is insensitive to mixing adsorbent with solution [1, 3, 7]. Steady state is thus achieved within the time frame of mixing protein solution and adsorbent.
7. Adsorption competition between two proteins is complex. There are 5 distinct solution-compositions that must be considered for two proteins of dissimilar MW [8, 9].
8. Mass adsorption from binary-protein solution near mg/mL concentrations exhibits two, pseudo-steady-state adsorption regimes connected by smooth transitions lasting 20–30 min, depending on binary-protein pair relative MW and concentration. As such, adsorption competition exhibits kinetics and is sensitive to solution mixing [8, 9].
9. Each different protein requires a separate calibration curve on the same gel to account for differences in staining density. Binary solutions are treated the same as single-protein solutions, multiplexing development of a band OD-concentration calibration curve. Calibration curves can be bi-modal over 0–10 mg/mL range [3, 4] with a 0.2 mg/mL lower limit of detection (LOD, [4]) using Coomassie blue gel stains.

Acknowledgments

This work was supported by National Institute of Health grant PHS 5R01HL069965. The authors appreciate the support from the Departments of Materials Science and Engineering and Bioengineering, The Pennsylvania State University.

References

1. Vogler EA (2012) Protein adsorption in three dimensions. Biomaterials 33:1201–1237
2. Anderson NL et al (2004) The human plasma proteome—a nonredundant list developed by combination of four separate sources. Mol Cell Proteomics 3(4):311–326
3. Noh H, Vogler EA (2006) Volumetric interpretation of protein adsorption: partition coefficients, interphase volumes, and free energies of adsorption to hydrophobic surfaces. Biomaterials 27:5780–5793
4. Noh H, Vogler EA (2006) Volumetric interpretation of protein adsorption: mass and energy balance for albumin adsorption to particulate adsorbents with incrementally-increasing hydrophilicity. Biomaterials 27:5801–5812
5. Noh H, Vogler EA (2007) Volumetric interpretation of protein adsorption: competition from mixtures and the vroman effect. Biomaterials 28:405–422
6. Noh H, Vogler EA (2008) Volumetric interpretation of protein adsorption: ion-exchange adsorbent capacity, protein pI, and interaction energetics. Biomaterials 29:2033–2048
7. Barnthip N et al (2008) Volumetric interpretation of protein adsorption: kinetic consequences of a slowly-concentrating interphase. Biomaterials 29:3062–3074
8. Barnthip N et al (2009) Volumetric interpretation of protein adsorption: kinetics of protein-adsorption competition from binary solution. Biomaterials 30:6495–6513
9. Barnthip N, Vogler EA (2012) Protein adsorption kinetics from single- and binary-solution. Appl Surf Sci 262:19–23
10. Parhi P et al (2010) Role of water and proteins in the attachment of mammalian cells to surfaces: a review. J Adhesion Sci and Tech 24:853–888

11. Parhi P et al (2009) Volumetric interpretation of protein adsorption: capacity scaling with adsorbate molecular weight and adsorbent surface energy. Biomaterials 30:6814–6824
12. Kao P et al (2010) Volumetric interpretation of protein adsorption: interfacial packing of protein adsorbed to hydrophobic surfaces from surface-saturating solution concentrations. Biomaterials 32:969–978
13. Sagiv J et al (1986) Self-assembling monolayers: a study of their formation, composition, and structure. In: Mittal KL, Bothorel P (eds) Surfactants in solution, vol 5. Plenum Press, New York, pp 965–978
14. Fadeev AY, McCarthy TJ (2000) Self-assembly is not the only reaction possible between alkyltrichlorosilanes and surfaces: monomolecular and oligomeric covalently attached layers of dichloro- and trichloroalkylsilanes on silicon. Langmuir 16:7268–7274
15. Wang M et al (2005) Self-assembled silane monolayers: fabrication with nanoscale uniformity. Langmuir 21(5):1848–1857
16. Oner D, McCarthy TJ (2000) Ultrahydrophobic surfaces. Effects of topography length scales on wettability. Langmuir 165:7777–7782
17. Gao L, McCarthy TJ (2009) Wetting 101. Langmuir 25(24):14105–14115

Chapter 13

Hyperspectral Microscopy for Characterization of Gold Nanoparticles in Biological Media and Cells for Toxicity Assessment

Christin Grabinski, John Schlager, and Saber Hussain

Abstract

Nanoparticles (NPs) are being implemented in a wide range of applications, and it is critical to proactively investigate their toxicity. Due to the extensive range of NPs being produced, in vitro studies are a valuable approach for toxicity screening. Key information required to support in vitro toxicity assessments include NP stability in biologically relevant media and fate once exposed to cells. Hyperspectral microscopy is a sensitive, real-time technique that combines the use of microscopy and spectroscopy for the measurement of the reflectance spectrum at individual pixels in a micrograph. This method has been used extensively for molecular imaging with plasmonic NPs as contrast agents (Aaron et al., Opt Express 16:2153–2167, 2008; Kumar et al., Nano Lett 7:1338–1343, 2007; Wax and Sokolov, Laser Photon Rev 3:146–158, 2009; Curry et al., Opt Express 14:6535–6542, 2006; Curry et al., J Biomed Opt 13:014022, 2008; Cognet et al., Proc Natl Acad Sci U S A 100:11350–11355, 2003; Sokolov et al., Cancer Res 63:1999–2004, 2003; Sönnichsen et al., Nat Biotechnol 23:741–745, 2005; Nusz et al., Anal Chem 80:984–989, 2008) and/or sensors (Nusz et al., Anal Chem 80:984–989, 2008; Ungureanu et al., Sens Actuators B 150:529–536, 2010; McFarland and Van Duyne, Nano Lett 3:1057–1062, 2003; Galush et al., Nano Lett 9:2077–2082, 2009; El-Sayed et al., Nano Lett 5:829–834, 2005). Here we describe an approach for using hyperspectral microscopy to characterize the agglomeration and stability of plasmonic NPs in biological media and their interactions with cells.

Key words Hyperspectral microscopy, Hyperspectral imaging, Light scattering, Plasmonic nanoparticles, Agglomeration, Cellular interaction

1 Introduction

Nanoparticles (NPs) are being implemented in a wide range of commercial, medical, military, and industrial applications, with public and private investments in nanotechnology projects reaching \$17.8 billion globally in 2010 [1]. With such a widespread implementation of NPs, exposure to humans and the environment is likely to occur at continually increasing levels [2]. Therefore, it is

Paolo Bergese and Kimberly Hamad-Schifferli (eds.), *Nanomaterial Interfaces in Biology: Methods and Protocols*, Methods in Molecular Biology, vol. 1025, DOI 10.1007/978-1-62703-462-3_13,

critical to proactively investigate NP toxicity in order to protect against human and environmental effects.

Due to the extensive variations in NPs being produced, in vitro studies are a necessary screening tool to compliment in vivo studies in assessing toxicity. Many in vitro studies have been published, demonstrating that NP toxicity is a function of size, agglomerate morphology, composition, crystallinity, surface area, surface chemistry, and surface charge [3–10]. However, the results are often contradictory across laboratories and provide little value for NP regulation due to lack of sufficient NP characterization [11–13].

For the majority of traditional in vitro toxicity studies, NPs are dispersed in biological media before exposure, which can result in agglomeration and adsorption of media components [9, 14]. Agglomeration and protein adsorption have been shown to affect NP uptake and toxicity [5, 15]. Additionally, NP agglomeration can result in a misinterpretation of dose, which is related to the NP agglomerate size and density distribution [13]. Therefore, NP stability must be characterized in the media used for dosing.

Common methods for investigating NP stability include dynamic light scattering and laser Doppler electrophoresis, which are used to measure hydrodynamic diameter and zeta potential, respectively. An additional method commonly employed for investigating the stability of plasmonic NPs is UV–Visible spectroscopy, which can be used for measuring shifts in plasmon resonance to estimate stability in various media [16]. However, none of these techniques are supplemented with visual data for characterizing NP interactions in biological media [17]. Additionally, UV–Visible spectroscopy only provides averaged data for the NP dispersion and is not useful for imaging localized NP interactions [17].

The fate and distribution of NPs once they are exposed to cells is also of critical importance. Many studies have shown that NPs accumulate in the endo-lysosomal system once internalized [3, 4, 7]. However, the extent of uptake and intracellular localization varies widely for different NPs, and this can have a direct effect on NP toxicity [9, 18]. Therefore, methods for detecting NPs in cells and tissues are critical for understanding their toxicity.

NP uptake and intracellular localization in vitro can be characterized using various microscopy techniques. For high resolution, transmission electron microscopy (TEM) is often used; however, this technique requires various steps for manipulating samples, including centrifugation and dehydration [19]. Although optical microscopy techniques can be used with minimal sample preparation, the major limitation is that fluorescent tagging is required to visualize the presence of NPs. Dark-field microscopy can be used to illuminate dense structures, such as cell membranes, organelles and NPs, without fluorescent tagging, but the identity of the structures cannot be confirmed using this technique alone [20].

Hyperspectral microscopy combines the use of microscopy and spectroscopy for the measurement of the reflectance spectrum at individual pixels in a micrograph. It can be used to image and track plasmonic NPs in complex biological environments and has been highly cited for molecular imaging using plasmonic NPs as contrast agents [17, 21–28] and/or local refractive index sensors [17, 29–32]. Recently, hyperspectral microscopy has been demonstrated as a useful technique for characterizing Au NP agglomeration, protein adsorption and cell uptake [18, 33].

Hyperspectral microscopy is particularly useful for analyzing plasmonic NPs, such as gold (Au), which absorb and scatter light in a predictive manner due to collective oscillations of electrons at the surface of a particle induced by an electromagnetic field. Light scattering by Au NPs is a function of size, composition, shape, and local refractive index [34–36]. Additionally, light scattering is affected by interparticle plasmon coupling, which decays over a distance on the order of the NP size, where coupled NPs exhibit a red-shifted and broadened plasmon peak [37]. These properties of Au NPs allow for spectral data to be analyzed for understanding NP agglomeration and protein adsorption in biological environments. Hyperspectral microscopy offers a combined qualitative and quantitative approach for dynamic imaging of NPs and live cells, requiring very little sample preparation. Therefore, it is an ideal approach for characterization of Au NP agglomeration and stability in biological media and interactions with cells to support toxicity assessment.

2 Materials

2.1 Hyperspectral Microscopy Instrument (See Note 1)

2.2 Consumables

1. Glass slides or coverslips for attaching Au NPs (*see* **Note 2**).
2. Glass slides or coverslips appropriate for cell culture.
3. Glass coverslips, cleaned in ethanol and dried.
4. Clear nail polish or silicon grease for sealing the coverslips to the slides.
5. Well-characterized Au NPs (*see* **Note 3**).
6. Cell-line or primary cells.
7. Cell culture media appropriate for the cells.
8. Buffer for rinsing cells.
9. Mounting media for refractive index matching of NP slides and NP exposed cell slides (optional).
10. Immersion oil for high magnification oil lens objectives (optional).

3 Methods

3.1 Protocol for Nanoparticle Characterization Using Hyperspectral Microscopy

1. Dilute Au NPs in the desired media (water, cell media with and without serum proteins) to desired concentration (recommended range: 1–100 mg/l).
2. Pipette 5–20 μl onto coated glass slide and incubate 10–120 s, then rinse with water and allow to dry (*see* **Note 4**).
3. Add mounting media to achieve desired refractive index (if applicable; *see* **Note 5**).
4. Add a clean glass coverslip and seal using silicon grease or clear nail polish (if applicable; *see* **Note 6**).
5. Allow to dry for at least 10 min at room temperature (*see* **Notes** 7 and **8**).
6. Image slides using an appropriate hyperspectral microscopy system (*see* **Note 1**).
7. Use software to collect average scattering spectra and localized spectra for pixels or regions of interest. *See* **Note 5** for data analysis. Examples for NP characterization are described in Figs. 1, 2, and 3. Data in the figures were collected using the CytoViva Hyperspectral Imaging System (CytoViva, Inc., Auburn, AL).

3.2 Protocol for Characterizing Nanoparticle Interactions with Cells Using Hyperspectral Microscopy

1. Plate cells on coverslips or cell culture slides and allow to adhere for desired time period (usually about 24 h).
2. Dilute Au NPs in the appropriate cell culture media to desired concentration (recommended range: 1–100 mg/l).
3. Expose NPs diluted in culture media to cells and incubate for desired exposure period (recommended: 0–48 h).
4. After the desired exposure period, rinse cells with buffer or fix cells (e.g., 4 % paraformaldehyde or cold methanol for 10 min at room temperature), then rinse with buffer.
5. Add Permount or media layer to achieve desired refractive index (if applicable; *see* **Note 5**).
6. For cells grown on chambered slides, remove chambers and add a clean glass coverslip; for cells grown on glass coverslips, transfer face down to a clean glass slide.
7. Seal cover glass with silicon grease or clear nail polish.
8. Allow to dry for at least 10 min (*see* **Notes** 7 and **8**).
9. Image slides using an appropriate hyperspectral microscopy system (*see* **Note 1**).
10. Use software to collect average scattering spectra and localized spectra for pixels or regions of interest. *See* **Note 5** for data analysis. Examples are described in Figs. 4, 5, and 6. Data in the figures were collected using the CytoViva Hyperspectral Imaging System (CytoViva, Inc., Auburn, AL).

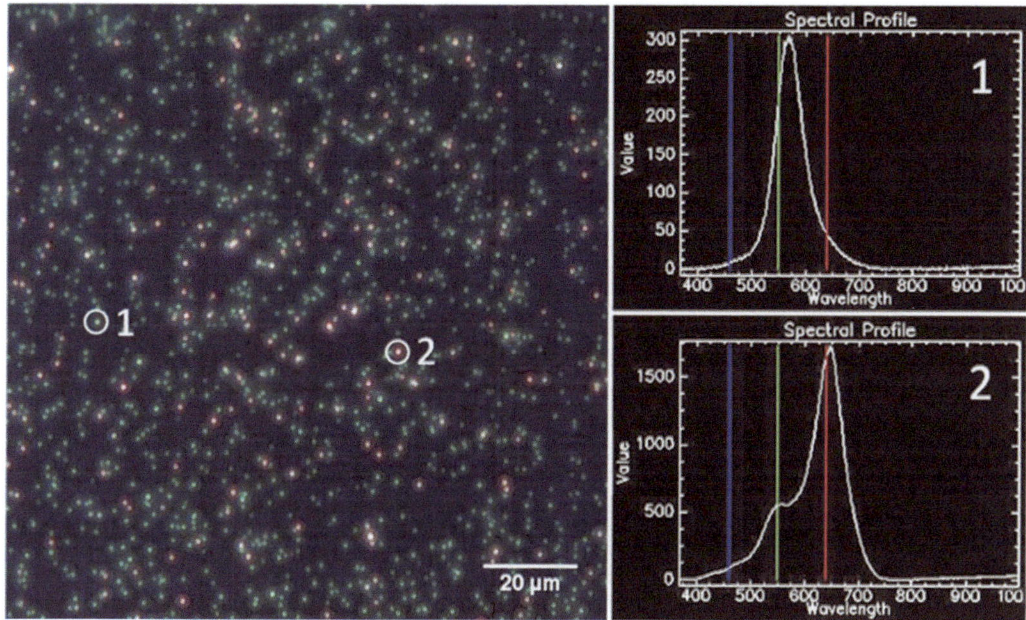

Fig. 1 Scattering properties of Au nanospheres from the National Institute of Standards and Technology (NIST) dispersed in water and stabilized in citric acid (TEM: 64.5 ± 8.7 nm). Au nanospheres were dried on a glass slide and coated with a mounting media with refractive index of 1.515. A glass coverslip was added and sealed using clear nail polish. Individual spherical Au NPs appeared *green*, while agglomerated Au NPs appeared *yellow* or *red* due to interparticle coupling. The scattering peak for *green* particles in the micrograph was ~569 nm, while the scattering peak for *red* particles was ~651 nm. The theoretical plasmon peak for an individual spherical Au nanosphere (65 nm) is 515, 538, and ~565 nm for local refractive index for air (~1.000), water (~1.333), and glass/immersion oil (~1.515), respectively.* *Theoretical values were determined using freely accessible software MiePlot and confirmed using published values [36]

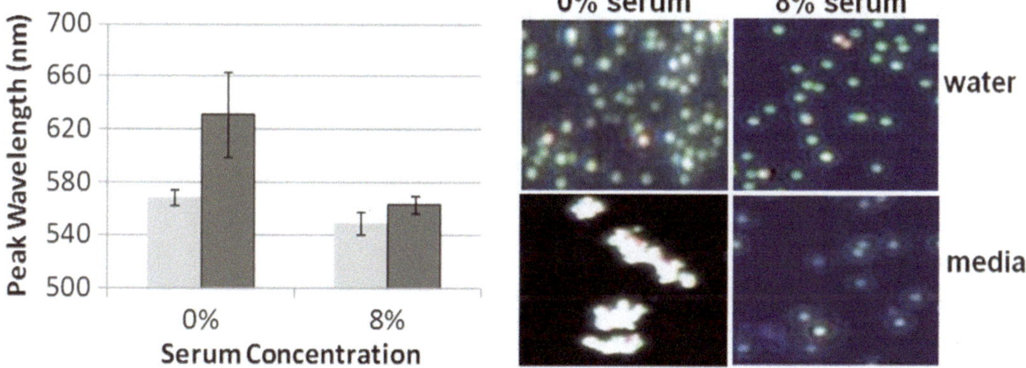

Fig. 2 Stability of Au nanospheres from NIST stabilized in citric acid (TEM: 64.5 ± 8.7 nm) in water or cell culture media, with and without addition of serum proteins. Au NPs were dispersed in the desired media, and then dried on a slide. A coverslip was added and sealed with clear nail polish. Au nanospheres were stable in water, agglomerated in cell exposure media, and were stable in both water and media when serum proteins were added at a concentration used typically for cell culture (8 % fetal bovine serum). The peak wavelengths (averaged over 50 pixels) demonstrated a significant right shifted plasmon peak in media with no serum, and a decreased peak in the presence of serum. Local refractive index effects may account for the small shifts between water and water plus serum and water plus serum versus media plus serum

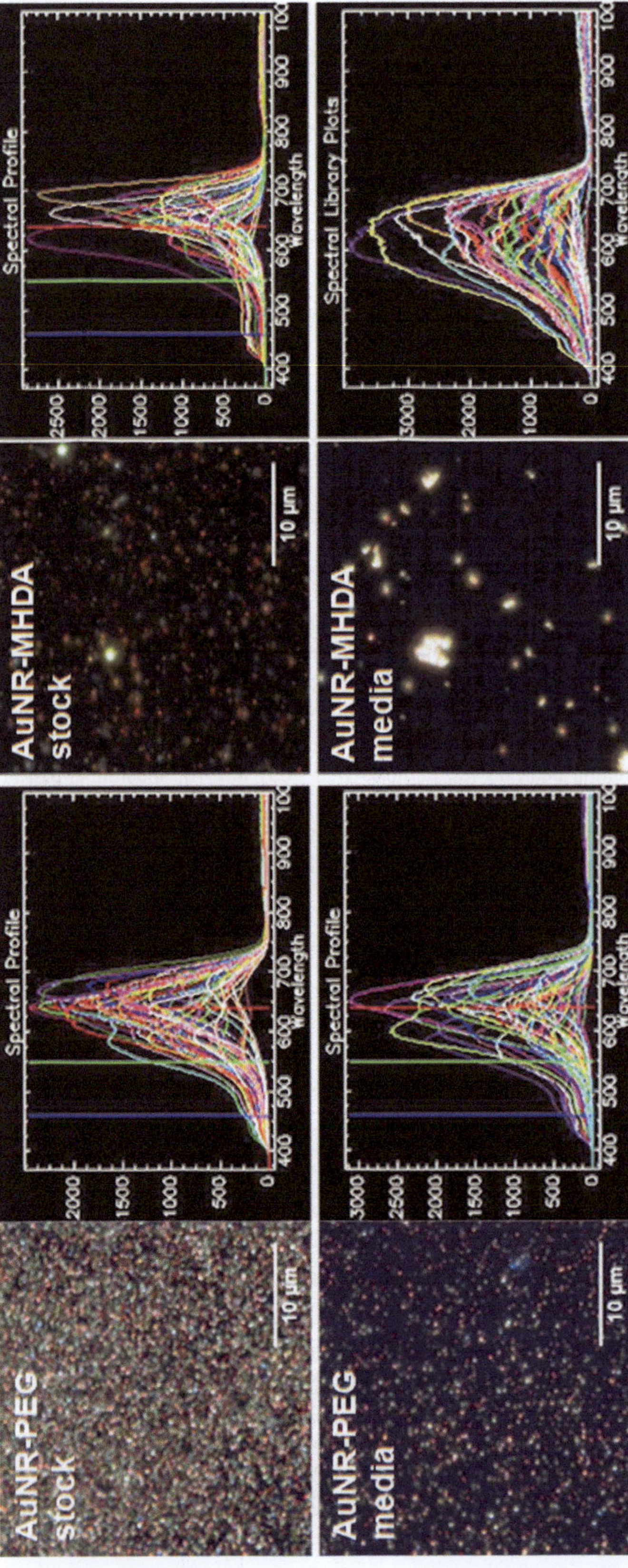

Fig. 3 Effect of surface chemistry on Au nanorod (aspect ratio ~4) agglomeration. Au nanorods functionalized with polyethylene glycol (PEG) did not agglomerate in media, while Au nanorods functionalized with mercaptohexadecanoic acid (MHDA) produced large agglomerates. This can be observed both in the dark-field images and the corresponding spectral plots shown to the *right* of each micrograph

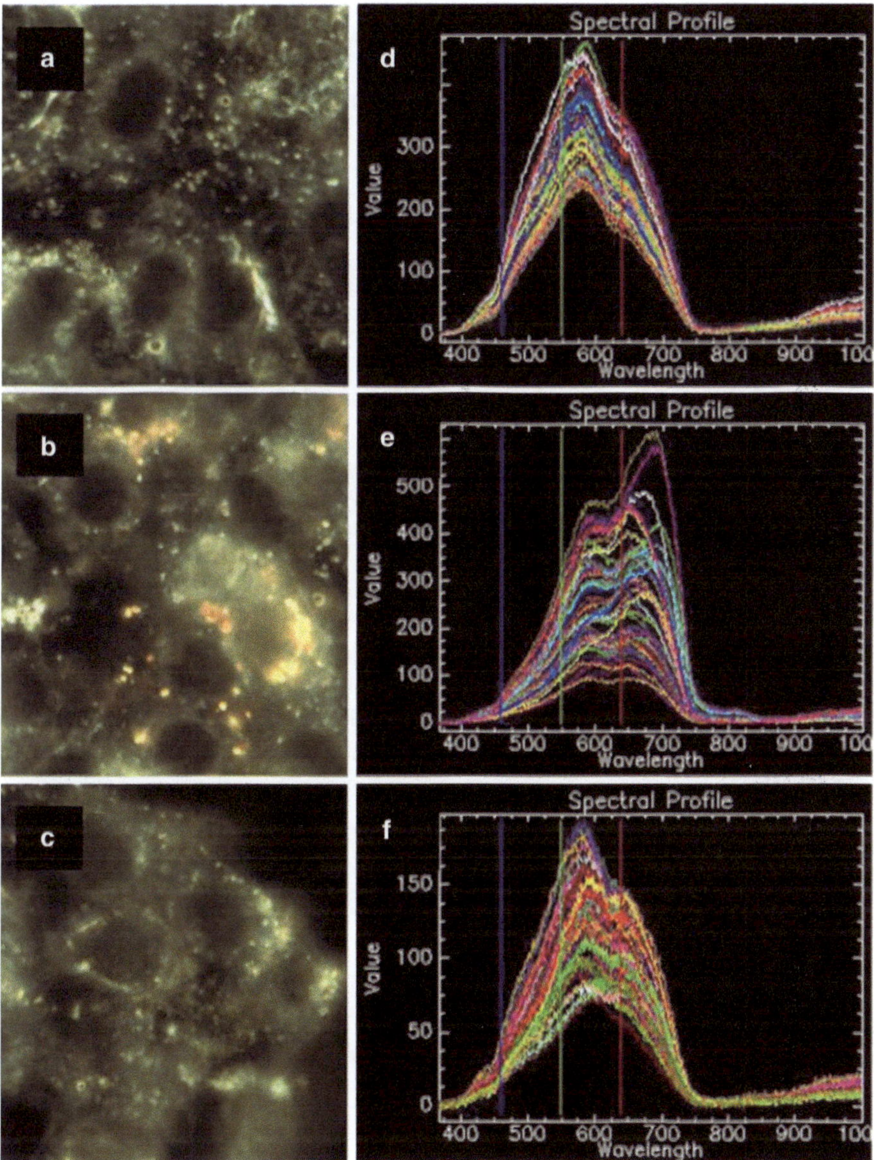

Fig. 4 Effect of stabilizing molecule on Au nanosphere (~10 nm) interaction with lung epithelial cells (A549; ATCC). (**a**, **d**) Control cells; (**b**, **e**) Au nanospheres stabilized in citrate; (**c**, **f**) Au nanospheres stabilized in tannic acid. Results showed *red* peaks for Au NP agglomerates in (**b**, **e**), indicating greater interaction of citrate stabilized Au nanospheres with cells. This was previously confirmed qualitatively using TEM and quantitatively using inductively coupled plasma-mass spectrometry [18]. Image reproduced from [9] with permission from Springer Science + Business Media

4 Notes

1. Hyperspectral microscopy is also referred to as either multispectral imaging or hyperspectral imaging, where hyper refers to high spectral resolution, yielding continuous spectral plots. Instruments capable of spectral imaging are marketed using any of the above

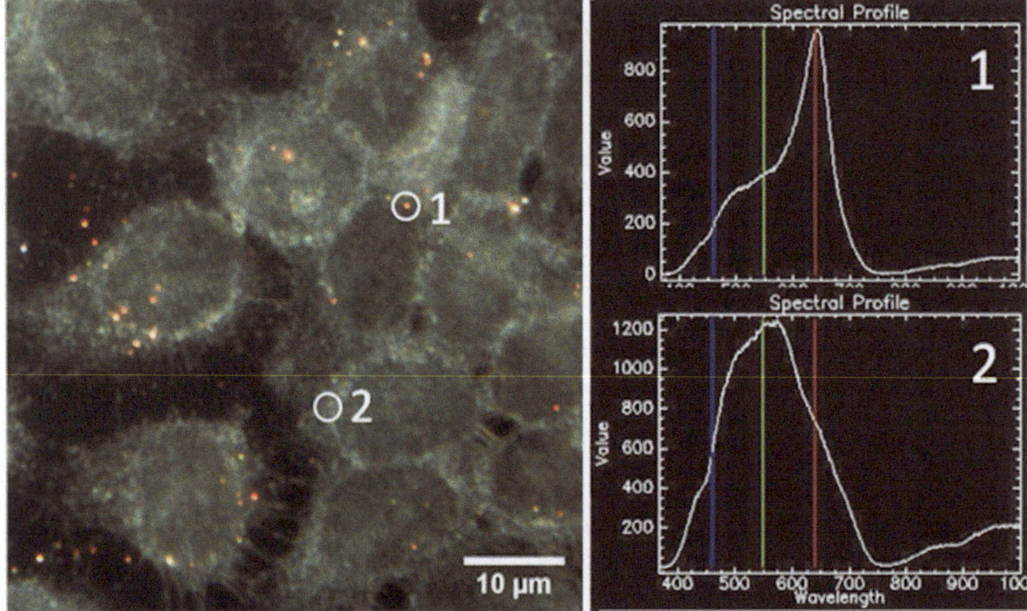

Fig. 5 PEGylated Au nanorod (aspect ratio ~4) interaction with HaCaT cells. PEGylated Au nanorods were easily identified in the cellular environment due to the *sharp red* peak and limited interparticle coupling due to the hydrophilic stable surface chemistry. Representative spectral plots are shown for both PEGylatedAu nanorods (*1*) and background cell scattering (*2*). Limited interaction of these particles with HaCaT cells was previously confirmed using TEM [9]

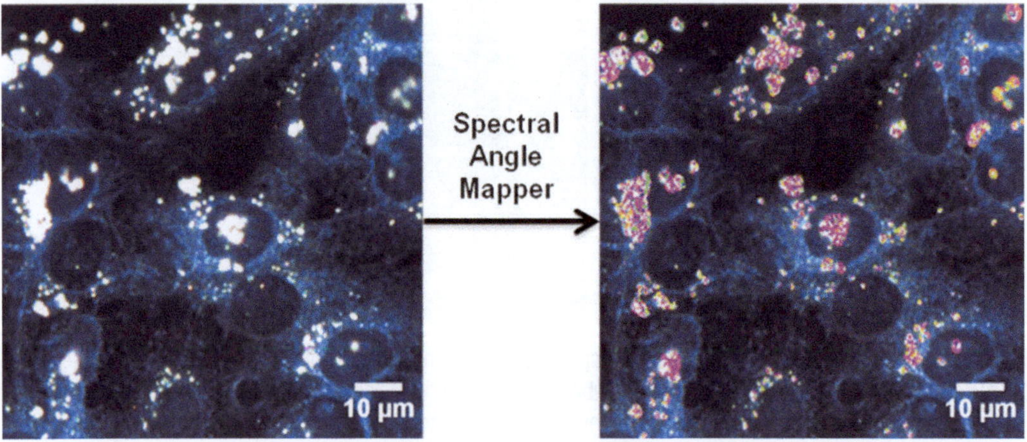

Fig. 6 MHDA functionalized Au nanorods (aspect ratio ~4) are identified in HaCaT cells using spectral angle mapper, which is automated approach embedded in software by ENVI that compares spectral data collected in the NP stability studies to pixel spectra in the cellular environment. Uptake of MHDA-Au nanorod agglomerates in HaCaT cells was previously confirmed using TEM [9]

names and generally consist of a research microscope equipped with a high resolution dark-field condenser, dispersive element (i.e., diffraction grating or prism), spectral detector, and imaging camera. The spectral range of the spectrophotometer is critical, especially for Au NPs, which scatter light in the near infrared region, such as high aspect ratio nanorods and nanocages.

For spherical Au NPs, which scatter in the visible region, it has been shown that it is possible to estimate the scattering spectra by analyzing the red/green ratio of NPs in dark-field images acquired using a color camera [29].

2. Coating the slides with a silane that will leave available thiol or amine groups, such as (3-mercaptopropyl)-trimethoxysilane (MPTMS), mercaptopropyltriethoxy silane (MPTES) or 3-aminopropyltriethoxy silane (APTES), can improve the attachment of Au NPs. The slides can be coated using published techniques [17, 29, 38], or pre-coated slides can be purchased.
3. The morphology and size distribution should be verified using TEM or atomic force microscopy. The stabilizing molecule or surface chemistry should be known (information from the manufacturer is sufficient).
4. Interparticle spacing is a function of concentration, droplet volume, and incubation time and should be optimized for adequate particle spacing.
5. Plasmon resonance shifts in the peak wavelength in the scattering spectra of Au NPs occur as a function of size, shape, electromagnetic coupling, local refractive index, and substrate effects. It is critical to understand these effects in order to properly prepare the samples and interpret the data acquired from hyperspectral microscopy experiments. Each of these effects is described in more detail below.
   (a) Size and shape: Au nanospheres scatter light from green to yellow when increased from 20 to 100 nm with the wavelengths increasing with size. The plasmon peak shifts to red as the particle becomes more oblate. Au nanorods also scatter increasing wavelengths with increasing aspect ratio, where the transverse peak ranges from green to yellow and the longitudinal peak ranges from red to the near infrared region. For Au nanorods, the scattering can change depending on the orientation of the rods, but this detail is not extremely critical for toxicity studies [39]. Stellated Au NPs (~100 nm) have also been shown to scatter red wavelengths [21].
   (b) Electromagnetic coupling: Particles that are close enough in proximity participate in electromagnetic coupling, which changes the scattering spectral response [28]. The shift in the plasmon resonance peak decreases with interparticle distance. Discrete dipole approximation was used to show a 15 nm shift in peak wavelength for 20 nm distance between two Au nanospheres (80 nm) and negligible for 100 nm distance [29]. Based on this type of approximation, shifts in plasmon peak can be used to assess aggregation of Au NPs in biological media and cells.
   (c) Local refractive index: Au NPs exhibit strong sensitivity to the local refractive index [40]. The peak wavelength shift has been found to increase linearly with refractive index

change [28, 41]. Dependence on refractive index increases for particles with higher aspect ratio [41]. For studies aiming to evaluate cellular interactions of NPs based on spectral data collected from NP slides, it is very important that the refractive index of the media or Permount used for preparation of slides for NP characterization matches that used for slides containing cells and NPs.

(d) Substrate effect: Hyperspectral microscopy data is collected in an asymmetric environment with a substrate underneath and either air or a media layer on top of the NPs. This results in a complex refractive index. Some studies have attempted to calculate an effective refractive index to account for this effect [24], but we find that when using objectives with an oil lens, the peak scattering wavelength is within 5 nm of the theoretical values for the refractive index matching that of immersion oil (*see* Fig. 1). For the application described here, this is sufficient.

6. A coverslip is only appropriate if imaging with a local refractive index of 1.515 (i.e., with an oil lens). The NPs can also be dried and imaged in air or covered with an immersion oil with the desired refractive index.
7. Allow the nail polish (or sealant) to dry completely before imaging with an immersion oil objective. If the coverslip is not sealed properly, movement of the objective near the edge of the coverslip may allow the immersion oil to seep under the coverslip, dislodging NPs from the slide.
8. NP slides can be imaged immediately or stored at room temperature or 4 °C. Live cells should be imaged immediately. Fixed slides can be stored at 4 °C for up to 6 months.

Acknowledgments

This work was supported by the Air Force Surgeon General and the Air Force Research Laboratory, Chief Scientist Seedling Program. Christin Grabinski receives a fellowship from the Oak Ridge Institute for Science and Education.

References

1. Lux Research (2011) Global nanotech spending. Presented at EuroNanoForum 2011 conference, 30 May 2011. http://www.euronanoforum2011.eu/wp-content/uploads/2011/09/enf2011_support-commercialisation_raje_fin.pdf. Accessed 14 Sept 2012
2. Reijnders L (2012) Human health hazards of persistent inorganic and carbon nanoparticles. J Mater Sci 47:5061–5073
3. Carlson C, Hussain SM, Schrand AM et al (2008) Unique cellular interaction of silver nanoparticles: size-dependent generation of reactive oxygen species. J Phys Chem B 112:13608–13619
4. Yu K, Grabinski CM, Schrand AM et al (2009) Toxicity of amorphous silica nanoparticles in mouse keratinocytes. J Nanopart Res 11:15–24
5. Murdock RC, Braydich-Stolle L, Schrand AM et al (2008) Characterization of nanomaterial

dispersion in solution prior to *in vitro* exposure using dynamic light scattering technique. Toxicol Sci 101:239–252
6. Grabinski CM, Braydich-Stolle LK, Lafdi KL et al (2007) Effect of particle dimension on biocompatibility of carbon-based nanomaterials. Carbon 45:2828–2835
7. Braydich-Stolle LK, Schaeublin NM, Murdock RC et al (2008) Crystal structure mediates mode of cell death in TiO_2 nanotoxicity. J Nanopart Res 11:1361–1374
8. Grabinski CM (2008) Biocompatibility of carbon-based nanomaterials. Master's thesis, University of Dayton
9. Grabinski C, Schaeublin N, Wijaya A et al (2011) Effect of gold nanorod surface chemistry on cellular interactions *in vitro*. ACS Nano 5:2870–2879
10. Schaeublin NM, Braydich-Stolle LK, Schrand AM et al (2011) Surface charge of gold nanoparticles mediates mechanism of toxicity. Nanoscale 3:410–420
11. Lapotko DO, Lukianova EK, Chizhik SA (2007) Methods for monitoring and imaging nanoparticles in cells. Proc SPIE-Int Soc Opt Eng 6447(644703):1–10
12. Hussain SM, Braydich-Stolle LK, Schrand AM et al (2009) Toxicity evaluation for safe use of nanoparticles: recent achievements and technical challenges. Adv Mater 21:1549–1559
13. Teeguarden JG, Hinderliter PM, Orr G et al (2007) Particokinetics in vitro: dosimetry considerations for *in vitro* nanoparticle toxicity assessments. Toxicol Sci 95:300–312
14. Nel AE, Madler L, Velegol D et al (2009) Understanding biophysicochemical interactions at the nano-bio interface. Nat Mater 8:543–557
15. Walkey CD, Olsen JB, Guo H et al (2011) Nanoparticle size and surface chemistry determine serum protein adsorption and macrophage uptake. J Am Chem Soc 134:2139–2147
16. Kah J, Zubieta A, Saavedra R et al (2012) Stability of gold nanorods passivated with amphiphilic ligands. Langmuir 28:8834–8844
17. Nusz GJ, Marinakos SM, Curry AC et al (2008) Label-free plasmonic detection of biomolecular binding by a single gold nanorod. Anal Chem 80:984–989
18. Mukhopadhyay A, Grabinski CM, Afrooz AN et al (2012) Effect of gold nanosphere surface chemistry on protein adsorption and cell uptake *in vitro*. Appl Biochem Biotechnol 167:327–337
19. Schrand AM, Schlager JJ, Dai K et al (2010) Preparation of cells for assessing ultrastructural localization of nanoparticles with transmission electron microscopy. Nat Protoc 5:744–757
20. Skebo JS, Grabinski CM, Schrand AM et al (2007) Assessment of metal nanoparticle agglomeration, uptake, and interaction using high-illuminating system. Int J Toxicol 26:135–141
21. Aaron J, de la Rosa E, Travis K et al (2008) Polarization microscopy with stellated gold nanoparticles for robust, in-situ monitoring of biomolecules. Opt Express 16:2153–2167
22. Kumar S, Harrison N, Richards-Kortum R et al (2007) Plasmonic nanosensors for imaging intracellular biomarkers in live cells. Nano Lett 7:1338–1343
23. Wax A, Sokolov K (2009) Molecular imaging and darkfield microspectroscopy of live cells using gold plasmonic nanoparticles. Laser Photon Rev 3:146–158
24. Curry A, Hwang WL, Wax A (2006) Epi-illumination through the microscope objective applied to darkfield imaging and microspectroscopy of nanoparticle interaction with cells in culture. Opt Express 14:6535–6542
25. Curry AC, Crow M, Wax A (2008) Molecular imaging of epidermal growth factor receptor in live cells with refractive index sensitivity using dark-field microspectroscopy and immunotargeted nanoparticles. J Biomed Opt 13:014022
26. Cognet L, Tardin C, Boyer D et al (2003) Single metallic nanoparticle imaging for protein detection in cells. Proc Natl Acad Sci U S A 100:11350–11355
27. Sokolov K, Follen M, Aaron J et al (2003) Real-time vital optical imaging of precancer using anti-epidermal growth factor receptor antibodies conjugated to gold nanoparticles. Cancer Res 63:1999–2004
28. Sönnichsen C, Reinhard BM, Liphardt J et al (2005) A molecular ruler based on plasmon coupling of single gold and silver nanoparticles. Nat Biotechnol 23:741–745
29. Ungureanu F, Wasserberg D, Yang N et al (2010) Immunosensing by colorimetric dark-field microscopy of individual gold nanoparticle-conjugates. Sens Actuators B 150:529–536
30. McFarland AD, Van Duyne RP (2003) Single silver nanoparticles as real-time optical sensors with zeptomole sensitivity. Nano Lett 3:1057–1062
31. Galush WJ, Shelby SA, Mulvihil MJ et al (2009) A nanocube plasmonic sensor for molecular binding on membrane surfaces. Nano Lett 9:2077–2082
32. El-Sayed H, Huang XH, El-Sayed MA (2005) Surface plasmon resonance scattering and absorption of anti-EGFR antibody conjugated gold nanoparticles in cancer diagnostics: applications in oral cancer. Nano Lett 5:829–834
33. White B, Strawbridge A, Grabinski CM et al (2012) Hyperspectral imaging (HSI) to evaluate the interaction of optically active nanoparticles in biological media and cells. Accepted to BIOS

34. Yguerabide J, Yguerabide EE (1998) Light scattering submicroscopic particles as highly fluorescent analogs and their use as tracer labels in clinical and biological applications: I. Theory. Anal Biochem 262:137–156
35. Mock J, Smith DR, Schultz S (2003) Local refractive index dependence of plasmon resonance spectra from individual nanoparticles. Nano Lett 3:485–491
36. Haiss W, Nguyen TKT, Aveyard J et al (2007) Determination of size and concentration of gold nanoparticles from UV-Vis spectra. Anal Chem 79:4215–4221
37. Kreibig U, Vollmer M (1995) Optical properties of metal clusters, vol 25. Springer, Berlin
38. Grabar KC, Smith PC, Musick MD et al (1996) Kinetic control of interparticle spacing in Au colloid-based surfaces: rational nanometer-scale architecture. J Am Chem Soc 118:1148–1153
39. Orendorf CJ, Sau TK, Murphy CJ (2006) Shape-dependent plasmon-resonant gold nanoparticles. Small 2:636–639
40. Lee KS, El-Sayed MA (2006) Gold and silver nanoparticles in sensing and imaging: sensitivity of plasmon response to size, shape, and metal composition. J Phys Chem B 110:19220–19225
41. Sönnichsen C, Geier S, Hecker NE et al (2000) Spectroscopy of single metallic nanoparticles using total internal reflection microscopy. Appl Phys Lett 77:2949–2951

Chapter 14

Immunocytochemistry, Electron Tomography, and Energy Dispersive X-ray Spectroscopy (EDXS) on Cryosections of Human Cancer Cells Doped with Stimuli Responsive Polymeric Nanogels Loaded with Iron Oxide Nanoparticles

Roberto Marotta, A. Falqui, A. Curcio, A. Quarta, and Teresa Pellegrino

Abstract

The cryosectioning technique is an alternative method for preparing biological material for Transmission Electron Microscopy (TEM). We have applied this technique to study the mechanism of cell internalization of stimuli-responsive polymeric nanogels exploited as cargo nanovectors. With respect to conventional TEM processing, cryosectioning technique better preserves the morphology of solvent-sensitive nanogels and enhances the visibility of membrane-bounded organelles inside the cell cytoplasm. In this chapter we describe the protocols we have established to perform Electron Microscopy (EM)-immunocytochemistry, Electron Tomography (ET), and Energy Dispersive X-ray Spectroscopy (EDXS) chemical analysis in Scanning TEM (STEM) on cryosections of HeLa cells treated with pH-responsive nanogels hosting short interference RNA (siRNAs) and iron oxide nanoparticles (IONPs).

Key words Tokuyasu cryosectioning technique, Immunocytochemistry, Electron tomography (ET), Energy dispersive X-ray spectroscopy (EDXS) chemical analysis, Scanning transmission electron microscopy (STEM), Stimuli-responsive polymeric nanoparticles, Iron oxide nanoparticles (IONPs) loaded nanogel, HeLa

1 Introduction

Tokuyasu cryosectioning technique of aldehyde-fixed biological material is a consolidated technique for preparing biological material for immuno Transmission Electron Microscopy (TEM) studies [1–4]. Briefly, the samples are chemically fixed and infiltrated with a cryoprotectant, plunge-frozen into liquid nitrogen, sectioned at low temperature, transferred to room temperature as frozen sections, and finally thawed and embedded in a thin layer of supporting and contrasting film [2, 3]. By this technique proteins and organic compounds remain in their natural aqueous environment and, because of the open surface structure, more antigens are

Paolo Bergese and Kimberly Hamad-Schifferli (eds.), *Nanomaterial Interfaces in Biology: Methods and Protocols*, Methods in Molecular Biology, vol. 1025, DOI 10.1007/978-1-62703-462-3_14,

accessible than in resin sections, giving a very high sensitivity in immunolabeling experiments. Moreover, the simplicity of sample preparation makes cryosection a convenient way to perform routine immunocytochemical localizations [5].

We have recently applied this technique to nanobiotechnology, with the aim to study the mechanism of cell internalization of stimuli-responsive solvent-sensitive polymeric nanogels loaded with inorganic nanoparticles and small oligonucleotides [6]. The soft nature of those polymeric nanogel materials which can undergo physicochemical modifications upon application of environment changes (such as pH or temperature) makes them appealing as drug carriers for the controlled release of drugs or payload elements. In a recent work, we showed how acidic pH-responsive nanogels made of polyvinyl pyridine-divinylbenzene allowed to host and transport siRNA oligonucleotides and superparamagnetic iron oxide nanoparticles (IONPs) both entrapped within the polymeric matrix [6]. Furthermore, thanks to the well distinct contrast of the superparamagnetic nanocrystals and of the nanogel under TEM, we could track the fate of the magnetic pH-responsive nanogels once administered to HeLa cells. We observed that upon endocytosis uptake, the IONPs and the functional siRNA were released into the cytoplasm under the action of the endogenous pH stimulus present inside the different endosomal compartments (*see* Fig. 1a). The IONPs may be also exploited for applying hyperthermia treatment to the cells, triggered by an external alternative magnetic field of appropriate frequency and field amplitude [7, 8]. In recent works IONPs have been combined with thermo-responsive polymers [9–12]. The increase in local temperature generated by the IONPs might also induce the conformational change of the polymer network, thus providing the external stimulus for shrinking the polymer networks and release the entrapped drug molecules [13].

The study and the characterization of such kind of inorganic–organic hybrid materials when different stimuli are applying either on cell cultures in vitro or once injected in animal models in vivo, are quite challenging and require particular care in order to preserve the natural conformation of the nanosystems in the different environmental conditions. For this reason we applied cryosectioning. Indeed, in comparison with conventional EM preparation methods, the advantages of cryosectioning in processing cells doped with solvent-sensitive polymeric nanogels are manifold. Since cryosectioning does involve neither the use of solvents during sample dehydration, nor the use of resin during the infiltration and the embedding steps, it better preserves the fine morphology of polymeric nanogels (*see* Figs. 1b and 2). Indeed, using cryosectioning we did not observe the typical artifacts shown by the polymeric nanogels when processed by conventional TEM, such as the swelling of the polymer that more than doubled its diameter and the

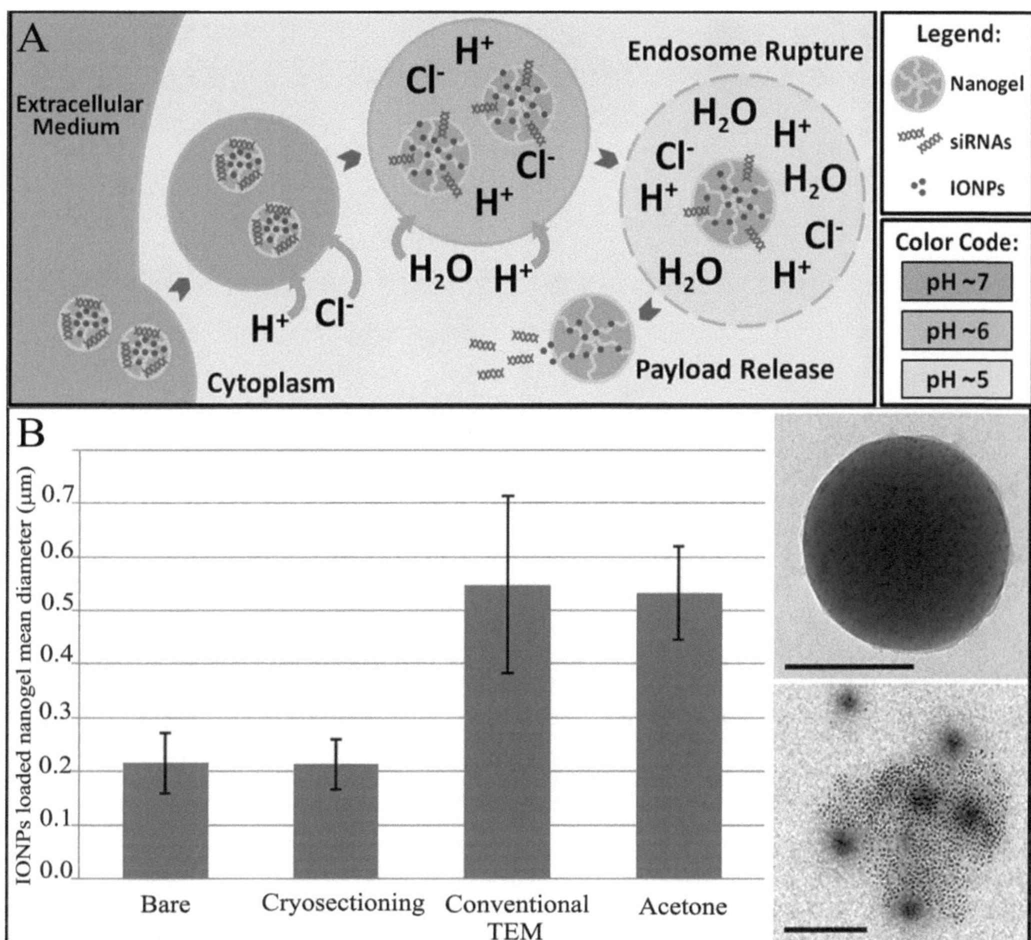

Fig. 1 Nanogel releasing mechanism. (**a**) Sketch showing the intracellular pathway and the pH-mediated release mechanism of the siRNA/IONPs loaded nanogels (NG): the endosome acidification promotes the NG swelling, due to the recall of counter ions and water (the "proton sponge" effect). This leads to the rupture of the endosomal membrane and to the escape of both siRNAs and IONPs from the endosome to the cytoplasm; (**b**) histogram showing the various diameters of the NGs when deposit from water (Bare), or when processed by cryosectioning, conventional TEM or exposed to acetone. Note that the diameter of the NGs processed by cryosectioning is similar to that of the bare NGs. Conventional TEM processing as well as incubation with acetone causes a remarkable increase of the NG diameter. The *insets* show TEM images of a bare NG (*up*) and of a NG after incubation with acetone (*bottom*). Scale bar are 200 nm

displacement towards the outside of the nanoparticles contained inside the nanogels (for a comparison between conventional processing and cryosectioning TEM characterizations applied to the same samples *see* Figs. 1b and 2). In addition, cryosectioning, enhancing the integrity and visibility of membranes, is particularly suited for studying nanoparticles uptake and trafficking. Indeed on cryosections it is easier the identification of membrane bounded cytoplasmic organelles (i.e., endosomes), which play a crucial role in the internalization processes (*see* Fig. 2 and ref. 6).

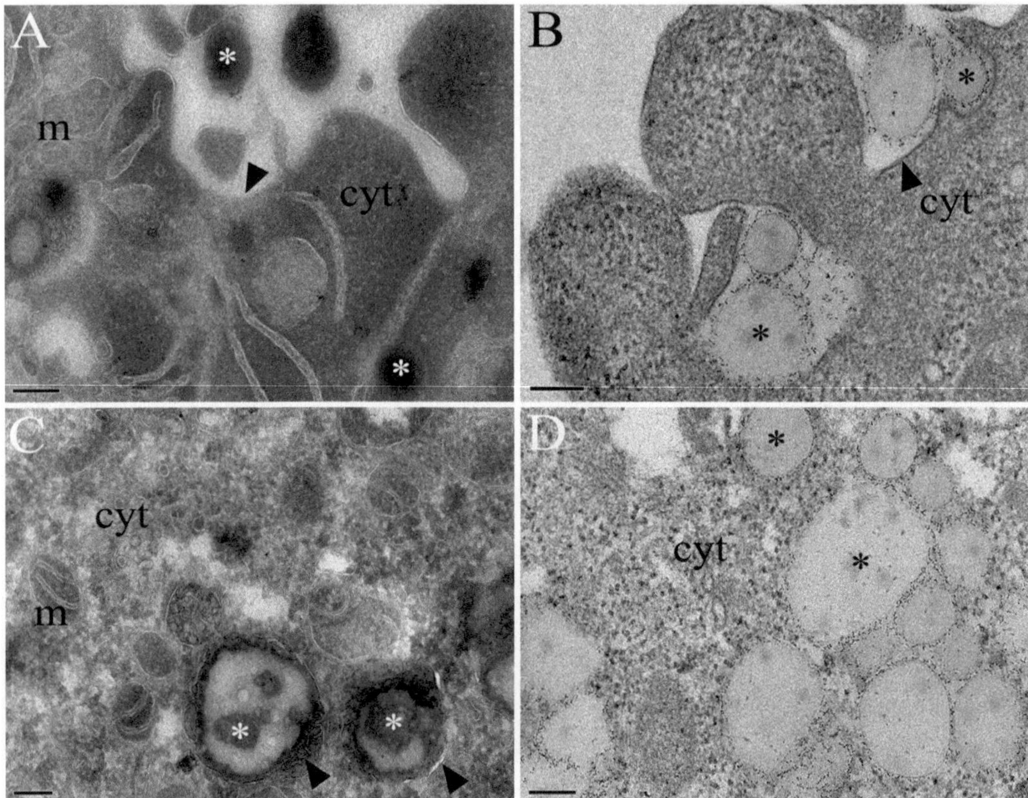

Fig. 2 TEM images of HeLa cells incubated with acidic pH-responsive nanogels loaded with IONPs using cryosectioning (**a**, **c**) and conventional TEM sample preparation (**b**, **d**). (**a**, **b**) The IONPs loaded nanogels (*asterisks*) are approaching the cell membrane (*arrowhead*). Note that in the sample processed using the cryosectioning technique the nanogels maintain their morphology. On the contrary, in the sample conventionally processed the nanogels appear dramatically swollen. (**c**, **d**) the IONPs loaded nanogels (*asterisks*) are inside the cell. Note that only in the samples processed using the cryosectioning technique the endosomes (*arrowheads*) are clearly visible as also the nanogels (*asterisks*) inside them. *cyt* cytoplasm, *m* mitochondria. Scale bars correspond to a length of 200 nm

Besides, cryosections represent an unexpectedly robust and flexible starting point on which to perform further EM-based analyses, such as immunocytochemistry, electron tomography (ET), and Energy Dispersive X-ray Spectroscopy (EDXS) chemical analysis. On this concern cryosectioning, involving the rapid freezing of the sample and avoiding liquid media, is a good preparation technique for successive microanalysis, avoiding the translocation of the elements of interest within cells or soft tissue that instead occur by chemical fixation [14]. Therefore, the application of cryosectioning to cells doped with solvent-sensitive polymeric nanoparticles is promising, especially given the wide and multidisciplinary EM characterizations further allowed with minimal structural artifacts.

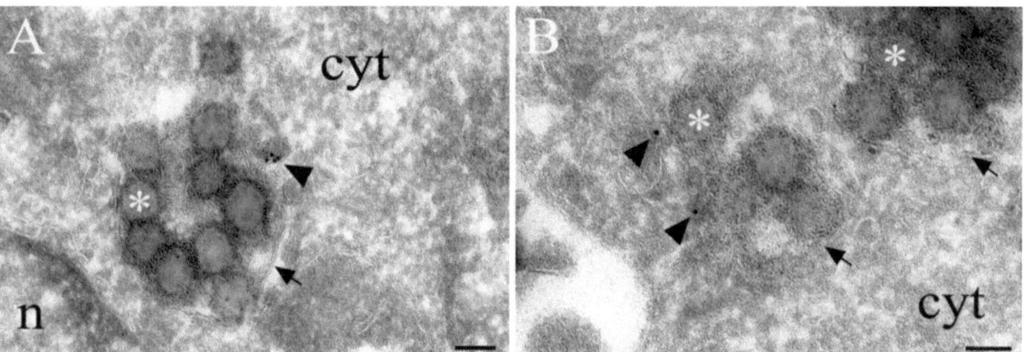

Fig. 3 Immunogold electron microscopy analysis of Rab7 in cryosection of HeLa cells doped with IONPs loaded nanogel. (**a**) Late endosome (*arrow*) containing IONPs loaded nanogels (*asterisk*). Note the cluster of gold particles (*arrowhead*) close to the endosomal membrane. (**b**) Late endosomes (*arrows*) containing IONPs loaded nanogels (*asterisks*). Note the gold nanoparticles (*arrowheads*) lining the endosomal membrane. *cyt* cytoplasm, *n* nucleus. Scale bars correspond to a length of 100 nm

The main aim of this chapter is to describe the protocols for preparing cryosections of cells doped with solvent-sensitive polymeric nanoparticles for EM-immunocytochemistry, ET and EDXS chemical analysis in Scanning TEM (STEM). As an example here we first describe the EM immunocytochemistry protocol we used to process cryosections of HeLa cells doped with IONPs loaded nanogels. Using a primary antibody against Rab7 (Sigma Aldrich), a late endosomal marker, and an appropriate secondary antibody conjugated to gold nanoparticles having their own contrast under TEM, we identified and localized the vesicular compartments containing the nanogels (*see* Fig. 3). Later, we provide an example of the ET protocol we have established to generate a tomogram on a semi-thin cryosection (of around 150 nm in thickness) of HeLa cells doped with IONPs loaded nanogels (*see* Fig. 4). ET is an electron microscopy technique based on the collection of several TEM images of the sample tilted in a wide angular range. The use of these projections allows to reconstruct the sample 3D-morphology [15]. Although ET should be in principle a useful tool in nanobiotechnology, allowing the study of nanoparticles in their sub-cellular context, so far relatively few studies dealing with ET on cells doped with nanoparticles are reported [16–20]. Moreover ET has been applied only seldom to labeled and unlabeled cryosections, despite the advantages of this technique (see above) [5, 21].

Finally, we end this chapter describing the STEM EDXS protocol we have established to perform STEM EDXS on cryosections of HeLa cells, allowing both the localization of the nanogels inside the cell cytoplasm and the identification of the IONPs inside the nanogels, via the iron EDXS mapping (*see* Fig. 5). EDXS is a well-established technique in biology to perform the chemical analysis of biological materials using the transmission electron microscope.

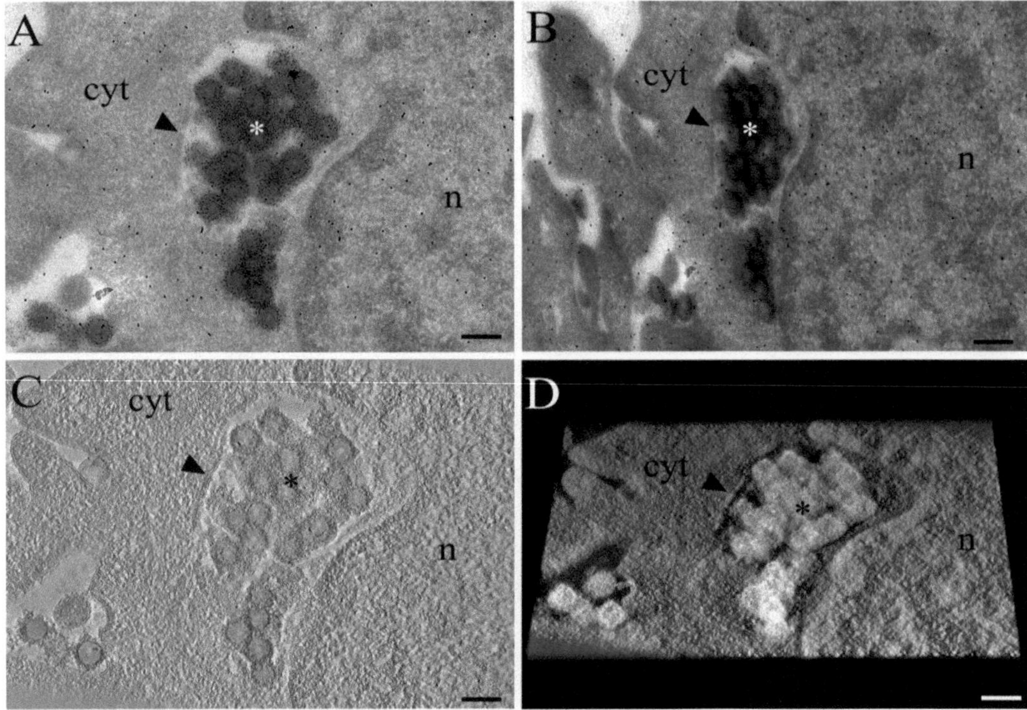

Fig. 4 Stages of data acquisition in electron tomography reconstruction. (**a**, **b**) a semi-thin (≈150 nm) cryosection of HeLa cells doped with IONPs nanogel is imaged at respectively 0° (**a**) and 60° of tilt angle (**b**) in a Jeol JEM 2200FS microscope operating at 200 kV of acceleration voltage. Small *black dots* are 15 nm gold particles used to align images. (**c**) A single tomogram slice displays fine detail of membranes and of IONPs decorating the nanogels (*asterisks*) inside the endosomes (*arrowheads*) not apparent in the original semi-thick section image. (**d**) A tomogram 3D visualization using the isosurface rendering as implemented in UCSF Chimera software, version 1.6.2. *cyt* cytoplasm, *n* nucleus. Scale bar correspond to a length of 200 nm

This technique allows for the identification of the chemical elements inside cells or tissues even if they are present in low atomic percentage (about 3–4 %) [22]. EDXS makes use of the X-ray spectrum emitted by the sample when irradiated with a high-energy electron beam. As the energy of the emitted X-rays is related to the elements constituting the sample, it can be used to obtain a spatially resolved chemical analysis. When EDXS is combined with STEM mode, it allows to measure elemental composition down to sub-micrometer scale in organic and biological thin sections [23]. In the latter method, the electron beam is focused in a way to obtain a very small electron probe (down to nanometer size). It is then raster scanned across the specimen and at each probe position, the corresponding X-ray emission spectrum is recorded. This spectrum may therefore be used to reconstruct a sample elemental map. Moreover, compared to EDXS in TEM mode, EDXS via STEM is more advantageous since the inelastic interaction that gives rise

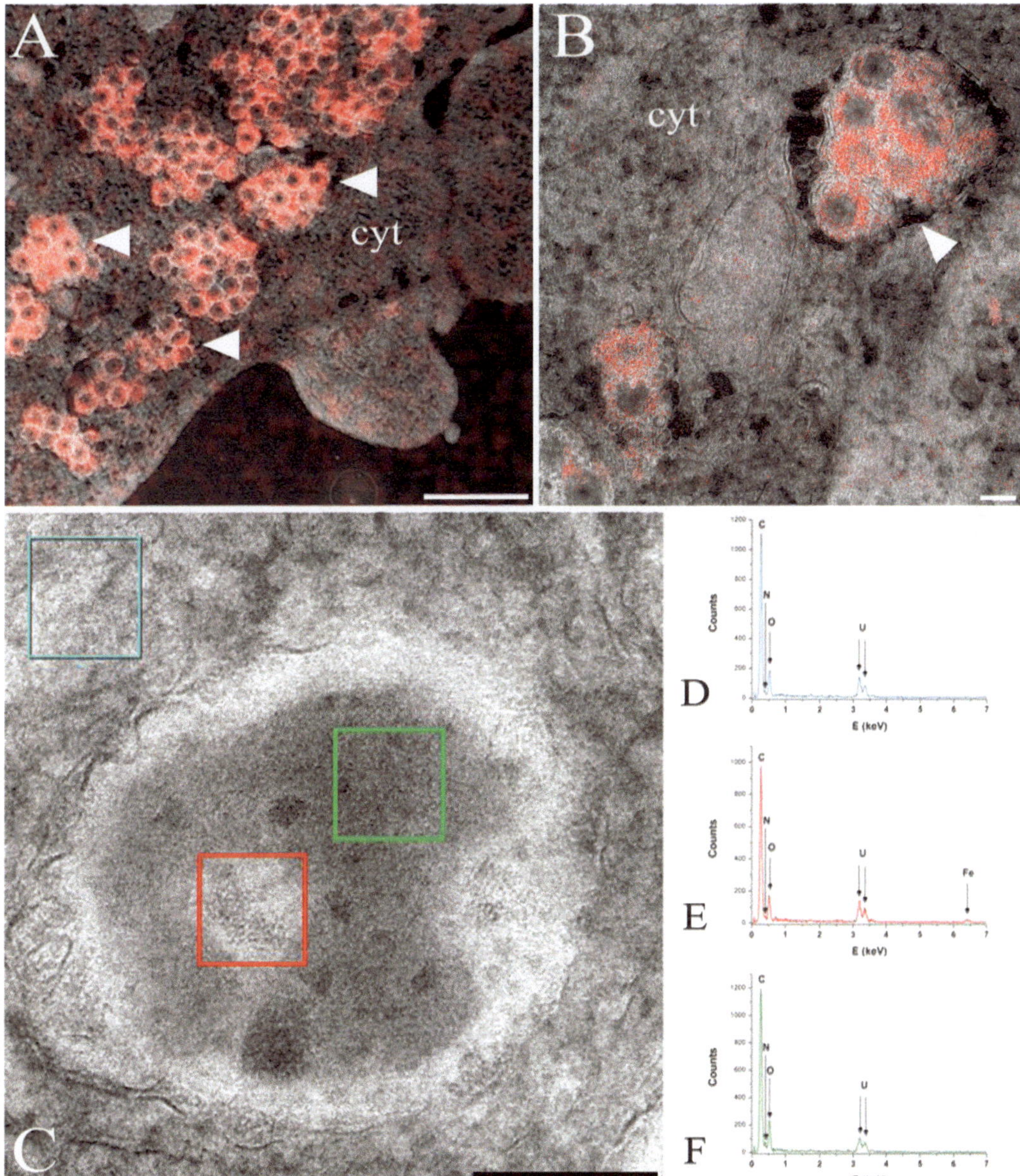

Fig. 5 EDXS analysis on cryosections of HeLa cells loaded with IONPs loaded nanogels. (**a**) Scanning transmission electron microscopy (STEM) image of several endosomes containing numerous IONP loaded nanogels (*arrowheads*) inside the cytoplasm of a cryosectioned HeLa cell. The superimposed EDXS iron map, corresponding to the IONPs inside the nanogels, is shown in *red*. (**b**) superimposed EDXS iron map on a higher magnification of a single endosome (*arrowhead*). Note that the IONPs are mainly around the edge of the nanogels trapped inside the endosome. (**c**) STEM image of an endosome containing IONPs loaded nanogels inside the cytoplasm of a cryosectioned HeLa cell. Three regions of interest, one outside the endosome (*blue square*) and two inside it (*green* and *red squares*) were selected for compositional mapping as shown in (**d–f**). Note that the *red square* in (**c**) encloses a IONPs loaded nanogel. *cyt* cytoplasm. Scale bars correspond to a length of 700 nm in (**a**), 100 nm in (**b**) and 150 nm in (**c**)

to X-Ray emission is effectively local [24]. In the last years this technique has been increasingly used in nanobiotechnology to characterize nanoparticles inside cells and cell compartments [25], although only recently it has been performed on cryosections [6].

2 Materials

2.1 Major Equipments

1. Transmission and Scanning Transmission Electron Microscope (TEM/STEM) equipped with a charge-coupled device (CCD) digital camera (*see* **Note 1**).
2. Energy Dispersive X-ray Spectroscopy detector (*see* **Note 2**).
3. Tomographic TEM specimen holder (*see* **Note 3**).
4. EDXS software (*see* **Note 4**).
5. Tomography acquisition software (*see* **Note 5**).
6. Software for tomogram reconstruction (*see* **Note 6**).
7. Software for tomogram 3D visualization (*see* **Note 7**).
8. Carbon vacuum evaporator (*see* **Note 8**)
9. Stereomicroscope with light source (*see* **Note 9**).
10. Adjustable volume pipette.
11. EM specimen grids (Nickel 200 mesh specimen grids Formvar coated and carbon sputtered).
12. Loops for the final drying step (*see* **Note 10**).
13. Fine forceps (e.g., Dumont Biologie #5 or #7).
14. Filter paper.
15. Parafilm (SPI).

2.2 Chemical and Solution

1. Colloidal gold solution of the appropriate size (*see* **Note 11**).
2. Phosphate buffered saline (PBS).
3. 2 % gelatin in PBS.
4. 0.15 % glycine in PBS.
5. Bovine serum albumin (BSA 1 % and 0.1 % in PBS).
6. Primary antibody with known antigen specificity.
7. Secondary antibody conjugated to colloidal gold nanoparticles.
8. 1 % glutaraldehyde in PBS.
9. 2 % uranyl acetate in distilled water, pH = 7.5.
10. Methyl cellulose saturated uranyl acetate, 9:1, pH = 4 (*see* **Note 12**).

3 Methods

In this paragraph we will describe in detail the various protocols we established for immunolabeling of cryosections (Subheading 3.1), ET (Subheading 3.2) and STEM EDXS chemical analysis (Subheading 3.3) performed on cryosections of HeLa cells doped with the magnetic pH-responsive nanogels. The preparation protocol of the pH-responsive nanogels loaded with iron oxide nanoparticles (IONPs) and the description of the doping protocol of HeLa cells are reported in ref. 6. A detailed description of the Tokuyasu cryosectioning technique can be found in refs. 1–4.

3.1 EM Immunolabeling of Cryosections of Doped Cells

The immunolabeling procedure is realized by floating the Formvar coated nickel grids (*see* **Note 13**) on top of different drops, cryosections facing the drops (*see* **Note 14**), using forceps (*see* **Note 15**). The drops are placed on a clean flat surface (*see* **Note 16**). Note that the backside of the grid should stay dry throughout the whole procedure (*see* **Note 17**).

3.1.1 Prepare the Sections for the EM Immunolabeling

1. Transfer the grids stored at 4 °C on the glass slide with sucrose/methyl cellulose to a cold 2 % gelatin plate. Then warm it to 40 °C and keep on the fluid gelatin for at least 20 min (*see* **Note 18**).
2. Transfer the grids in 10–20 mM glycine in PBS for 5 min (*see* **Note 19**).
3. Transfer the grids in 1 % BSA in PBS for 2 min (*see* **Note 20**).

3.1.2 EM Immunolabeling and Fixation of the Immunoreagents

1. Dilute primary antibody in 1 % BSA/PBS at a final concentration of 2–20 μg/ml (*see* **Note 21**).
2. Incubate the grids at 37 °C, on 10 μl drops of the specific primary antibody solution (*see* **Note 22**) for more than 20 min (*see* **Note 23**). For the immunolabeling shown in Fig. 3, as a marker for late endosomes we used an anti-Rab7 polyclonal primary antibody (Sigma-Aldrich) produced in rabbit, diluted 1:50 in the blocking solution.
3. Rinse with 0.1 % BSA/PBS, four changes, 2 min each to wick away the unbounded primary antibody.
4. Transfer the grid on 10 μl drops of the suitable secondary antibody (*see* **Note 24**) solution conjugated with colloidal gold nanoparticles of appropriate size (*see* **Note 25**). Dilute the secondary antibody in 0.1 % BSA/PBS and incubate at room temperature for more than 20 min.
5. Rinse with 0.1 % BSA/PBS, four changes, 2 min each; then rinse with plain PBS, four changes, 2 min each to wick away the unbounded secondary antibody. For the immunolabeling

shown in Fig. 3, we used an anti-rabbit secondary antibody conjugated with 15 nm colloidal gold (Aurion, Delta Microscopie), diluted 1:100 in the blocking solution.

6. Transfer the grid on 1 % glutaraldehyde in PBS for 5 min (*see* **Notes 26** and **27**).
7. Rinse abundantly with distilled water, more than six changes, 1 min each (*see* **Note 28**).

3.1.3 Contrast Enhancement and Final Embedding

1. Transfer the grids on a solution of 2 % uranyl acetate, pH = 7.5 for 5 min (*see* **Note 29**).
2. Rinse briefly with distilled water, two changes.
3. Transfer quickly the grids over two drops of cold uranyl acetate/methyl cellulose (1:9), and leave the sections floating for 5–10 min on a third drop of that mixture. All uranyl acetate/methyl cellulose drops should be on ice (*see* **Note 30**).

3.1.4 Drying

1. Pick up the grid in a wire loop (*see* **Note 31**). Push the loop in the uranyl acetate/methyl cellulose drop at some distance from the grid, bring it underneath and lift the grid from the uranyl acetate/methyl cellulose drop. The grid is then in the center of the loop with excess uranyl acetate/methyl cellulose hanging underneath.
2. Tilt the loop and grid to an angle of 45–60° (the excess uranyl acetate/methyl cellulose downwards) and touch the side of the loop to well absorbing filter paper with the sections facing the filter paper. The amount of excess uranyl acetate/methyl cellulose should be enough to make contact and disappear into the filter paper, leaving behind a thin even film on the surface of the grid, which remains in the loop center (*see* **Note 32**).

3.2 Electron Tomography

Semithin cryosections are coated (on one side) with 10 or 15 nm gold particles that serve as fiducial markers in the fine alignment process during tomogram assembly (*see* **Note 33**). To maintain well preserved the cell ultrastructure is very important that only the Formvar membrane supporting the cryosections but not the cryosections themselves contacts the colloidal gold solution (*see* **Note 34**).

3.2.1 Preparing Cryosections for Electron Tomography

1. Transfer the grid face up into a drop of colloidal gold solution of the appropriate size for 1 min. Previously dilute the colloidal gold solution 1:3 in bidistilled water (*see* **Note 35**).
2. Wash the grid face up three changes to wick away the excess of gold solution.
3. Carefully dry the grids with filter paper (*see* **Note 36**).

3.2.2 Adjust the Eucentric Height of Your Specimen

1. Identify on your sample the region of interest (ROI) on which perform tomography.
2. Gradually tilt the stage over a range of ±30° (*see* **Note 37**). If necessary re-center the ROI changing the *Z* coordinate.
3. Repeat **step 2** until the ROI does not move significantly.

3.2.3 Data Collection

This step has become more routine with the aid of computer controlled microscopes.

1. Before the data acquisition, check if your ROI can be tilted over an angular range of ±60° or more, without appearing of shadows due to the TEM grid.
2. Acquire your data while your specimen is tilted over the chosen angular range. TEM images are acquired every 1° or 2° to complete a tilt series.
3. Before capturing each image always center your ROI and check the focus (*see* **Note 38**).

3.2.4 Tomogram Reconstruction

1. The tilt series can be collected and/or finally recorded as a single image stack file (*see* **Note 39**), containing all the projections of the sample taken in correspondence of the different tilting angles (*see* **Note 40**).
2. Use the software interface Etomo in IMOD version 4.5 [26] (or similar) to generate a one-axis tomogram from the image stack (*see* **Note 41**). Some of the major steps include: (a) aligning the images in the stack for cross correlation generating a coarse aligned stack; (b) creating a fiducial model based on the position of the fiducial gold particles; (c) using the fiducial model, generating a finely aligned stack (*see* **Note 42**); and finally (d) generating the tomogram by weighted back projection (WBP) or simultaneous iterative reconstruction technique (SIRT).

   This process can be repeated on adjacent serial sections to capture 3D structural data through a greater volume of the cell.
3. Scan through the entire tomogram with the “Viewing mode” present in 3dmod, version 4.5, as implemented in IMOD. This is the first way to understand the volume of data, and to grasp the 3D arrangement of the structures of interest within the tomogram.

3.2.5 3D Reconstruction Based on Isosurface Rendering

Manual and computer assisted segmentation allows the isolation, identification and visualization of the ROIs inside a tomogram. Although manual segmentation, i.e., the hand drawing of contours in each slice of a tomogram, can be influenced from several biases (*see* **Note 43**), it is widely used with biological samples due to the complex shape and low contrast characteristic of biological structures [28]. Thanks to the good membrane contrast present in tomograms generated from cryosections, it is possible to generate 3D reconstruction of complex structures avoiding manual segmentation, using the less time consuming isosurface rendering. By this procedure the segmentation occurs from gray level volumes using a visually estimated gray level threshold. As an example, we used the isosurface rendering as implemented in the UCSF Chimera software, version 1.6.2 [27] (*see* **Note 44**) to the 3D visualization of the tomogram generated from a 150 nm thick cryosection of HeLa cells doped with IONPs loaded nanogels (*see* Fig. 4).

1. Opening your tomogram with Chimera will automatically start "Volume Viewer" window.
2. In "Volume Viewer" select the rendering mode to use for 3D tomogram visualization using the "Style" option (*see* **Note 45**). For the 3D rendering shown in Fig. 4d we used the solid mode, in which the data are shown as a semitransparent solid.
3. Set the "step" setting on the "Volume Viewer" window to value below 4 (*see* **Note 46**).
4. Finely tune the gray-level thresholds dragging horizontally or vertically the small squares and connecting lines on the gray level histogram in the "Volume Viewer" window to optimize the 3D rendering.
5. Under the "Volume Viewer" window, use the "Planes" section to display a slab of the tomogram, normal to a specified axis (default *Z* axis). Change the slab thickness by editing the "Depth" value.
6. Set the "step" setting on the "Volume Viewer" window to 1 for a full resolution 3D rendering.

3.3 Energy Dispersive X-ray Spectroscopy (EDXS) Analysis in STEM: HAADF Mode

One of the advantages of using STEM for EDXS analysis is that the electron probe can be used to irradiate a very small area of the specimen, whose the size is very close to that of the electron probe used. Thus, the X-rays obtained can be spatially resolved at a resolution corresponding to the probe size.

3.3.1 Preparing Cryosections for EDXS Analysis

During EDXS analysis the samples are heavily irradiated: for this reason the cryosections should be adequately protected against beam damage (*see* **Note 47**).

1. Sputter coat the cryosections with 20–30 nm of carbon (*see* **Note 48**). Indeed carbon coating has a protective effect reducing mass loss due to electron beam irradiation [29].
2. Fill the liquid nitrogen cold trap of the microscope before starting with the analysis to minimize the microscope column contamination.
3. Lowering the temperature of the specimen, if possible, can reduce its sensitivity to structural damage and mass loss [29].

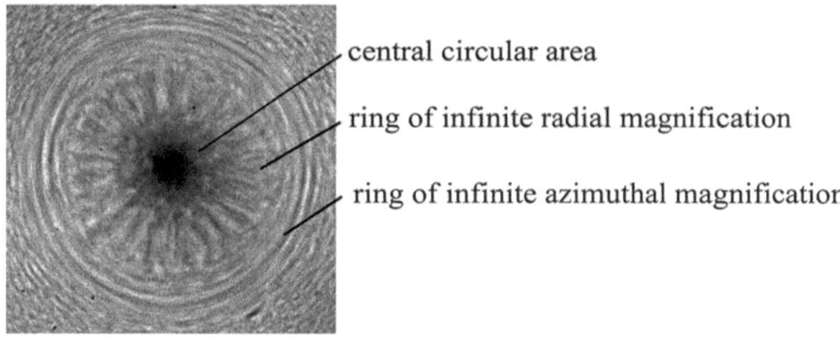

Fig. 6 Ronchigram from an amorphous specimen. Modified from [23]

3.3.2 Adjust the Microscope Alignment in STEM: HAADF Mode

In STEM mode, the electron optics that determines the spot size and focus is located above the specimen, together with the scan coils. The relevant stigmators for STEM mode are thus the condenser stigmators, while the STEM camera length is controlled by the electron optics located below the specimen.

1. Without condenser apertures, select "spot mode" for a stationary electron probe. Then optimize the spot size, mainly considering that reducing the spot size increases the resolution, but decreases the signal to noise ratio (*see* **Note 49**).
2. At high magnification (more than 500 k), in standard focus condition, with the electron probe not scanning, use the *Z*-height control to adjust the axial coordinate *Z* of the sample until you see the Ronchigram, a sort of fish-eye view of the specimen that provides the best way to perform a fast STEM alignment (*see* Fig. 6). Center the Ronchigram using the projector lens shift.
3. If the microscope is well aligned, the Ronchigram from an amorphous specimen should be circularly symmetric (*see* Fig. 6 and **Note 50**). If it is elongated, correct the condenser astigmatism using the corresponding stigmators.
4. Insert the High Angle Annular Dark Field (HAADF) or Bright Field (BF) detector (the corresponding camera length should be already selected), to form the STEM image with only the electrons scattered at high angle or transmitted, respectively. Finally, center the appropriate condenser aperture with respect to the Ronchigram.
5. Start the beam scanning, and carefully adjust the image brightness and contrast (*see* **Note 51**).

3.3.3 Operating the Analytical Transmission Electron Microscope

1. Insert the EDXS detector inside the microscope column using the suitable EDXS software.
2. Search the specimen at low magnification; and try to not pre-expose the region you want analyze at higher magnification.

3. Find an interesting area of your sample and set it to the eucentric height (*see* Subheading 3.2.2).
4. Use suitable spot size and condenser aperture to get reasonable X-ray counts and dead time (*see* **Note 52**). If necessary tilt the specimen towards the EDX detector to increase the counts per second (*see* **Note 53**).

3.3.4 Spectrum Acquisition and Qualitative Chemical Analysis

The purpose of qualitative analysis is to uncover the chemical elements composing a specimen by identifying the lines in the X-ray spectrum. The possible ambiguities in assigning the element corresponding to a given line can be resolved by taking into account additional lines.

1. Set—in your suitable EDXS software—the appropriate energy range depending on the atomic mass of the elements you are looking for in your sample. For light elements (B, NA, etc.) use low maximum energy (such as 10 or 20 KeV) to increase resolution, as usually the number of usable channels is fixed and independent from maximum energy.
2. Set—in your suitable EDXS software—a spectrum acquisition time to get reasonable X-ray counts. If possible, set the acquisition time as "live time" so that the acquisition will stop after the dead time-corrected acquisition time is elapsed.
3. Start a spectrum acquisition from the selected ROI using the suitable EDXS software (*see* **Note 54**). As an internal control, move into a hole and obtain a spectrum: EDXS counts from vacuum should be normally very low (dead time <2 %).
4. Assign all spectrum peaks present in the X-ray spectrum to characteristic X-rays peaks of chemical elements possibly contained in the sample. If a good match between the displayed line markers corresponding to the expected elements and the spectrum peaks can be achieved, be confident that the identification is correct. In case of doubt, a subsequent quantification run should clarify the situation (*see* **Note 55**).

3.3.5 Elemental Mapping Acquisition

During elemental mapping acquisition, a point-by-point X-ray spectrum acquisition is performed to determine the localization (and possible amount) of the chemical elements in the sample. Each resulting element map represents the two-dimensional distribution of the regarding element over the sample area scanned by the electron probe.

1. Observe the input count rate and check if appropriate. Set up image/map resolution and dwell time.
2. Capture a STEM image of the sample, so that the program can recognize the used analytical conditions (spatial coordinates, magnification, etc.). Eventually select on the image the area to be mapped.

3. Define the chemical elements for which maps are to be recorded. If no prior knowledge of the sample composition is available, the elements can be identified acquiring an average spectrum of the mapping area. In principle, the chemical maps could be reconstructed even after the end of the localized EDXS spectrum acquisition.
4. Start to acquire the map. After a measurement is started, the mapped images should gradually improve in signal to noise ratio.

4 Notes

1. We use a Jeol JEM 2200FS microscope, equipped with a field-emission gun (FEG) and operating at 200 kV of acceleration voltage. The images were recorded using a 1 k×1 k Gatan charge-coupled device (CCD) camera.
2. We use a Bruker Nano's XFlash 5060T spectrometer (Bruker) equipped with a 60 mm^2 silicon drift detector (SDD).
3. We use an ultra-narrow gap tomography holder (Fischione).
4. We use the QuanTax Esprit 1.9 software (Bruker).
5. We use the Jeol Recorder tomography acquisition software, version 2.26 (Jeol).
6. We use Etomo in IMOD, version 4.5 [26].
7. We use UCSF Chimera software, version 1.6.2 [27].
8. We use the Emitech K950X carbon vacuum evaporator (Emitech).
9. We use a Leica MZ6 (Leica).
10. The loop, 3.5 mm in diameter, should be made from copper or platinum wire 0.2 mm diameter and mounted on Eppendorf tips.
11. We use 15 nm Gold Sol (Aurion).
12. Add to 90 ml of the methyl cellulose solution 10 ml of 4 % uranyl acetate and mix gently. Centrifugate the solution for 95 min at 90,000×g (4 °C). The supernatant is poured into 50 ml vials with screwcap and can be stored at 4 °C in the dark for about 3 months.
13. For a detailed description on how to make Formvar coated nickel grids please refer to refs. 1, 3.
14. Drops of 100–200 µl, which can carry up to five grids at once, are used for the majority of rinsing steps. Use drops of 5–10 µl for each grid for antibody and immuno-gold solutions.
15. The grids may be transferred also using wire loops. But using forceps washing is more efficient since less of the incubation fluid is transferred to the next drop.

16. The flat surface is obtained by using a parafilm sheet that is adhered to a glass plate by some drop of distilled water.
17. If a grid accidentally sinks in a drop, wash it in distilled water. Then dry the back side of the grid carefully wiping with filter paper. Before reentering the incubation procedure, let the grid float on clean distilled water for a short while.
18. The aim of this procedure is to make assessable the section to the incubation solutions removing the dried mixture of sucrose and methyl cellulose from the section surface.
19. The aim of this procedure is to quench free aldehyde groups. When non-specific binding becomes a problem, you can extend this step.
20. The aim of this procedure is to block other sites where the immuno-reagents may stick non specifically. When non-specific binding becomes a problem, you can extend this step.
21. When non-specific binding becomes a problem, the antibody may be further diluted (~10 times).
22. To avoid that the drop dry out easily, it is important to cover the drops by a small petri dish in a whet chamber.
23. The duration of the immuno-incubation is not very critical. It is important to consider that molecules that do not react with aldehydes (like membrane lipids) and soluble proteins may escape easily when the fixation is week: in this case is better to keep the incubation short (i.e., 20 min). On the contrary, proteins that are integrated in membranes, or bound to the cytoskeleton will not easily escape, and labeling of these may be favored by longer incubation periods.
24. Remember that the secondary antibody should be against the species that primary antibody is raised. For example, if the primary antibody is raised in rabbit, an anti-rabbit secondary antibody should be used.
25. It is extremely important to carefully select the appropriate size for the gold nanoparticles conjugated to the secondary antibody. Indeed, the gold nanoparticles should be clearly distinguishable from other possible nanoparticles present inside the sections. As an example we used a secondary antibody conjugated with 15 nm colloidal gold nanoparticles on cryosections of HeLa cells doped with nanogels containing 10 nm iron oxide nanoparticles (*see* Fig. 3).
26. Do this step in a fume-hood since the glutaraldehyde is toxic.
27. The aim of this step is an extra fixation of the immuno-reagents on the sections before the rinsing and contrast -staining part of the procedure.
28. The aim of this task is to remove the phosphate or other ions thoroughly from the sections in order to prevent uranyl

precipitates during the final staining and embedding. Use fresh and good quality distilled water.

29. The purpose of this step is to enhance contrast and stabilize membrane lipids.
30. The purpose of this step is to stain the sections for contrast and to support them by polymers (methyl cellulose) in order to prevent drying artifacts.
31. The loop should have an inner diameter between 3.5 and 4 mm.
32. Be careful with the heavy metal uranium in this part of the procedure. Discard all remnants properly. Think of the pieces of filter paper with absorbed UA/MC.
33. It is extremely important to carefully select the appropriate size for the gold nanoparticles used as fiducial markers. Indeed, to avoid mistakes during the tomogram alignment, the gold nanoparticles should be clearly distinguishable from other possible nanoparticles present inside the sections. As an example we used 15 nm colloidal gold nanoparticles as fiducial markers on cryosections of HeLa cells doped with nanogels containing 10 nm iron oxide nanoparticles (*see* Fig. 4).
34. The cryosection ultrastructure will be deeply deteriorated if the cryosections come in contact with the colloidal gold solution since they are not embedded in resin.
35. To find out the optimal dilution for the colloidal gold solution, try different dilution on your sections, considering that for the fine alignment of your image stack during the tomogram generation procedure a number of gold nanoparticles between 20 and 40 should be present.
36. Remove almost all the solution with filter paper by carefully blotting the edges of each grids and then allow them to dry for a few minutes.
37. During eucentric height adjustment tilt your specimen at small angular steps to avoid that your ROI is shifted outside the area you are observing.
38. The vast majority of tomographic software have a function that automatically center (autotracking) and focus (autofocus) your ROI.
39. The file containing the projections of your sample has the following interchangeable file extension (i.e., tmg, st, or mrc) depending on the tomographic software you are using.
40. Consider that using the programs TIF2MRC or DM2MRC as implemented in Imod 4.5 you can easily convert a series of tilted images (in TIF or DM3 format) in a single file (*see* http://bio3d.colorado.edu/imod/doc/program_listing.html).

41. At the IMOD home page (http://bio3d.colorado.edu/imod/#Guides) you can find useful tutorials, including also videos that describe in details how to generate tomograms using ETOMO.
42. The program Tiltalign uses the position of the gold particles in the fiducial model and a variable metric minimization approach to solve for displacements, rotations, tilts, and magnification differences in the tilted projections.
43. Since in manual segmentation the contours are drawn from the operator, this approach is time consuming and, to a certain degree, also subjective.
44. UCSF Chimera is freely available at: http://www.cgl.ucsf.edu/chimera/
45. The rendering modes available in Volume Viewer are "surface," "mesh," or "solid". See the Chimera User's Guide (freely available at http://www.cgl.ucsf.edu/chimera/current/docs/UsersGuide/framecore.html) for further details.
46. The step setting controls sampling density. Consider that lower step setting values mean better resolution but higher computational time. See the Chimera User's Guide for further details.
47. Consider that the uranyl acetate used to stain the cryosections enhances their resistance to irradiation.
48. Usually we perform several depositions of carbon using low evaporation time (1,000–1,500 ms), thus avoiding damaging the cryosections.
49. Using a spot size between 0.7 and 1 nm was found to be a good compromise between high X-Ray cont rate and spatial resolution.
50. Looking at the Ronchigram (*see* Fig. 6), you should be able to see (1) a central circular area containing a normal shadow image of the sample at rather high magnification; (2) a ring where everything is streaked out in the radial direction, the ring of infinite radial magnification; and (3) at higher radius, another ring where everything is streaked out into a circular pattern, the ring of infinite azimuthal magnification.
51. A line scan can be used to help adjust the brightness and contrast. The line scan is a plot of the image density on a line across the sample. The image density should vary across the full dynamic range of the monitor, without white or black areas, that indicate that the image is saturated at high or low intensity. Consider that tuning the contrast you are expanding or decreasing the amount of variation in intensity. On the contrary with "brightness" you are increasing or decreasing the overall signal level. Thus, maximize the amount of contrast so that the full range of intensities is used, but without saturating

the image. Then adjust the brightness so that the overall image intensity is correct.

52. With high spot size and large aperture the counts per seconds increases, but the resolution of the image decreases.
53. Usually tilting the specimen towards the EDX detector of 10–20° is recommended in order to maximize the X-Ray signal and to avoid shadow effects from the specimen grid.
54. With QuanTax Esprit 1.9 to start a spectrum acquisition click the acquire bottom in the Spectra workspace.
55. In the majority of EDX software there is a "finder options" for the automatic identification of the unknown peaks of the spectrum.

References

1. Tokuyasu KT (1973) A technique for ultracryotomy of cell suspensions and tissues. J Cell Biol 57:551–565
2. Liou W, Geuze HJ, Slot JW (1996) Improving structural integrity of cryosections for immunogold labeling. Histochem Cell Biol 106: 41–58
3. Webster P (1999) The production of cryosections through fixed and cryoprotected biological materials and their use in immunocytochemistry. In: Nasser Hajibagheri MA (ed) Electron microscopy methods and protocols, vol 117. Humana Press, Totowa, NJ, pp 49–76
4. Posthuma G, van Donselaar E, Griffith J, Oorschot VMJ, van Dijk S, Slot JW (eds) (2006) Ultrathin cryo-sectioning and immuno – gold labeling. A practical introduction. Cell Microscopy Center, Department of Cell Biology, University Medical Center Utrecht, The Netherlands
5. Murk J, Postuma G, Koster AJ, Guuze HJ, Verkleij AJ, Kleijmeer MJ, Humbel BM (2003) Influence of aldehyde fixation on the morphology of endosomes and lysosomes: quantitative analysis and electron tomography. J Microsc 212:81–90
6. Curcio A, Marotta R, Riedinger A, Palumberi D, Falqui A, Pellegrino T (2012) Magnetic pH-responsive nanogels as multifunctional delivery tool for small interfering RNA molecules and iron oxide nanoparticles. Chem Commun 28(48):2400–2402
7. Laurent S, Dutz S, Häfeli U, Mahmoudi M (2011) Magnetic fluid hyperthermia: focus on superparamagnetic iron oxide nanoparticles. Adv Colloid Interface Sci 166:8–23
8. Gazeau F, Lévy M, Wilhelm C (2008) Optimizing magnetic nanoparticle design for nanothermotherapy. Nanomedicine 3(6): 831–844
9. Pradhan P, Giri J, Rieken F, Koch C, Mykhaylyk O, Döblinger M, Banerjee R, Bahadur D, Plank C (2010) Targeted temperature sensitive magnetic liposomes for thermo-chemotherapy. J Control Release 142(1):108–121
10. Liu T, Liu K, Liu D, Chen S, Chen I (2009) Temperature-sensitive nanocapsules for controlled drug release caused by magnetically triggered structural disruption. Adv Funct Mater 19(4):616–623
11. Deka S, Quarta A, Di Corato R, Riedinger A, Cingolani R, Pellegrino T (2011) Magnetic nanobeads decorated by thermo-responsive PNIPAM shell as medical platforms for the efficient delivery of doxorubicin to tumour cells. Nanoscale 3(2):619–629
12. Chen S, Li Y, Guo C, Wang J, Ma J, Liangs X, Yang L, Liu H (2007) Temperature-responsive magnetite/PEO-PPO-PEO block copolymer nanoparticles for controlled drug targeting delivery. Langmuir 23(25):12669–12676
13. Louguet S, Rousseau B, Epherre R, Guidolin N, Goglio G, Mornet S, Duguet E, Lecommandoux S, Schatz C (2012) Thermoresponsive polymer brush-functionalized magnetic manganite nanoparticles for remotely triggered drug release. Polym Chem 3:1408–1417
14. Morgan AJ, Winters C, Stürzenbaum S (1999) X-ray microanalysis techniques. In: Hajibagheri MAN (ed) Electron microscopy methods and protocols, vol 117. Humana Press, Totowa, NJ, pp 245–276
15. Frank J (2005) Electron tomography. Methods for three-dimensional visualization of structures in the cell, 2nd edn. Springer, Albany, NY

16. Porter AE, Muller K, Skepper J, Midgley P, Welland M (2006) Uptake of C60 by human monocyte macrophages, its localization and implications for toxicity studied by high resolution electron microscopy and electron tomography. Acta Biomater 2:406–419
17. Sousa AA, Aronova MA, Kim YC, Dorward LM, Zhang G, Leapman RD (2007) On the feasibility of visualizing ultrasmall gold labels in biological specimens by STEM tomography. J Struct Biol 159:507–522
18. Uchida M, Willits DA, Muller K, Willis AF, Jackiw L, Jutila M, Young MJ, Porter AE, Douglas T (2008) Intracellular distribution of macrophage targeting ferritin–iron oxide nanocomposite. Adv Mater 21(4):458–462
19. Cai X, Chen HH, Wang CL, Chen ST, Lai SF, Chien CC, Chen YY, Kempson IM, Hwu Y, Yang CS, Margaritondo G (2011) Imaging the cellular uptake of tiopronin-modified gold nanoparticles. Anal Bioanal Chem 401: 809–816
20. Nair BJ, Fukuda T, Mizuki T, Hanajiri T, Maekawa T (2012) Intracellular trafficking of superparamagnetic iron oxide nanoparticles conjugated with TAT peptide: 3-dimensional electron tomography analysis. Biochem Biophys Res Commun 421:763–767
21. Vicidomini G, Gagliani MC, Cortese K, Krieger J, Buescher P, Bianchini P, Boccacci P, Tacchetti C, Diaspro A (2010) A novel approach for correlative light electron microscopy analysis. Microsc Res Tech 73:215–224
22. Sigee DC, Morgan AJ, Sumner AT, Warley A (eds) (1993) X-ray microanalysis in biology. Experimental techniques and applications. Cambridge University Press, Cambridge
23. Laquerriere P, Banchet V, Michel J, Zierod K, Balossier G, Bonhomme P (2001) X-ray microanalysis of organic thin sections in TEM using an UTW Si(Li) detector: comparison of quantification methods. Microsc Res Tech 52: 231–238
24. D'Alfonso AJ, Freitag B, Klenov D, Allen LJ (2010) Atomic-resolution chemical mapping using energy-dispersive x-ray spectroscopy. Phys Rev 81:100101(R)
25. Busch W, Bastian S, Trahorsch U, Iwe M, Kuhnel D, Meißner T, Springer A, Gelinsky M, Richter V, Ikonomidou C, Potthoff A, Lehmann I, Schirmer K (2011) Internalisation of engineered nanoparticles into mammalian cells in vitro: influence of cell type and particle properties. J Nanopart Res 13:293–310
26. Kremer JR, Mastronarde DN, McIntosh JR (1996) Computer visualization of three-dimensional image data using IMOD. J Struct Biol 116(1):71–76
27. Pettersen EF, Goddard TD, Huang CC, Couch GS, Greenblatt DM, Meng EC, Ferrin TE (2004) UCSF Chimera – a visualization system for exploratory research and analysis. J Comput Chem 13:1605–1612
28. Donohoe BS, Mogelsvang S, Staehelin LA (2006) Electron tomography of ER, Golgi and related membrane systems. Methods 39: 154–162
29. Egerton RF, Li P, Malac M (2004) Radiation damage in the TEM and SEM. Micron 35: 399–409

Part III

Implementing Bio–Nano Interfaces

Chapter 15

Zwitterion Siloxane to Passivate Silica Against Nonspecific Protein Adsorption

Zaki G. Estephan and Joseph B. Schlenoff

Abstract

Passivating surfaces against protein adsorption is important for many biotechnological applications. Current approaches have been exploiting the use of zwitterions instead of poly(ethylene glycol) (PEG). Commonly used zwitterions are polymeric and are grafted onto surfaces using specialized polymerization techniques. Here we describe the synthesis of a monomeric zwitterion siloxane and its covalent attachment to silica surfaces (nanoparticle and planar) in a one-pot one-step aqueous method requiring no catalyst.

Key words Silane, Silica nanoparticle, Silicon wafer, Non-fouling, Protein repellent surface

1 Introduction

Modifying surfaces with zwitterions, instead of poly(ethylene glycol), to repel proteins and prevent bacterial adhesion (Fig. 1) has gained interest lately. The effectiveness of different zwitterionic chemical functionalities, i.e., sulfobetaine, carboxybetaine, or phosphobetaines, against protein or bacterial adhesion has been studied [1, 2]. Both carboxybetaines and sulfobetaines have shown interesting resistive properties and have been therefore extensively used to modify nanoparticles for dual colloidal stability and protein resistivity [3]. Most approaches have focused on polymeric zwitterions. For planar substrates, surface passivation has been as simple as dipping the substrate in the polymeric zwitterion or using thiol self assembled monolayers (SAMs) in the case of gold substrates. For example, Holmlin et al. used zwitterionic SAMs to moderate the adsorption of proteins onto gold passivated surfaces [4]. In contrast, Salloum et al. used the layer-by-layer technique to coat glass coverslips with poly(allylamine hydrochloride), PAH, and poly(acrylic acid)-*co*-poly(3-[2-(acrylamido)-ethyl dimethylammonio] propane sulfonate), PAA-*co*-PAEDAPS, to control cell adhesion by changing the ratio of the zwitterion in the random

Paolo Bergese and Kimberly Hamad-Schifferli (eds.), *Nanomaterial Interfaces in Biology: Methods and Protocols*, Methods in Molecular Biology, vol. 1025, DOI 10.1007/978-1-62703-462-3_15, © Springer Science+Business Media New York 2013

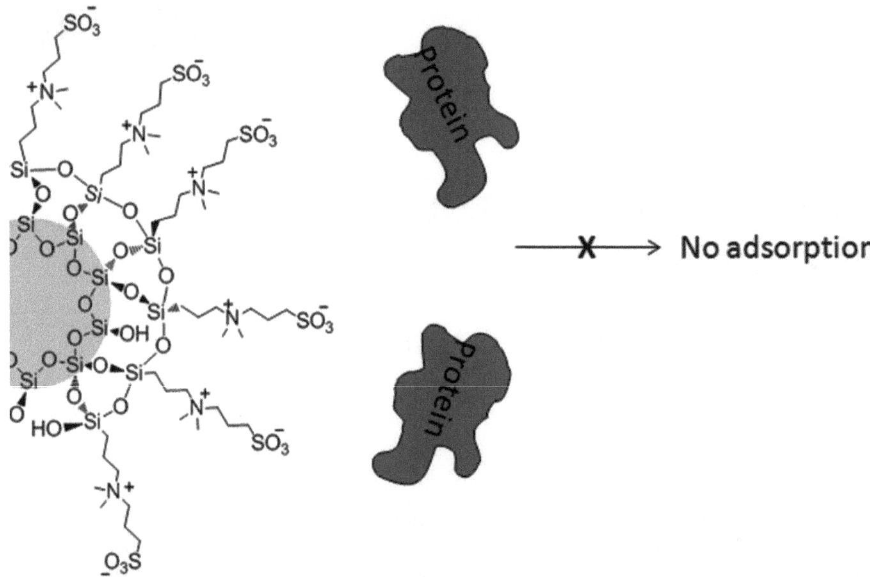

Fig. 1 Scheme of zwitterion passivation of silica nanoparticles against nonspecific protein adsorption

copolymer [5]. Other techniques involve surface initiated atom transfer polymerization (ATRP) onto an initiator covered gold or glass substrates [6, 7]. Most of these approaches have been extended to nanoparticles. For example, Rouhana et al. prepared zwitterated gold nanoparticles by exchanging the citrate ligand with a zwitterion disulfide [8], whereas Jia et al. and Matsuura et al. used surface initiated ATRP to graft zwitterionic polymer to gold and silica nanoparticles respectively [9, 10].

The authors have previously reported a method for passivating silica surfaces against nonspecific protein interaction [11, 12]. In this chapter we provide a step-by-step procedure to functionalize silica nanoparticles and planar silicon wafers with a zwitterion siloxane. The reaction is based on covalent attachment of the siloxane group of the zwitterion to the surface silanols of the substrate.

2 Materials

Prepare all solution with ultrapure water (18 MΩ) at room temperature. Follow local and federal waste disposal regulations when disposing of materials.

2.1 Zwitterion Siloxane Synthesis Components

1. HPLC grade acetone.
2. Propane sultone from TCI America.
3. (*N*,*N*-dimethyl-3-aminopropyl)trimethoxysilane from Gelest.

2.2 Nanoparticle Zwitteration Components

1. Ludox® TM-40 colloidal silica from Sigma Aldrich.
2. Dialysis was performed using Spectra/Pro dialysis tubing with molecular weight cutoff (MWCO) of 3,500.

2.3 Planar Surface Zwitteration Components

1. RCA solution: mix 1 ml of 28 % NH_4OH with 1 ml of 30 % H_2O_2 with 5 ml H_2O.
2. Piranha solution (Caution: strong oxidizer, not to be stored in closed containers): mix 7 ml of concentrated H_2SO_4 with 3 ml of 30 % H_2O_2. Let the solution cool for 15 min before use.

3 Methods

3.1 Synthesis of Zwitterion Siloxane: 3-(Dimethyl(3-(Trimethoxysilyl)Propyl)-Ammonio)Propane-1-Sulfonate (Fig. 2)

The following procedure is performed under inert conditions at room temperature:

1. Dissolve 4.45 g of propane sultone in 37 ml of acetone.
2. Add 7.5 g of (*N*,*N*-dimethyl-3-aminopropyl)trimethoxysilane (*see* **Note 1**).
3. Let the solution stir for 6 h. The solution turns turbid after few minutes and eventually precipitates a white solid.
4. Wash the solution extensively with acetone. The precipitate is first dispersed in acetone and then poured into a suction filtration flask. The process is repeated three times or until the acetone remains colorless after filtration.
5. Let the precipitate dry overnight under Ar.

3.2 Zwitteration of Silica Nanoparticles

The following procedure is performed for Ludox TM-40 silica suspension (40 wt% suspension in water) but can be applied to other silica nanoparticles.

1. Dilute the silica nanoparticle to a final concentration of 10 wt% by adding 3.75 g of Ludox TM-40 to around 10 g of water.
2. Disperse the proper amount of zwitterion siloxane (Fig. 1) in around 1 g of water (*see* **Notes 2** and **3**).
3. Add the dissolved zwitterion siloxane to the silica nanoparticles.
4. Bring the final mass of the solution to 15 g by addition of water.
5. Shake the solution and introduce to a heated bath preset at 80 °C.

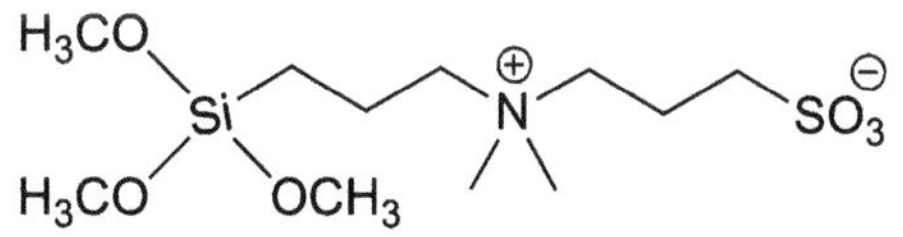

Fig. 2 Chemical structure of zwitterion siloxane: 3-(dimethyl(3-(trimethoxysilyl)propyl)-ammonio)propane-1-sulfonate

6. Let the reaction proceed for 6 h.
7. After cooling to room temperature dialyze the system against 4 l of water for 24 h, changing the dialysate at 6 h intervals.
8. Measure the solids content after dialysis by drying around 1 ml of the dialyzed solution under vacuum at 25–30 °C.

3.3 Zwitteration of Planar Silicon Wafer

1. Immerse the silicon wafer into an RCA solution for 10 min.
2. Rinse extensively with water.
3. Immerse the wafer into cold piranha for 10 min.
4. Wash with water.
5. Insert the silicon wafer into a sealable flask containing 100 mM of zwitterion siloxane (0.165 g zwitterion siloxane in 5 ml of H_2O).
6. Seal the flask tightly to avoid the evaporation of water and heat the solution at 80 °C for 15 h.
7. After cooling to room temperature, wash the silicon wafer with water.

4 Notes

1. Equivalent number of moles of propane sultone and (*N*,*N*-dimethyl-3-aminopropyl)trimethoxysilane are used. Each corresponds to 1 M concentration.
2. The zwitterion should readily dissolve in water to give a colorless solution. The presence of color indicates impurities or unreacted components from synthesis.
3. For the silica nanoparticles used in our laboratory (specific surface area of 140 m^2/g) 0.19 g of zwitterion siloxane was needed to saturate the surface. In our calculations it was assumed that 1 mol of zwitterion siloxane reacts with 3 mol of surface silanol to form a monolayer. Surface saturation was observed close to the theoretical monolayer as defined here. The recommended amount of siloxane is thus $8.9\times10^{-4}mS$ where m is the mass of nanoparticlesing and S is their specific surface area in m^2/g.

References

1. West SL, Salvage JP, Lobb EJ, Armes SP, Billingham NC, Lewis AL, Hanlon GW, Lloyd AW (2004) The biocompatibility of crosslinkable copolymer coatings containing sulfobetaines and phosphobetaines. Biomaterials 25:1195–1204
2. Emmenegger CR, Brynda E, Riedel T, Sedlakova Z, Houska M, Alles AB (2009) Interaction of blood plasma with antifouling surfaces. Langmuir 25:6328–6333
3. Zhang Z, Chao T, Chen SF, Jiang SY (2006) Superlow fouling sulfobetaine and carboxybetaine polymers on glass slides. Langmuir 22:10072–10077
4. Holmlin RE, Chen XX, Chapman RG, Takayama S, Whitesides GM (2001)

Zwitterionic SAMs that resist nonspecific adsorption of protein from aqueous buffer. Langmuir 17:2841–2850

5. Salloum DS, Olenych SG, Keller TCS, Schlenoff JB (2005) Vascular smooth muscle cells on polyelectrolyte multilayers: hydrophobicity-directed adhesion and growth. Biomacromolecules 6:161–167
6. Zhang Z, Chen SF, Chang Y, Jiang SY (2006) Surface grafted sulfobetaine polymers via atom transfer radical polymerization as superlow fouling coatings. J Phys Chem B 110: 10799–10804
7. Cheng G, Li GZ, Xue H, Chen SF, Bryers JD, Jiang SY (2009) Zwitterionic carboxybetaine polymer surfaces and their resistance to long-term biofilm formation. Biomaterials 30: 5234–5240
8. Rouhana LL, Jaber JA, Schlenoff JB (2007) Aggregation-resistant water-soluble gold nanoparticles. Langmuir 23:12799–12801
9. Jia G, Cao Z, Xue H, Xu Y, Jiang S (2009) Novel zwitterionic-polymer-coated silica nanoparticles. Langmuir 25:3196–3199
10. Matsuura K, Ohno K, Kagaya S, Kitano H (2007) Carboxybetaine polymer-protected gold nanoparticles: high dispersion stability and resistance against non-specific adsorption of proteins. Macromol Chem Phys 208:862–873
11. Estephan ZG, Jaber JA, Schlenoff JB (2010) Zwitterion-stabilized silica nanoparticles: toward nonstick nano. Langmuir 26: 16884–16889
12. Estephan ZG, Schlenoff PS, Schlenoff JB (2011) Zwitteration as an alternative to PEGylation. Langmuir 27:6794–6800

Chapter 16

Preparation and Characterization of DNA Block Copolymer Assemblies Loaded with Nanoparticles

Xi-Jun Chen, Robert J. Hickey, and So-Jung Park

Abstract

We have recently developed a universal procedure to functionalize inorganic nanoparticles with a dense layer of DNA through the self-assembly of DNA block copolymers and nanoparticles. This functionalization strategy allows one to combine the useful physical properties of inorganic nanoparticle with the enhanced DNA binding properties that originate from the high surface DNA density. In particular, the hybrid nanostructures exhibit orders of magnitude higher binding constants than regular DNA strands. This chapter presents a detailed protocol for the preparation and characterization of DNA block copolymer assemblies loaded with nanoparticles.

Key words DNA, Block copolymer, Nanoparticles, Self-assembly

Abbreviations

CPG	Controlled pore glass
DLS	Dynamic light scattering
DMF	*N*,*N*-Dimethylformamide
DNA-*b*-PS	Block copolymer of DNA and polystyrene
FAM	Fluorescein
FRET	Förster/Fluorescent Resonance Energy Transfer
MNP	Magnetic nanoparticles
MNP@PS@DNA	DNA-*b*-polystyrene assemblies loaded with magnetic nanoparticles
PBS	Phosphate buffer saline
PS	Polystyrene
PS-MNP	Polystyrene modified magnetic nanoparticles
PS@DNA	Self assembly of DNA-*b*-polystyrene
PTFE	Polytetrafluoroethylene
TEM	Transmission electron microscope
THF	Tetrahydrofuran

Paolo Bergese and Kimberly Hamad-Schifferli (eds.), *Nanomaterial Interfaces in Biology: Methods and Protocols*, Methods in Molecular Biology, vol. 1025, DOI 10.1007/978-1-62703-462-3_16, © Springer Science+Business Media New York 2013

1 Introduction

Bio-conjugated nanoparticles have been actively exploited in various biological and medical applications for the past couple of decades to take advantage of the useful physical properties of inorganic nanoparticles in detecting, studying, and manipulating biological systems [1–6]. In particular, DNA-modified gold nanoparticles have shown great promise in many biomedical applications such as DNA detection, drug delivery, and gene regulation owing to the unique optical properties of gold nanoparticles and the densely immobilized DNA strands on the nanoparticle surface [7, 8]. We have recently developed a new DNA-functionalization method based on the self-assembly of DNA block copolymers and nanoparticles [9]. In this approach, nanoparticles are encapsulated in DNA block copolymer assemblies by the slow addition of water to the mixture of DNA block copolymers and nanoparticles in a polar organic solvent. The resulting assemblies are composed of nanoparticles embedded in the hydrophobic polymer core and the hydrophilic DNA shell at the exterior. Since this approach is based on self-assembly, it can be applied to virtually any types of nanoparticles. It is also important to note that this methodology leads to a densely packed DNA layer on the surface because every polymer strand is conjugated to DNA. Due to the high density, the DNAs on the assemblies show dramatically enhanced binding properties. For example, they can recognize complementary DNA at low salt concentrations where regular DNA strands do not form duplex structures. While DNA block copolymers have been previously synthesized for other purposes [10–13], this work was the first to demonstrate that the self-assembly of nanoparticles and DNA block copolymers can be used to prepare DNA-functionalized nanoparticles with enhanced binding properties. This enhanced binding property in combination with the capability to load nanoparticles and other small molecules makes the DNA assemblies an excellent material for DNA detection and delivery applications.

Here, we present detailed procedures on (1) the synthesis and characterization of DNA block copolymers, (2) the self-assembly of DNA block copolymers and nanoparticles, and (3) the characterization of nanoparticle-loaded DNA block copolymer assemblies. Specifically, this protocol presents detailed procedures for the assemblies of iron oxide magnetic nanoparticles (MNP) and DNA block copolymers composed of an oligonucleotide and polystyrene (DNA-*b*-PS).

2 Materials

All aqueous solutions are prepared using ultrapure water (deionized Type I water purified by Barnstead NanoPure DIamond, Model D11901, with resistivity of 18 MΩ at room temperature). All reagents are used without further purification unless otherwise indicated.

2.1 Synthesis of DNA-b-PS

2.1.1 Synthesis of Phosphoramidite-Terminated PS

1. Hydroxyl-terminated PS, 5 g (Mn = 10.4 kg/mol, Polymer Source, Canada) (*see* **Note 1**).
2. Dichloromethane (anhydrous, ≥99.8 %).
3. 2-Cyanoethyl *N,N*-diisopropylchlorophosphoramidite, 1 g (Sigma-Aldrich) (*see* **Note 2**).
4. Diisopropylethylamine (purified by redistillation, 99.5 %, Sure/Seal™, Sigma-Aldrich).
5. Argon gas (high purity).
6. Reaction apparatus: 100-mL three-neck round bottom flask, Schlenk line, Schlenk line gas adapter, two rubber stoppers, PTFE-encased magnetic stir bar, magnetic stirrer with heat plate.
7. Deuterated chloroform for NMR: Chloroform-d (99.8 atom % D).
8. Bruker DMX 360 MHz NMR (DMX360) for ^{31}P NMR spectroscopy.

2.1.2 Solid State Synthesis of Oligonucleotides

1. ABI 391 DNA/RNA synthesizer (Applied Biosystems).
2. Beta-Cyanoethyl (CE) phosphoramidite bases (Glen Research, Sterling, VA, USA) for a 10 μmol scale synthesis of oligonucleotide sequence: 5′-FAM A_{10} ATC CTT ATC AAT ATT-3′ (*see* **Note 3**). Each nucleoside phosphoramidite is dissolved in anhydrous acetonitrile to a concentration of 0.1 M (*see* **Note 4**).
3. dT-CE phosphoramidite controlled pore glass (CPG) beads, 1,000 Å (Glen Research).
4. Synthesis support column (10 μmol, crimp style caps, with Teflon filters, Applied Biosystems).
5. Activator solution: 0.45 M tetrazole in acetonitrile (Glen Research).
6. Capping mix A: tetrahydrofuran (THF)/pyridine/acetic anhydride (Glen Research).
7. Capping mix B: 15 % 1-methylimidizole in THF (Glen Research).
8. Detritylation solution: 3 % trichloroacetic acid/dichloromethane (Glen Research).

9. Oxidizing solution: 0.1 M iodine in THF/pyridine/H_2O (Glen Research).
10. Dichloromethane (stabilized with amylene, peptide synthesis, ≥99.9 %, Acros Organics).
11. Acetonitrile (anhydrous, DNA synthesis).
12. Argon (Ultrahigh Purity).

2.1.3 Solid State Coupling of Phosphoramidite-Terminated PS to Oligonucleotides

1. Activator solution: 0.45 M tetrazole in acetonitrile (Glen Research).
2. Detritylation solution: 3 % trichloroacetic acid/dichloromethane (Glen Research).
3. *N,N*-Dimethylformamide (HPLC grade).
4. Reaction apparatus: three-neck round bottom flask (100 mL), Schlenk line, PTFE-encased magnetic stir bar, magnetic stirrer with heat plate (Fisher Scientific).
5. Oxidizing solution: 0.1 M iodine in THF/pyridine/H_2O (Glen Research).
6. Acetonitrile (anhydrous, DNA synthesis).
7. Ammonium hydroxide solution (25 %) (*see* **Note 5**).
8. 20 mL Scintillation vial.
9. Water bath.
10. Filter paper (grade 1, Whatman).

2.1.4 Characterization of DNA-b-PS

1. Fluorometer (Fluorolog3, Jobin Yvon Horiba).
2. UV–Vis Spectrometer (Model 8453, Agilent).
3. Micro quartz cuvettes: 2-clear window for UV–Visible spectroscopy measurements and 3-clear window for fluorescence spectroscopy.
4. Zetasizer Nano Series (Malvern Instruments Ltd.).
5. Agarose (Ultrapure, Invitrogen).
6. 1× Tris-acetate-EDTA (TAE) casting and running buffer.
7. Nucleic acid sample loading buffer, 5× (Bio-Rad).
8. Horizontal nucleic acid electrophoresis system, basic power supply (Bio-Rad).

2.2 Synthesis of MNP and the Self-Assembly of DNA-b-PS and MNP

2.2.1 Synthesis of Oleic Acid Stabilized MNP

1. Iron(III) acetylacetonate.
2. 1,2-Hexadecanediol.
3. Oleic acid.
4. Oleylamine.
5. Diphenyl ether (*see* **Note 6**).
6. Acetone.

7. Hexanes.
8. Chloroform.
9. Reaction apparatus: spatula, PTFE-encased magnetic stir bar, magnetic stirrer with heat plate, two-neck 100 mL round bottom flask with 14/20 necks, glass condenser with 14/20 neck, glass 90° angle adapter with 14/20 neck, clamps, Schlenk line, vacuum pump, nitrogen gas.
10. Thermocouple with a 14/20 joint adapter (stainless steel, J-1/16-U-24, J-KEM Scientific).
11. Heating controller with connecting cords (Model 210/Timer, J-KEM Scientific).
12. Heating mantle (soft shell, 100 mL, OS-100, J-KEM Scientific).
13. Benchtop centrifuge (Allegra 64RCentrifuge, No: 367586, Beckman Coulter) with 50 mL centrifuge tube rotor (F0685 Rotor, Beckman Coulter).
14. Centrifuge tubes (50 mL).
15. Transmission Electron Microscope (TEM) (JEM-1400, JEOL).
16. TEM grids (CF200-Cu, Electron Microscopy Sciences).
17. TEM grid tweezers.

2.2.2 Preparation of PS-Modified MNP (PS-MNP) via Ligand Exchange

1. Carboxyl-terminated PS, Mn = 11.2 kg/mol (Polymer Source).
2. Acetone.
3. *N*,*N*-Dimethylformamide (DMF).
4. 1.5 mL microcentrifuge tube.
5. Benchtop centrifuge (Allegra 64RCentrifuge, No: 367586, Beckman Coulter) with 2 mL centrifuge tube rotor (F2400H Rotor, Beckman Coulter).

2.2.3 Preparation of MNP-Incorporated DNA-b-PS Assemblies (MNP@PS@DNA)

1. *N*,*N*-Dimethylformamide (DMF).
2. Scintillation vial (20 mL).
3. PTFE-encased magnetic stir bar.
4. Magnetic stirrer and heat plate.
5. Automatic micropipettes (Eppendorf Research, 1–10 μL).
6. Timer/Stopper.
7. Regenerated cellulose dialysis tubing (nominal MWCO 6,000–8,000 Da, pore size 1.8 nm, FisherBrand).
8. Dialysis clips.
9. 800 mL beaker.

2.3 Characterization of MNP@PS@ DNA

1. TEM (FEI-Technai T12).
2. TEM grids (CF200-Cu, Electron Microscopy Sciences).
3. TEM grid tweezers.
4. Zetasizer Nano Series (Malvern Instruments, Ltd.).
5. Plastic disposable cuvette, 4.5 mL (Brandtech).

2.4 DNA Hybridization Monitored by the Förster Resonance Energy Transfer (FRET)

1. Cy3-labeled complementary DNA (Sequence: 3′-Cy3 T_{10} TAG GAA TAG TTA TAA-5′) (Integrated DNA Technologies, USA): Dissolve the DNA in water at a concentration of 1×10^{-4} M (Extinction coefficient of DNA at 260 nm: 268,456 M^{-1} cm^{-1}) [14].
2. 1.5 mL microcentrifuge tube.
3. Vacufuge concentrator (Eppendorf).
4. 0.6 M phosphate buffer saline (PBS, 0.6 M NaCl, 20 mM phosphate buffer, pH 7).
5. Micro quartz cuvettes: 3-clear window for fluorescent spectroscopy.
6. Fluorometer (FluoroLog3, Jobin Yvon Horiba).
7. Temperature controller (TLC50, Quantum Northwest).
8. Cy5-labeled complementary DNA (Sequence: 5′-Cy5 A_{10} ATC CTT ATC AAT ATT-3′) (Integrated DNA Technologies, USA).

3 Methods

3.1 Synthesis, Purification, and Characterization of DNA-b-PS

The following procedures describe the synthesis of one specific example of DNA block copolymer, DNA-*b*-PS, with the DNA sequence 5′-FAM A_{10} ATC CTT ATC AAT ATT-3′. The polymer is attached at the 5′ end of DNA. The DNA block copolymer is prepared by solid-state synthesis by standard phosphoramidite chemistry. Other types of DNA block copolymers can be synthesized in a similar fashion using different hydroxyl-terminated polymers (*see* **Note 7**).

3.1.1 Synthesis of Phosphoramidite-Terminated PS

1. Weigh out 4.16 g (0.4 mmol) of hydroxyl-terminated PS (Mn = 10.4 kg/mol) and add it to a 3-neck round bottom flask equipped with a PTFE-coated magnetic stir bar, a Schlenk line adapter and two rubber stoppers.
2. Vacuum and then purge the reaction flask containing the polymer with argon, and repeat the procedure three times.
3. Under a gentle flow of argon, add 15 mL of anhydrous dichloromethane to the reaction flask with stirring and allow the polymer to dissolve completely (*see* **Note 8**).

4. Add 0.378 mL (1.68 mmol) of chlorophosphoramidite (*see* **Note 9**) and 0.290 mL (1.68 mmol) of anhydrous diisopropylethylamine to the reaction flask.
5. Allow the solution to stir for 3 h under a blanket of argon (*see* **Note 10**).
6. Analyze the crude product by ^{31}P NMR. The signal for phosphoramidite-terminated PS is at $\delta = 147$ ppm (*see* **Note 11**).

3.1.2 Solid State Synthesis of DNA

1. Prepare phosphoramidite solutions by adding 18.77 mL, 10 mL, 8 mL, and 1 mL of anhydrous acetonitrile to crimped vials containing dA-CE (1.0 g), dT-CE (0.5 g), dC-CE (0.5 g), and 6-fluorescein (100 μmol) phosphoramidites, respectively. Swirl the vials gently to make sure that the reagents are completely dissolved.
2. Prepare a solid state synthesis column for a 10 μmol synthesis by weighing out 0.33 mg of dT-CE CPG beads (1,000 Å). Tightly secure the beads into the support column and make sure there are no leaks.
3. Prepare the DNA synthesizer by loading the dT-CE CPG into the support column and screwing in the CE-phosphoramidite bottles to their appropriate dedicated spots on the instrument.
4. Enter the DNA sequence (5′-A_{10} ATC CTT ATC AAT ATT-3′), select 10 μmol synthesis method, choose the option to keep the trityl end group "on," and start the synthesis.
5. When the synthesis is complete, couple a fluorescein dye (FAM) tag at the 5′ end of the oligonucleotide by selecting the 10 μmol synthesis method again, and entering DNA sequence of 5′<u>**X**</u>A3′, where <u>**X**</u> denotes the position of the bottle containing the dye-phosphoramidite on the DNA synthesizer. Select the option to keep the trityl end group "on." Start the synthesis (*see* **Note 12**).

3.1.3 Synthesis of DNA-b-PS

The synthetic scheme for DNA-*b*-PS is presented in Fig. 1 (*see* **Note 13**).

1. Detritylate the DNA on CPG beads manually by pushing detritylation solution through the supporting column containing the beads using a syringe until no orange color is observed.
2. Dry the beads by blowing a stream of argon gas through the column.
3. Pour the dried CPG beads containing the synthetic oligonucleotides prepared in the previous section into the round bottom flask equipped with a stir bar and Schlenk line adapter and two rubber stoppers.

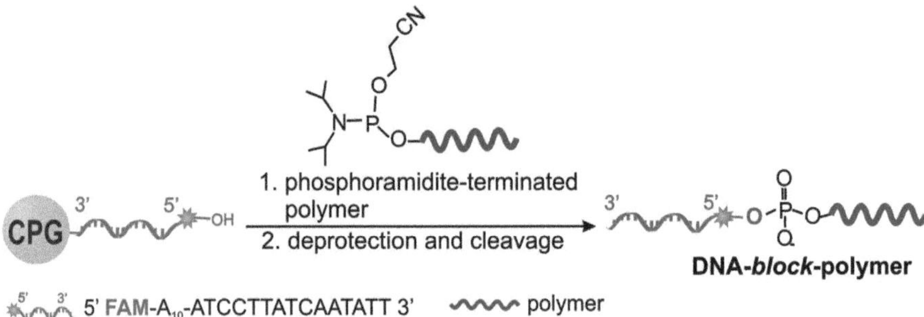

Fig. 1 Schematic description for the synthesis of DNA block copolymers

4. Add 3 mL of anhydrous dichloromethane to the CPG beads to avoid static cling while the flask is purged with a gentle stream of argon for 10 min.
5. Transfer the phosphoramidite-terminated PS to the flask containing the DNA on CPG beads as fast as possible by cannula to reduce the exposure to oxygen.
6. Add 15 mL of dimethylformamide to make a 1:1 DMF–CH_2Cl_2 solution (*see* **Note 14**).
7. Add 2.7 mL of activator solution to the flask at 3:1 molar ratio of activator: phosphoramidite-terminated PS.
8. Stir the reaction for a few hours or overnight under a blanket of argon.
9. Stop the stirring of the reaction and allow the CPG to settle to the bottom.
10. Decant the supernatant and wash the beads with five aliquots of 10 mL chloroform to remove unreacted polymer.
11. Allow the beads to air dry.
12. Collect the beads and pack them back into the solid support column once again. Oxidize the phosphoramidite to phosphate by passing the oxidation solution through the column with a syringe drop wise for 2 min.
13. Rinse the beads with acetonitrile until no yellow color from the iodine in the oxidation solution is observed on the beads.
14. Dry the beads and transfer them to a scintillation vial. To deprotect and cleave the DNA from the CPG beads, add 10 mL of concentrated ammonium hydroxide to the dried beads and leave the solution undisturbed for 6–12 h at 55 °C (*see* **Note 15**).
15. Separate the supernatant from the CPG beads by gravity filtration (*see* **Note 16**). This fraction will contain mainly uncoupled DNA strands and a small amount of DNA block copolymers.

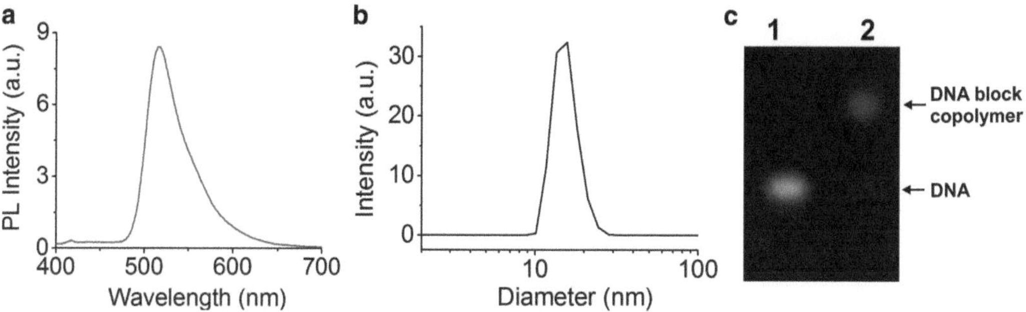

Fig. 2 (**a**) An emission spectrum of FAM-labeled DNA-*b*-PS dispersed in water. (**b**) DLS data of simple micelles of DNA-*b*-PS (PS@DNA) in water. (**c**) Gel electrophoresis result for (*1*) DNA and (*2*) PS@DNA

16. Wash the beads with two 5 mL aliquots of DMF, and collect the DMF solution in a separate vial. This fraction will contain mainly DNA block copolymers.
17. Collect the emission spectrum of DNA-*b*-PS in water with the excitation wavelength at 490 nm to check the presence of the FAM dye in the DNA block-copolymers (Fig. 2a).
18. Determine the concentration of FAM-modified DNA-*b*-PS in DMF using the extinction coefficient of FAM in DMF at 518 nm (1.3×10^4 M^{-1} cm^{-1}) using UV–Vis spectroscopy (*see* **Note 17**).
19. Determine the hydrodynamic diameter of simple micelles of DNA-*b*-PS in water (PS@DNA) by dynamic light scattering (DLS). The average of three separate measurements was 15 ± 4 nm (Fig. 2b).
20. Prepare a 3 % agarose gel by dissolving 1.5 g of agarose with 50 mL of 1× TAE buffer with heating in a microwave oven for 2 min. Cast the gel by pouring the solution into the gel tray and allow it to cool and solidify. Add 2 μL of loading buffer to 10 μL DNA samples (10 pmol). Add enough 1× TAE buffer to cover the gel. Load the sample into the well and run the gel at 60 V for 1 h. Image the gel under UV lamp (Fig. 2c).

3.2 Self-Assembly of DNA-b-PS and MNP

Oleic acid stabilized MNPs were synthesized by the thermal decomposition method [15].

3.2.1 Synthesis of Oleic Acid Stabilized MNPs

1. Weigh out 0.71 g of iron(III) acetylacetonate, 2.58 g of 1,2-hexadecanediol, and 1.69 g of oleic acid.
2. Add all three chemicals to the two-neck 100 mL round bottom flask with 14/20 necks.
3. Measure out 2 mL of oleylamine and 20 mL of diphenyl ether.
4. Add these two chemicals to the two-neck 100 mL round bottom flask with 14/20 necks.

5. Assemble the reaction flask and connect it to the Schlenk line (*see* **Note 18**).
6. Add the magnetic stir bar to the round bottom flask.
7. First attach the glass condenser to the round bottom flask, and then the 90° angle adapter to the glass condenser.
8. Attach the thermocouple with a 14/20 joint adapter to the other neck of the round bottom flask (*see* **Note 19**).
9. Secure the round bottom flask within the heating mantle on the magnetic stirrer by using metal clamps.
10. Connect both the thermocouple and the heating mantle to the heating controller using the connector cords.
11. Make sure all the glass and adapter fittings are tight and secure.
12. Connect the secured reaction flask, assembled with the heating apparatus, to the Schlenk line.
13. Turn on both vacuum pump and nitrogen gas.
14. To make sure the reaction proceeds under inert conditions without water and oxygen, evacuate the round bottom flask and refill it with nitrogen (*see* **Note 20**).
15. To do this, first, slowly open the vacuum valve of the Schlenk line and apply vacuum to the flask.
16. After about 30 s of applying vacuum, close the vacuum valve of the Schlenk line and then open the nitrogen gas valve of the Schlenk line.
17. When opening the nitrogen gas valve of the Schlenk line, open it slowly to make sure that no air enters the flask.
18. Repeat **steps 15–17** two more times.
19. On the last cycle, keep the nitrogen gas valve open, and keep it open during the duration of the reaction.
20. Next, turn on the heating controller and set the temperature to 200 °C.
21. Heat the reaction flask to 200 °C and hold it at that temperature for 30 min.
22. After 30 min, increase the temperature to 265 °C.
23. Hold at 265 °C for another 30 min.
24. After 30 min, cool the reaction to room temperature by detaching the heating mantle from the heating controller and removing the round bottom flask from the heating mantle (*see* **Note 21**).
25. Once the temperature goes down to room temperature, close the nitrogen gas valve and pour the reaction mixture into two 50 mL centrifuge tubes in equal amounts.
26. The reaction mixture should be dark-brown and cloudy.

27. Add 35 mL of acetone to each centrifuge tubes and centrifuge at 8,000 rpm (7,012 × *g*) for 10 min (*see* **Note 22**).
28. Decant the supernatant and keep the brown precipitate in the centrifuge tubes.
29. The brown precipitate is the iron oxide nanoparticles.
30. Add 5 mL of hexane to each centrifuge tube to redisperse the particles.
31. Repeat **steps 28–30** two more times.
32. After the last centrifugation step, air-dry the particles in the hood for 30 min and redisperse the particles in 5 mL of chloroform for a total of 10 mL solution of particles.
33. Characterize the particles with TEM.
34. To prepare the TEM sample, place a single droplet of nanoparticle solution to the grid using a glass pipette (*see* **Note 23**). The size distribution of nanoparticles synthesized using this method is typically 4.5 ± 0.4 nm.

3.2.2 Preparation of PS-Modified MNP (PS-MNP)

1. Mix 50 μL (in chloroform) of the 5 nm oleic acid stabilized MNP solution (50 μg/mL) in a 1.5 mL microcentrifuge tube with 100 μL of 10 mg/mL of carboxyl-terminated PS (in chloroform). The polymer will be in large excess compared to the number of oleic acid strands on the nanoparticle surface.
2. Let it stand at room temperature overnight.
3. Clean off excess PS by precipitating the nanoparticles with 500 μL of acetone.
4. Centrifuge the mixture at 7,000 rpm (4,492 × *g*) for 15 min. Pipette off the supernatant. Resuspend the precipitates of PS-MNP in 100 μL of DMF. Add 500 μL of acetone to reprecipitate the nanoparticles.
5. Repeat the centrifugation, decantation and resuspension procedure three more times (*see* **Note 24**).
6. After the cleaning step, resuspend the final solution of PS-MNP in 100 μL of DMF.

3.2.3 Preparation of MNP@PS@DNA (Fig. 3a)

1. Prepare a 300 μL of 10 μM solution of DNA-*b*-PS in DMF, and mix it with 10 μL of 50 μg/mL PS-MNP (in DMF) prepared in the previous section in a scintillation vial equipped with a magnetic stir bar.
2. Add an additional 1 mL of DMF to the solution.
3. With stirring, slowly add 10 μL of water to the solution at 30-s intervals for a period of 15 min (*see* **Note 25**).
4. Allow the solution to stir overnight.
5. Add an additional 1 mL of water to the solution, drop by drop.

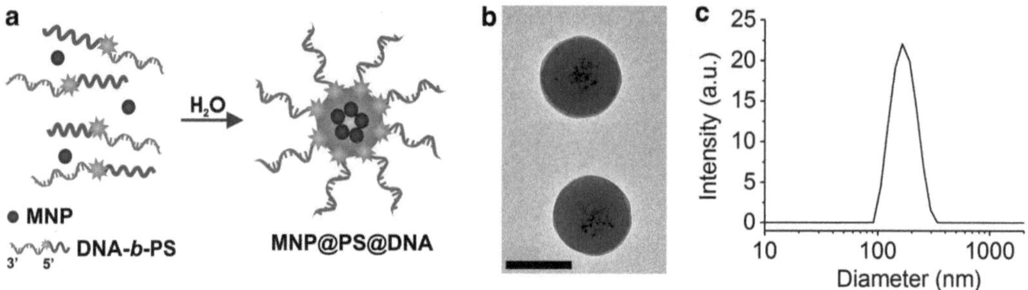

Fig. 3 (**a**) Schematic description for the self-assembly of DNA-*b*-PS and MNP. (**b**) A TEM image of MNP@PS@DNA (Scale bar = 100 nm). (**c**) DLS data of MNP@PS@DNA

6. Wet a piece of regenerated cellulose dialysis tubing, clip one end of the tubing and pipette the solution into the other end and clip that end with another clamp.
7. Dialyze the assembly solution against water overnight in an 800 mL beaker filled with ultrapure water.
8. Pipette the assembly solutions into 1.5 mL microcentrifuge tubes, and centrifuge at 14,000 rpm (17,968 × *g*) for 1 h.
9. Carefully use the pipette to remove the supernatant.
10. Add 100 μL of ultrapure water to the precipitate and vortex the precipitate to disperse the assemblies in the solution.

3.3 Characterization of MNP@PS@ DNA

3.3.1 Transmission Electron Microscope (TEM)

Prepare TEM samples by dropping 7.5 μL of the MNP@PS@DNA sample prepared above onto a TEM grid and wick away the excess liquid using a Kimwipe. Repeat this procedure three times. The diameter of the MNP@PS@DNA assemblies based on the TEM measurement was 113 ± 15 nm (Fig. 3b).

3.3.2 Dynamic Light Scattering (DLS)

The hydrodynamic diameter of the assemblies is determined by light scattering using Malvern Zetasizer by measuring the MNP@PS@DNA solution in water with a disposable cuvette. The hydrodynamic diameter of the MNP@PS@DNA assemblies was determined to be 161 ± 8 nm from three separate measurements (Fig. 3c).

3.4 DNA Hybridization Properties of DNA Block Copolymer/Nanoparticle Assemblies

3.4.1 DNA Melting Study by Förster Resonance Energy Transfer (FRET)

The FRET between FAM on the assemblies and Cy3 on the target DNA was used to monitor DNA hybridization of MNP@PS@DNA. Hybridization of the FAM- and Cy3-modified DNA results in energy transfer from FAM to Cy3 at an excitation wavelength that can excite the donor (FAM) but not the acceptor (Cy3). Thus, the increase in the Cy3 emission and the decrease in FAM emission can be used to monitor DNA hybridization. The FRET efficiency (E_{FRET}) was calculated by the equation: $E_{FRET} = 1 - (F_{DA}/F_D)$, where F_D is the emission intensity of the donor in the absence of the acceptor, and F_{DA} is the emission intensity of the donor in the presence of the acceptor [16].

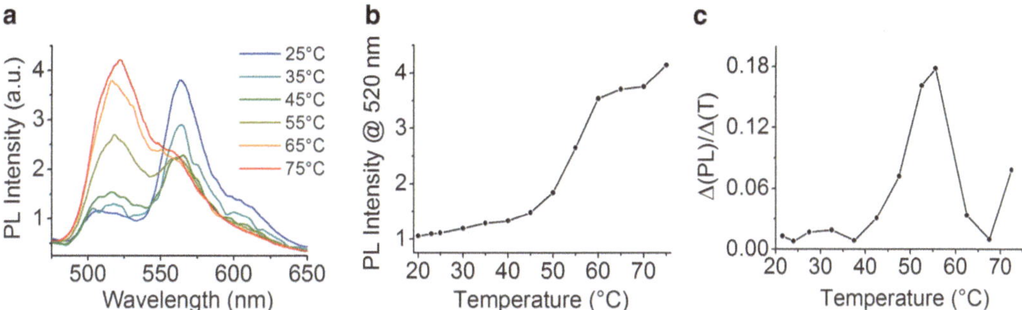

Fig. 4 Thermal denaturation monitored by FRET between MNP@PS@DNA and complementary Cy3-DNA in 0.3 M PBS. (**a**) Fluorescence spectra of MNP@PS@DNA mixed with Cy3-DNA in 0.3 M PBS collected with excitation wavelength of 430 nm at varying temperatures. (**b**) DNA melting curve constructed from the PL intensity at 520 nm. (**c**) The first derivative of the melting curve in (**b**)

1. Place 800 pmol of Cy3-labeled complementary DNA (Cy3-DNA, Sequence 3′-Cy3 T_{10} TAG GAA TAG TTA TAA-5′) in a 1.5 mL microcentrifuge tube and evaporate off the water with a Vacufuge concentrator.
2. Add 400 μL of MNP@PS@DNA solution containing 200 pmol of PS-DNA to the dried Cy3-labeled complementary DNA.
3. Add 400 μL of 0.6 M PBS solution to the solution prepared above containing the Cy3-labeled complementary DNA and MNP@PS@DNA to make a solution with a final NaCl concentration of 0.3 M and a total volume of 800 μL.
4. Equilibrate the solution at room temp for 6 h.
5. Pipette the solution containing MNP@PS@DNA and Cy3-DNA into the 3-clear window quartz cuvette with a cap to prevent evaporation.
6. Set the temperature controller to 25 °C and equilibrate the solution for 3 min.
7. Increase the temperature by 5 °C intervals and equilibrate for 3 min between each increase. Collect an emission spectrum of the sample with an excitation wavelength of 430 nm at each interval after equilibration (Fig. 4a).
8. Read the fluorescence intensity at 520 nm (emission maximum of FAM) for each spectrum collected. Plot a graph of the emission intensity versus temperature to construct thermal denaturation curves (Fig. 4b) (*see* **Note 26**).
9. Take the first derivative of the emission intensity versus temperature graph to determine the melting temperature of the sample (Fig. 4c).

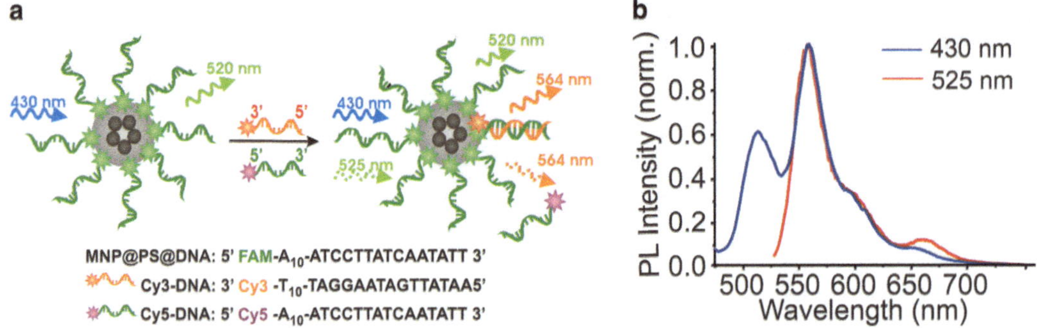

Fig. 5 (**a**) Schematic description of the "competition" experiment in water with MNP@PS@DNA, Cy3-labeled target strands (Cy3-DNA) and Cy5-labled competition strands (Cy5-DNA). (**b**) Fluorescence spectra of the competition experiment collected at 20 °C. Excitation wavelength of 430 nm (*blue*) was used to monitor the binding of target Cy3-DNA to MNP@PS@DNA by probing the energy transfer for the FAM-Cy3 FRET pair. E_{FRET}(FAM-Cy3) was determined to be 36 %. The fluorescence spectrum taken with the excitation wavelength of 525 nm (*red*) confirms that the FRET efficiency between Cy5-DNA and Cy3-DNA is very low (E_{FRET}(Cy3-Cy5) < 1 %) and that the Cy5-DNA strands do not bind to Cy3-DNA in water

3.4.2 "Competition" Binding Experiment

Perform a preferential binding experiment to examine the enhanced binding capability of the MNP@PS@DNA. In this experiment, Cy5-modified DNA with the same sequence as the DNA on MNP@PS@DNA is added to the solution of MNP@PS@DNA and Cy3-DNA as a competition strand (Fig. 5a). The degree of hybridization of the Cy3-target DNA with the DNA block copolymer assemblies and the competition DNA was monitored by the energy transfer between FAM (donor) to Cy3 (acceptor) and Cy3 (donor) to Cy5 (acceptor), respectively, at excitation wavelengths for the donor dye.

1. Mix 800 pmol of Cy3-DNA and 200 pmol of Cy5-labeled "competition" DNA (sequence: 5′Cy5 A_{10} ATC CTT ATC AAT ATT-3′) in a 1.5 mL microcentrifuge tube and evaporate off the water with a Vacufuge concentrator.
2. Add 400 μL of MNP@PS@DNA stock solution containing 200 pmol of PS-DNA to the dried Cy3-labeled complementary DNA and Cy5-labeled competition DNA.
3. Add 400 μL of ultrapure water to the solution prepared above (i.e., the solution containing the Cy3-DNA, MNP@PS@DNA, and Cy5-DNA).
4. Equilibrate the solution at room temp for 6 h.
5. To measure the degree of hybridization between the complementary Cy3-DNA and MNP@PS@DNA and the degree of hybridization between the Cy3-DNA and competition Cy5-DNA, take emission spectra at excitation wavelengths of 430 nm and 525 nm (Fig. 5b), respectively.

4 Notes

1. It is important that the hydroxyl-terminated polymer used for the chlorophosphoramidite synthesis is completely dry. To achieve this, the polymer can be dried under vacuum (<100 mTorr) overnight to remove residual water molecules.
2. (a) 2-Cyanoethyl *N,N*-diisopropylchlorophosphoramidite is stored in the freezer (−20 °C) as recommended by the manufacturer. (b) Integrity of the chlorophosphoramidite can be checked using ^{31}P NMR (chemical shift at 180 ppm). It is stable up to 6 months if stored properly.
3. "A_{10}" denotes ten adenine bases in the sequence.
4. For the synthesis of other oligonucleotides sequences, the appropriate amount of each nucleoside phosphoramidite and anhydrous acetonitrile needed to make a 0.1 M solution can be found on the Glen Research website [17].
5. *Caution*: Ammonium hydroxide used in the deprotection and cleaving steps is highly basic and has a very pungent smell. Take caution in handling it and always open it in a well-ventilated hood.
6. Diphenyl ether is a solid at room temperature. Before starting the synthesis, it is helpful to have the diphenyl ether sit in hot water. It should take about 20 min to melt.
7. *Important*: The length of the entire synthesis of the DNA block copolymer including preparation time can take up to 10–12 h. It is preferred that all the steps are carried out on the same day due to the instability of the phosphoramidite-terminated polymer. Therefore, we suggest to first set up the DNA synthesis in the morning, and then set up the phosphoramidite reaction of the hydroxyl-terminated polymer when there is about 4–5 h left in the DNA synthesis.
8. During the 3 h reaction of the chlorophosphoramidite with the hydroxyl-terminated polymer, make sure the flow of argon is gentle enough to provide a blanket of the gas over the solution, but not too high that it can evaporate the dichloromethane solvent.
9. The chlorophosphoramidite purchased from the manufacturer does not come in a Sure/Seal™ bottle. Additional steps are needed to avoid water condensation and oxidation. Before use, place the cold chlorophosphoramidite in the desiccator to allow it to warm up to room temperature. Degas the reagent by measuring out a larger amount (~5 times excess) of the chlorophosphoramidite than the procedure calls for, and inject it into a vacuum/purge vial capped with a rubber stopper. Subsequently, bubble the solution with argon for 5 min.

10. After the 3-h reaction time, the solution becomes light yellow in color.
11. (a) Given that phosphoramidite-terminated polymer is highly prone to oxidation to become phosphate, no purification was carried out to avoid the possibility of exposure to oxygen. (b) ^{31}P NMR can be calibrated by adding an internal standard trioctylphosphine (TOP), with a ^{31}P NMR peak at −30 ppm [18]. (c) The oxidized product (i.e., polymer phosphate) has a peak at ~0–18 ppm. The unreacted chlorophosphoramidite has a peak at 180 ppm.
12. The synthesis of a dye-modified DNA can be performed in one step instead of the two-steps described in the procedure by programming in the sequence (5′-$\underline{\mathbf{X}}A_{10}$ ATC CTT ATC AAT ATT-3′), where $\underline{\mathbf{X}}$ denotes the position of the FAM phosphoramidite.
13. Solid state coupling of phosphoramidite-terminated PS to the oligonucleotides on CPG beads is not performed using the DNA synthesizer due to the poor solubility of the PS in acetonitrile, the typical solvent used in DNA synthesis. Other acetonitrile soluble polymers can be coupled using the DNA synthesizer [19].
14. A solvent mixture of DMF–CH_2Cl_2 (1:1) was used to resolve the solubility difference between DNA and PS.
15. During this step, make sure to wrap the vial with foil to protect the dye from photobleaching. *Caution*: During this step, ammonia is produced as a by-product from the degradation of ammonium hydroxide. Therefore, be cautious when opening the vial after the reaction. Cool down the solution to room temperature before opening the vial, and open the vial in a well-ventilated hood.
16. Filtration will go faster by first pipetting the supernatant into the filter paper and then pipette the beads into the filter paper.
17. The extinction coefficient of FAM in DMF was determined by creating a calibration curve by preparing a series of dilutions from a stock solution of FAM in DMF. The extinction coefficient provided is the average of three measurements.
18. Make sure that the glass and adapter fittings are snug and the round bottom flask necks are clean of all chemicals. If there is some residue, wipe the necks with Kimwipes.
19. Make sure the tip of the thermocouple is touching the reaction mixture but not the spinning magnetic stir bar.
20. The magnetic stir bar should be spinning while using Schlenk line techniques.
21. Be cautious after the last heating step. The heating mantle is extremely hot. Wear heat resistant gloves when handling the mantle.

22. When purifying the particles, make sure to notice the solution turning cloudy as the acetone is added to the hexane solution. This cloudiness indicates flocculation of particles.
23. When preparing the TEM grid, use a diluted solution (10× dilution of the as synthesized particles) of nanoparticles. If the solution is too concentrated, the particles will completely cover the grid, which makes it difficult to distinguish individual particles.
24. One can check if PS is present in the solution by pouring the supernatant into water. If PS is present, the solution will turn white and turbid.
25. In the preparation of MNP@PS@DNA, the solution should look a bit turbid after the slow water addition. However, there should be no visible precipitates.
26. The changes in the emission intensity of FAM are used to monitor the hybridization process instead of Cy3 due to the temperature-dependent fluorescence intensity of Cy3 dyes.

Acknowledgments

This work was supported by the NSF Career Award (DMR-087646) and NSF-IGERT Fellowship (DGE-0221664).

References

1. Bardhan R, Lal S, Joshi A, Halas NJ (2011) Theranostic nanoshells: from probe design to imaging and treatment of cancer. Acc Chem Res 44:936–946
2. Della Rocca J, Liu D, Lin W (2011) Nanoscale metal-organic frameworks for biomedical imaging and drug delivery. Acc Chem Res 44:957–968
3. Xie J, Liu G, Eden HS, Ai H, Chen X (2011) Surface-engineered magnetic nanoparticle platforms for cancer imaging and therapy. Acc Chem Res 44:883–892
4. Erathodiyil N, Ying JY (2011) Functionalization of inorganic nanoparticles for bioimaging applications. Acc Chem Res 44:925–935
5. Yoo D, Lee J-H, Shin T-H, Cheon J (2011) Theranostic magnetic nanoparticles. Acc Chem Res 44:863–874
6. Kievit FM, Zhang M (2011) Surface engineering of iron oxide nanoparticles for targeted cancer therapy. Acc Chem Res 44:853–862
7. Chen X-J, Sanchez-Gaytan BL, Qian Z, Park S-J (2012) Noble metal nanoparticles in DNA detection and delivery. Wiley Interdiscip Rev Nanomed Nanobiotechnol 4:273–290
8. Rosi NL, Mirkin CA (2005) Nanostructures in biodiagnostics. Chem Rev 105:1547–1562
9. Chen X-J, Sanchez-Gaytan BL, Hayik SEN, Fryd M, Wayland BB, Park S-J (2010) Self-assembled hybrid structures of DNA block-copolymers and nanoparticles with enhanced DNA binding properties. Small 6:2256–2260
10. Li Z, Zhang Y, Fullhart P, Mirkin CA (2004) Reversible and chemically programmable micelle assembly with DNA block-copolymer amphiphiles. Nano Lett 4:1055–1058
11. Alemdaroglu FE, Ding K, Berger R, Herrmann A (2006) DNA-templated synthesis in three dimensions: introducing a micellar scaffold for organic reactions. Angew Chem Int Ed 45:4206–4210
12. Jeong JH, Park TG (2001) Novel Polymer-DNA hybrid polymeric micelles composed of hydrophobic Poly(D, L-lactic-co-glycolic Acid) and hydrophilic oligonucleotides. Bioconjug Chem 12:917–923
13. Chien M-P, Rush AM, Thompson MP, Gianneschi NC (2010) Programmable shape-shifting micelles. Angew Chem Int Ed 49: 5076–5080
14. Northwestern University. oligonucleotide properties calculator used to calculate extinction coefficient of DNA. http://www.basic.northwestern.edu/biotools/OligoCalc.html

15. Sun S, Zeng H (2002) Size-controlled synthesis of magnetite nanoparticles. J Am Chem Soc 124:8204–8205
16. Clegg RM, David M.J.L.a.J.E.D. (1992) Fluorescence resonance energy transfer and nucleic acids. Methods Enzymol, vol. 211, Academic, p 353–388
17. The information is available at the Glen Research website. Glen Research. http://www.glenresearch.com/
18. Kuno M, Lee JK, Dabbousi BO, Mikulec FV, Bawendi MG (1997) The band edge luminescence of surface modified CdSe nanocrystallites: probing the luminescing state. J Chem Phys 106:9869–9882
19. Zimmermann J, Kwak M, Musser AJ, Herrmann A (2011) Amphiphilic DNA block copolymers: nucleic acid-polymer hybrid materials for diagnostics and biomedicine. Methods Mol Biol 751:239–266

Chapter 17

Polyaspartic Acid Coated Iron Oxide Nanoprobes for PET/MRI Imaging

Taku Cowger and Jin Xie

Abstract

Iron oxide nanoparticles, due to their exceptional magnetic property, biocompatibility, and biodegradability, have long been studied as contrast agents for magnetic resonance imaging (Xie et al., Curr Med Chem 16(10):1278–1294, 2009; Xie et al., Adv Drug deliv Rev 62(11):1064–1079, 2010). While previous applications mostly target reticuloendothelial system (RES) organs such as liver and lymph nodes, recent efforts have been made to impart targeting peptides or antibodies onto particle surface to enable site-specific targeting after systemic administration (Xie et al., Adv Drug Deliv Rev 62(11):1064–1079, 2010; Cai and Chen, Small 3(11):1840–1854, 2007; Corot et al., Adv Drug Deliv Rev 58 (14):1471–1504, 2006; Xie et al., Acc Chem Res 44(10):883–892). Moreover, other imaging functionalities can be loaded onto nanoparticles to achieve multimodality imaging probes (Cai and Chen, Small 3(11):1840–1854, 2007; Lee et al., J Nucl Med Soc Nucl Med 49(8):1371–1379, 2008). In this protocol, we describe the procedure of constructing an iron oxide nanoparticle (IONP)-based probe with high affinity towards integrin $\alpha_v\beta_3$ for positron emission tomography (PET) and magnetic resonance imaging (MRI) dual modality imaging. The related characterizations and validation experiments, including particle concentration determination, Prussian blue staining, animal model preparation, and in vivo PET/MRI imaging will also be discussed.

Key words IRON oxide nanoparticles, Integrin $\alpha_v\beta_3$, Multimodal imaging, Positron emission tomography, Magnetic resonance imaging, Tumor angiogenesis

1 Introduction

Nanomaterials, especially nanoparticles, have been extensively studied in the recent decades for their potentials in improving the current clinic diagnostic and therapeutic techniques [2, 3]. These materials usually have unique physical or physiochemical properties, and when conjugated with targeting molecules, become "smart" gadgets that can pinpoint and/or kill diseased tissues in a living subject [7]. One of the most studied nanomaterials in the field has been IONPs [4, 5]. With strong magnetic property, biocompatibility, and biodegradability, IONPs are candidate nanoplatforms to construct functional nanogadgets for both imaging

Paolo Bergese and Kimberly Hamad-Schifferli (eds.), *Nanomaterial Interfaces in Biology: Methods and Protocols*, Methods in Molecular Biology, vol. 1025, DOI 10.1007/978-1-62703-462-3_17,

and therapy purposes [2, 3]. One of the primary applications of IONPs is to work as contrast agents for MRI. Several IONP formulations have been approved by the Federal Drug Administration (FDA) and used in the clinics to improve imaging quality [1].

While MRI provides high resolution and good anatomical details, its signal-to-background ratio, even with contrast agents, is suboptimal. In fact, none of the current imaging techniques are perfect and there are always trade-offs, for instance, between resolution and sensitivity. To address this issue, efforts have been made to develop multimodality imaging systems and imaging probes that allow simultaneous interrogation with two or even more imaging methods [6, 8–10]. Such an approach can greatly improve the diagnosis quality and lower misdiagnosis rate. For instance, positron emission tomography/computed tomography (PET/CT) systems have been developed and become an important diagnostic tool. PET/MRI is another promising technique under intensive studies and is expected to be implemented nationwide soon. Engineering IONPs and loading onto them radioisotopes is a common way to achieve PET/MRI dual functional imaging probes.

In the current protocol, we describe the detailed procedure for preparing tumor targeting IONPs with both PET and MRI imaging functionalities. IONPs are made from a high temperature thermal decomposition method. They participate in a ligand exchange with polyaspartic acid (PASP), a biocompatible and biodegradable polymer with high affinity toward IONP surface. Subsequently, a cyclic peptide, c(RGDyK), and a macrocyclic chelating agent, 1,4,7,10-tetreaazacyclododecane-*N*, *N′*, *N″*, *N‴*-tetraacetic acid (DOTA), are both covalently conjugated onto the PASP coating. Here c(RGDyK) works as targeting a motif for its high affinity toward integrin $\alpha_v\beta_3$, a cell adhesion mediating receptor that is highly expressed on many types of tumor cancer cell surface, and more importantly, on tumor vasculature [10, 11]. This makes integrin $\alpha_v\beta_3$ a tumor biomarker. Imaging probes and drug delivery vehicles using RGD or its derivatives as tumor targeting motifs have been widely reported [12–15]. Due to the fact that multiple RGD can be loaded onto a single particle surface, RGD-nanoparticle conjugates usually show even higher affinity toward integrin $\alpha_v\beta_3$, the so-called multivalency effect [15]. DOTA is introduced onto IONP surface as a metal chelator. DOTA can form serum-stable complexes with many types of transition metals and is widely used in PET or single-photon emission computed tomography (SPECT) imaging to assist radioisotope labeling [16]. In the current protocol, we load ^{64}Cu, a radioisotope commonly used for PET imaging, onto IONPs.

2 Materials

2.1 Reagents

1. Iron chloride.
2. Sodium oleate.
3. Ethanol.
4. Hexane.
5. 1-octadecene.
6. DOTA.
7. 1-ethyl-3-[3-(dimethylamino)propyl]carbodiimide.
8. *N*-hydroxysulfonosuccinimide (SNHS, Sigma).
9. Polyaspartic acid (PASP) (SPECTRUM).
10. O-[*N*-(3-Maleimidopropionyl)aminoethyl]-O′-[3-(*N*-succinimidyloxy)-3-oxopropyl]heptacosaethyleneglycol(NHS-PEG-MAL, Sigma).
11. 4-Maleimidobutyric acid *N*-succinimidyl ester (Sigma).
12. PBS buffer.
13. RGD peptide (c(RGDyK); Peptides International).
14. *N*-Succinimidyl *S*-acetylthioacetate (SATA, Pierce Biotechnology Inc.).
15. DMSO.
16. Acetonitrile.
17. Trifluoroacetic acid.
18. Hydroxylamine hydrochloride.
19. Tris(2-carboxyethyl)phosphine hydrochloride (TCEP.HCl; Pierce Biotechnology Inc.).
20. Sodium hydroxide.
21. Hydrochloric acid.
22. U87MG human glioblastoma cells.
23. MCF-7 human breast cancer cells.
24. Low-glucose DMEM (GIBCO).
25. Minimum essential medium (GIBCO).
26. Fetal bovine serum (FBS; GIBCO).
27. Cell-binding buffer (20 mM Tris, 150 mM NaCl, 2 mM $CaCl_2$, 1 mM $MnCl_2$, 1 mM $MgCl_2$, 0.1 % (wt/vol) bovine serum albumin; pH 7.4).
28. ^{125}I-Echistatin (GE Healthcare).
29. Isoflurane (RxElite Inc.).
30. U87MG tumor-bearing athymic nude mice (Harlan).

2.2 Equipment

1. HPLC (Dionex).
2. Analytical HPLC column (Vydac).
3. Semi-preparative HPLC column (Vydac).
4. Freeze-dry system (Labconco Corp.).
5. PD-10 column (GE Healthcare, GE Healthcare).
6. Centrifuge (Allegra X-15R, Beckman-Coulter).
7. Refrigerated microcentrifuge (Eppendorf).
8. Rotary vaporizer (Buchi, RII).
9. 96-well plate (Thermolab Systems).
10. Vacuum manifold (Millipore).
11. Vacuum pump (VWR).
12. Dry bath incubator (Fisher Scientific).
13. Polystyrene culture test tube (Fisher Scientific).
14. γ-counter (Packard).
15. GraphPad Prism (GraphPad Software Inc.).
16. 4-well chamber slides (Thermo-Fisher).
17. ICP-MS (VG Elemental).
18. Rodent anesthesia system (Summit Anesthesia Solutions).
19. 7 T small animal MRI system (Varian).
20. microPET R4 rodent scanner (Siemens Medical Solutions).

3 Methods

3.1 Preparation of Iron-Oleate Precursor

1. 10.8 g of iron chloride ($FeCl_3 \cdot 6H_2O$, 40 mmol) and 36.53 g of sodium oleate (120 mmol) are dissolved in a mixture solvent composed of 80 ml ethanol, 60 ml distilled water, and 140 ml hexane.
2. The resulting solution is heated up to 80 °C and kept at that temperature for 4 h.
3. The organic (top) layer is washed three times with 30 ml distilled water and collected by a separatory funnel.
4. Use a rotary evaporator to remove hexane and ethanol. The final product is in a waxy solid form.

3.2 Synthesis of IONPs

1. 9.0 g iron-oleate and 1.58 ml oleic acid is dissolved in 35 ml 1-octadecene with magnetic stirring (*see* **Note 1**).
2. The solution is heated stepwise at 150 °C for 45 min, 200 °C for 45 min, and finally to reflux at ~320 °C for 45 min (*see* **Note 2**).

3. Cool down the mixture to room temperature. Add 30 ml ethanol to precipitate IONPs. Collect IONPs by centrifuge at 8,000 rpm (7,700 × *g*) (*see* **Note 3**).
4. Decant the supernatant and redissolve the nanoparticles in 15 ml hexane with sonication. Add 30 ml ethanol to precipitate the nanocrystals. Repeat the washing step for three times (*see* **Note 4**).

3.3 Preparation of PASP Coated IONPs

1. 20 mg IONPs are dissolved in 5 ml hexane and added onto 5 ml 25 mg/ml PASP solution in water to form a two phase mixture.
2. Sonicate the mixture for 5 min and then centrifuge. Wash the aqueous phase three times with hexane.
3. Collect the aqueous phase and run it through a PD-10 column for purification and solvent exchange. Briefly, load the solution onto a column and wait until all the solution is in. Add 3.5 ml borate buffer (50 mM, pH 8.5) and collect only the deepest-colored fractions at the bottom. The product is PASP-coated IONPs (PASP-IONPs).

3.4 Preparation of Thiolated RGD

1. Mix c(RGDyK) (5 mmol) in 1 ml of borate buffer (pH 8.5) with 100 ml DMSO solution containing 6 mmol of SATA. Incubate at room temperature with gentle shaking.
2. Monitor the reaction by analytical RP-HPLC (reverse phase-high performance liquid chromatography). The mobile phase is changed from 95 % solvent A (0.1 % TFA in water) and 5 % solvent B (0.1 % TFA in acetonitrile) (0–2 min) to 35 % solvent A and 65 % solvent B at 32 min. The flow rate is 1 ml/min and the UV absorbance is monitored at 218 nm.
3. When the reaction comes to completion, quench with 100 ml of 2 % TFA in water.
4. Freeze-dry the crude mixture. Redissolve the product in 1 ml water.
5. Add 100 ml of 0.5 M hydroxylamine solution. Adjust the pH to 6.0 with 0.5 M sodium hydroxide (*see* **Note 5**).
6. Use RP-HPLC to monitor the reaction. When the reaction comes to completion, purify the thiolated c(RGDyK) (RGD-SH) by semi-preparative RP-HPLC. The gradient is the same as described above. The flow rate is changed to 5 ml/min. Collect the fractions containing pure RGD-SH and freeze-dry (Fig. 1).
7. Store RGD-SH under acidic condition (pH 3–4) in water to prevent disulfide formation (*see* **Note 6**).

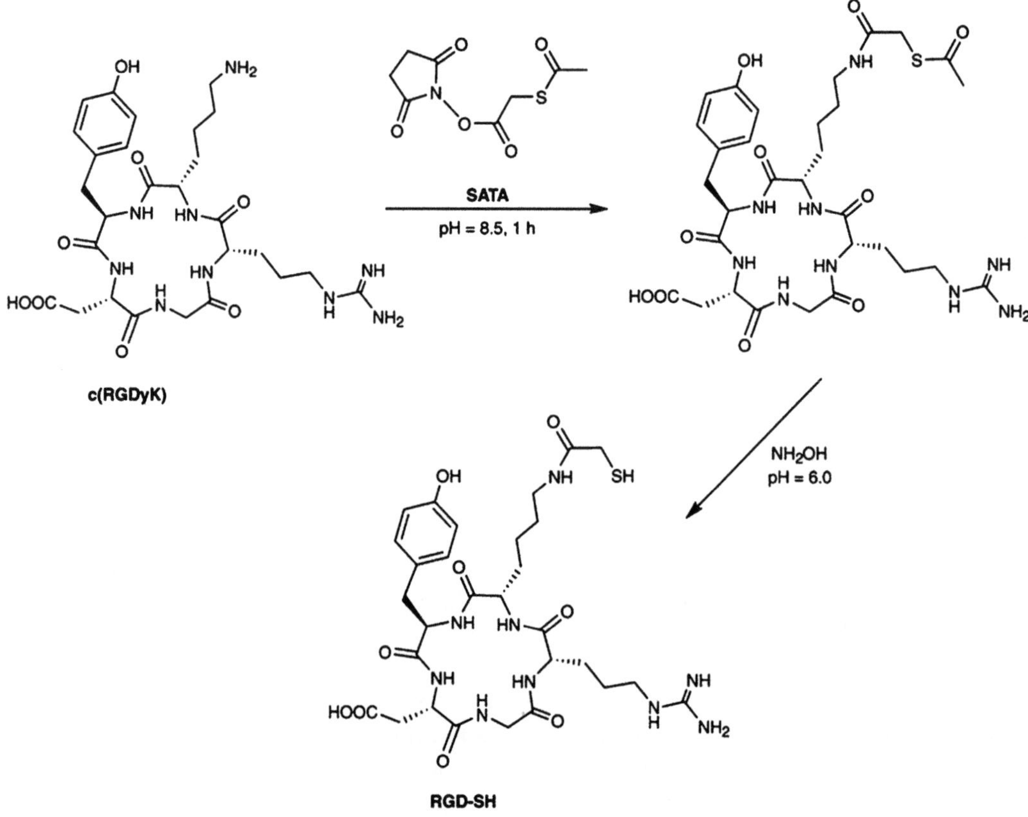

Fig. 1 Converting c(RGDyK) to RGD-SH. Modified with permission from ref. 16

3.5 Conjugation of DOTA and RGD onto PASP-IONPs

1. DOTA is activated by 1-ethyl-3-[3-(dimethylamino)propyl] carbodiimide (EDC) and *N*-hydroxysulfonosuccinimide (SNHS) at pH 5.5 for 30 min with a DOTA:EDC:SNHS molar ratio of 10:5:4.
2. 0.8 μmol activated DOTA and 1.2 μl NHS-PEG-MAL (4.1 mg) are added to a 600 μl PASP-IONP solution in borate buffer (2.5 mg Fe/ml, pH = 8.5). Incubate the mixture at 4 °C for 1 h.
3. Add 1.0 mg RGD-SH (1.5 μmol) to the solution. Adjust the pH to 7.0 and incubate the mixture overnight at 4 °C (*see* **Note 7**).
4. Run the raw product through a PD-10 column to remove unreacted species. Briefly, load the solution onto a column and wait until all the solution is in. Add PBS buffer (pH 7.4) as eluent and collect only the deepest-colored fractions at the bottom (*see* **Note 8**).
5. The product, RGD and DOTA modified IONPs (RD-IONPs), is stored at 4 °C and is stable for up to 4 weeks (Fig. 2).

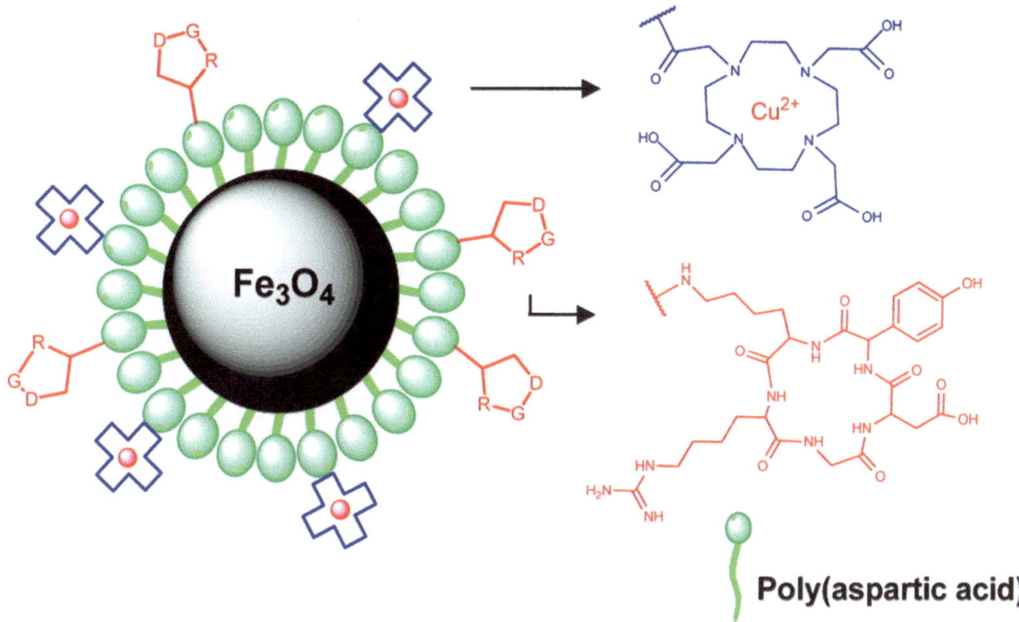

Fig. 2 Schematic diagram of the ^{64}Cu loaded RD-IONPs for PET and MRI imaging. Modified with permission from ref. 7

3.6 Determination of IONP Concentration

Take 20 μl of the particle stock solution and decompose the particles in 10 % HNO_3 with heating. Add water to adjust the final volume to 10 ml. The iron content is to be analyzed by ICP-MS. The results will be converted to obtain the iron concentration in the original solution (*see* **Note 9**).

3.7 Cell Culture

U87MG cells are cultured in DMEM (low glucose) supplemented with 10 % (vol/vol) FBS at 37 °C. MCF-7 cells are cultured in MEM supplemented with 10 % (vol/vol) FBS at 37 °C. The cells are used when they reach 70–85 % confluency.

3.8 Binding Affinity Assay

1. Prepare ^{125}I-echistatin solution in cell-binding buffer (20 mM Tris, 150 mM NaCl, 2 mM $CaCl_2$, 1 mM $MnCl_2$, 1 mM $MgCl_2$, 0.1 % (wt/vol) bovine serum albumin; pH 7.4; 0.5 μCi/ml). It is estimated that about 50 μl solution is needed for each well (*see* **Notes 10** and **11**).
2. Prepare three stock solutions of different concentrations of RD-IONPs in cell-binding buffer. Parallel analyses with free RGD peptide can be carried out as a comparison. Triplicate samples are recommended.
3. Add 0.02–0.03 μCi of ^{125}I-echistatin and appropriate amount of the RD-IONP stock solution to each well of a 96-well plate. The final concentrations should range from 1.0 μg Fe/ml to 1.0 mg Fe/ml.

4. Prepare a U87MG stock in cell-binding buffer at two million cells/ml concentration. Add 1×10^5 U87MG cells (50 μl) to each well and adjust the total volume to 200 μl per well with cell-binding buffer. Incubate for 2 h at room temperature.
5. Use a vacuum manifold to remove the incubation buffer from the 96-well plate and wash three times with the cell-binding buffer (100 μl per well).
6. Heat-dry the 96-well plate in a dry bath incubator. It usually takes 15 min to dry the membrane.
7. Collect the membranes from each well and put them into polystyrene culture test tubes.
8. Measure the radioactivity on each membrane with a γ-counter.
9. Fit the data by nonlinear regression and calculate the best-fit IC_{50} (inhibitory concentration of 50 %) values. RD-IONPs should have much lower IC_{50} values (higher binding affinity) than the unconjugated RGD peptide due to the multivalent efficiency.

3.9 Cell Staining

1. Seed U87MG (integrin $\alpha_v\beta_3$-positive) or MCF-7 (integrin $\alpha_v\beta_3$-negative) cells (1×10^5 cells per well) onto 4-well cell culture chambers and incubate overnight.
2. Aspirate the cell culture medium from the dish. Wash the cells twice with PBS.
3. Fix the cells with ice-cold 95 % EtOH for 15 min.
4. Wash the cells 2–3 times with 1 ml cell-binding buffer (3–5 min each time).
5. Add PASP-IONP or RD-IONP solutions into each well. Adjust the final particle concentration to 0.02 mg Fe/ml. Incubate at 37 °C for 1 h with gentle shaking (*see* **Note 12**).
6. Remove the particles and wash the cells three times with PBS buffer.
7. Incubate the cells with Prussian blue staining solution, which is a 1:1 mixture of 20 % hydrochloric acid and 10 % potassium ferrocyanide solution, for 10 min at room temperature. Wash cells with PBS twice.
8. Incubate cells with fast red nuclear staining solution for 10 min.
9. Consecutive dehydrations with 70, 90, and 100 % EtOH.
10. Remove the chamber and mount the slide for microscopy.

3.10 Preparation of U87MG Xenograft Models

The animal models will be prepared by injecting 5×10^6 U87MG cells subcutaneously into the right flank of athymic nude mice. Imaging studies will be performed when the tumor size reaches ~500 mm^3, which usually takes 3–4 weeks. The tumor size will be calculated as $a \times b^2/2$, where a represents the longer dimension and b represents the shorter dimension of the tumor.

3.11 Loading ^{64}Cu onto RD-IONPs

1. Before in vivo PET imaging, incubate RD-IONPs with ^{64}Cu (185 GBq/g Fe) in 0.1 N sodium acetate (pH 6.5) buffer for 45 min at 40 °C (*see* **Note 13**).
2. Run the mixture through a PD-10 column with PBS buffer (pH 7.4) as the mobile phase. Collect the fractions with the deepest color.

3.12 In Vivo PET Studies

1. Prepare U87MG xenograft models as above mentioned. Wait until the tumor size reaches ~500 mm^3.
2. Label IONPs with ^{64}Cu by following the above mentioned procedure.
3. Anesthetize the animals using the rodent anesthesia system with isoflurane (2 % (vol/vol) isoflurane in 0.2 l/min O_2 flow).
4. Inject 100–200 μl ^{64}Cu labeled RD-IONPs intravenously into the U87MG tumor-bearing mice at a dose of 10 mg Fe/kg. For the control group, free c(RGDyK) (10 mg/kg) is co-injected with the particle probes.
5. Perform PET scans on a microPET R4 rodent scanner at selective time points post injection.
6. For each scan, 3-dimensional ROIs are drawn over the tumor and organs on decay-corrected whole-body coronal images. The average radioactivity is obtained from the mean pixel values within the ROI volume, and is converted to counts per milliliter per minute by using a predetermined conversion factor.
7. Given a tissue density of 1 g/ml, the counts per milliliter per minute are converted to counts per gram per minute, and the values are divided by the injected dose to obtain the imaging-ROI-derived percentage injected dose per gram (% ID/g).
8. If needed, harvest the tumor and major organs. Take PET images again with the organs. Subsequent ex vivo tissue staining can be performed when after the reactivity is decayed.

3.13 In Vivo MRI Studies

1. Prepare U87MG xenograft models as described above. Wait until the tumor size reaches ~500 mm^3.
2. Anesthetize the animals using the rodent anesthesia system with isoflurane (2 % (vol/vol) isoflurane in 0.2 l/min O_2 flow).
3. RD-IONPs are intravenously injected through the tail vein (10 mg Fe/kg per mouse). For the control group, c(RGDyK) (10 mg/kg) is co-injected with the particles.
4. T_2-weighted fast spin-echo images are acquired on a 7.0 T small animal MRI system, before and at selective time points post injection. The following are suggested parameters: TE 40 ms, TR 3,000 ms, thickness 1 mm, FOV 6×6, NEX 1.0, Echo 1/1 (*see* **Note 14**).

4 Notes

1. Iron-oleate is a very viscous compound and can prove difficult to handle. It is best to keep the spatula slightly moist with hexane to help transfer the metal oleate into the reaction flask.
2. Any smoke observed during the refluxing period is due to the lack of a closed system and the solvent is being removed. In that case, cease the heating and repair any leak in the system.
3. IONPs can be split into multiple tubes to ensure good purification.
4. Small portion of aggregations may be observed. To remove these, centrifuge at low speed (2,500 rpm (460 × *g*)) and dispose the aggregates at the bottom.
5. If the pH is too low, it may result in incomplete de-protection of the thiol. If the pH is too high, disulfide bond may form when the thiol group is released.
6. RGD-SH can be stored for up to a few months under an acidic condition.
7. It is highly recommended to check the purity of the RGD-SH by analytical HPLC before conjugation. If significant amount of disulfide already formed, some TCEP·HCl (1 mg or less) can be added.
8. NAP-5 columns can be used when dealing with smaller amount of samples.
9. Remember that 2 % HNO_3 is commonly used in ICP analysis as the eluent. Estimate the iron content and make sure it lies within the range of the standards.
10. It is imperative to obtain appropriate training and abide by all regulatory rules when handling radioactivity. The radioactive waste needs to be collected and disposed under the guidance of the institutional radiation safety office.
11. The presence of certain metal ions (e.g., Mn^{2+} and Mg^{2+}) is essential for integrin $\alpha_v\beta_3$ binding. Binding buffer without these ions will result in much lower readings.
12. Aggregation may form during particle aging. The aggregates may stick onto the slides and interfere with the staining. Use fresh samples when possible. Sonication prior to the staining may also help.
13. It is imperative to obtain appropriate training and abide by all regulatory rules when handling radioactivity. The radioactive waste needs to be collected and disposed under the guidance of the institutional radiation safety office.

14. A monitoring system should be used to monitor physiological signals of mice such as their heart rate, respiration, and body temperature. Respiration rate is usually obtained from a small pneumatic pillow sensor placed next to the animal's abdomen. Heart rate is monitored using ECG by placing three leads with subdermal needle electrodes. Temperature is monitored with a rectal thermometer or a fiber optic sensor.

Acknowledgments

This work was supported by NIH grant R00 4R00CA153772.

References

1. Xie J, Huang J, Li X et al (2009) Iron oxide nanoparticle platform for biomedical applications. Curr Med Chem 16(10):1278–1294
2. Xie J, Lee S, Chen X (2010) Nanoparticle-based theranostic agents. Adv Drug Deliv Rev 62(11):1064–1079
3. Cai W, Chen X (2007) Nanoplatforms for targeted molecular imaging in living subjects. Small 3(11):1840–1854
4. Corot C, Robert P, Idee JM et al (2006) Recent advances in iron oxide nanocrystal technology for medical imaging. Adv Drug Deliv Rev 58(14):1471–1504
5. Xie J, Liu G, Eden HS et al (2011) Surface-engineered magnetic nanoparticle platforms for cancer imaging and therapy. Acco Chem Res 44(10):883–892
6. Lee HY, Li Z, Chen K et al (2008) PET/MRI dual-modality tumor imaging using arginine-glycine-aspartic (RGD)-conjugated radiolabeled iron oxide nanoparticles. J Nucl Med 49(8):1371–1379
7. Lee S, Xie J, Chen X (2010) Peptides and peptide hormones for molecular imaging and disease diagnosis. Chem Rev 110(5): 3087–3111
8. Chen K, Xie J, Xu H et al (2009) Triblock copolymer coated iron oxide nanoparticle conjugate for tumor integrin targeting. Biomaterials 30(36):6912–6919
9. Xie J, Chen K, Huang J et al (2010) PET/NIRF/MRI triple functional iron oxide nanoparticles. Biomaterials 31(11):3016–3022
10. Cai W, Chen X (2008) Multimodality molecular imaging of tumor angiogenesis. J Nucl Med 49(Suppl 2):113S–128S
11. Cai W, Niu G, Chen X (2008) Imaging of integrins as biomarkers for tumor angiogenesis. Curr Pharm Des 14(28):2943–2973
12. Chen X, Park R, Shahinian AH et al (2004) 18F-labeled RGD peptide: initial evaluation for imaging brain tumor angiogenesis. Nucl Med Biol 31(2):179–189
13. Li ZB, Cai W, Cao Q et al (2007) (64) Cu-labeled tetrameric and octameric RGD peptides for small-animal PET of tumor alpha(v)beta(3) integrin expression. J Nucl Med 48(7):1162–1171
14. Liu Z, Li ZB, Cao Q et al (2009) Small-animal PET of tumors with (64)Cu-labeled RGD-bombesin heterodimer. J Nucl Med 50(7):1168–1177
15. Xie J, Chen K, Lee HY et al (2008) Ultrasmall c(RGDyK)-coated Fe3O4 nanoparticles and their specific targeting to integrin alpha(v) beta3-rich tumor cells. J Am Chem Soc 130(24):7542–7543
16. Cai W, Chen X (2008) Preparation of peptide-conjugated quantum dots for tumor vasculature-targeted imaging. Nat Protoc 3(1):89–96

Chapter 18

Ligand Synthesis and Passivation for Silver and Large Gold Nanoparticles for Single-Particle-Based Sensing and Spectroscopy

Daniel Montiel, Emma V. Yates, Li Sun, Marissa M. Sampias, John Malona, Erik J. Sorensen, and Haw Yang

Abstract

Silver and large gold nanoparticles are more efficient scatterers than smaller particles, which can be advantageous for a variety of single-particle-based sensing and spectroscopic applications. The increased susceptibility to surface oxidation and the larger surface area of these particles, however, present challenges to colloid stability and controllable bio-conjugation strategies. In this chapter, ligand syntheses and particle passivation procedures for yielding stable and bio-conjugatable colloids of silver and large gold nanoparticles are described.

Key words Gold nanoparticles, Silver nanoparticles, Bio-conjugation, Nanoparticle passivation

1 Introduction

Materials on the nanometer scale can exhibit optical properties different than the corresponding bulk materials. For noble metal nanoparticles, the source of the size-dependent optical response is the collective oscillation of conduction-band electrons in response to an applied electromagnetic field, i.e., the particle's plasmon resonance [1–3]. In the near-field region (distance to particle surface $\ll$wavelength), nanoparticles can enhance the local field leading to emission enhancement for emitters present in the field. The plasmon resonance and near-field distributions can be tuned using differently shaped nanoparticles or hierarchical structures composed of multiple particles.

Noble metal nanoparticles, with plasmon resonances in the visible spectrum, have attracted a lot of attention in the literature for a variety of applications. The synthesis of these particles typically involves the reduction of the metal ions out of solution leading to

Paolo Bergese and Kimberly Hamad-Schifferli (eds.), *Nanomaterial Interfaces in Biology: Methods and Protocols*, Methods in Molecular Biology, vol. 1025, DOI 10.1007/978-1-62703-462-3_18, © Springer Science+Business Media New York 2013

Ostwald ripening of seed particles [4]. Particles are typically ripened to a desired size and then stabilized with a surfactant.

The advantageous optical properties and the known surface chemistry of these nanoparticles have led experimenters to conjugate them to biological materials for a variety of applications. One can attach small molecules to the nanoparticle surface to prevent nonspecific adsorption of biomolecules or attach molecules with an orthogonal chemistry for hierarchical construction. Applications of these hybrid particles range from analyte detection [5], electron microscopy contrast agents [6], photo-thermal therapies [7, 8], drug delivery systems [9], bio-barcoding [10, 11], among others [12–16].

For certain applications, e.g., single-particle-based sensing and spectroscopy, using particles with increased scattering cross-sections such as larger-diameter gold nanoparticles may be desirable. Utility of any hybrid nano-material relies on the ability to produce stable colloids resistant to aggregation and nonspecific adsorption of biomolecules under physiologically relevant conditions. As the particle size becomes greater, however, the increased susceptibility to surface oxidation and the larger surface area per particle present challenges to robust incorporation of these particles into the optical toolbox of nano-materials [17–20]. The passivation of bare metal nanoparticle surfaces with organic ligands is important for colloid stability and minimizing nonspecific adsorption of biomolecules. The attachment chemistry to the metal nanoparticle [21], coverage efficiency [22, 23], ligand length [24], and terminal group are all important control parameters [25].

Thiol groups are commonly used for attachment of small molecules to noble metal nanoparticles because chemical bonds are formed with the metal surface [26]. To reduce ligand lability and reduce susceptibility to oxidative cleavage, dithiol groups have been used to increase coordination to the gold surface [21, 27, 28].

The ligand's terminal group can serve multiple roles. One of these is to increase steric and electrostatic repulsion between particles to increase colloid stability. Zwitterionic ligands, although uncharged, form a charged double layer around the nanoparticle and have been shown to yield stable colloids [29–33] under physiologically relevant conditions.

Another role for the terminal moiety is to provide an orthogonal chemistry for rational construction of biological or inorganic nanoparticles. Streptavidin, a 53 kDa tetrameric protein, binds strongly to biotin [34] and has been used to construct bioinorganic composites for various applications [35–37].

For self-assembled monolayers on flat surfaces there exist techniques such as ellipsometry [38] and quartz crystal microbalance [39] measurements to quantify surface coverage. Nanoparticles typically display particle-to-particle variation; the extent of surface coverage [23; 40] and nonspecific adsorption [41] need to be characterized to understand the scope of a new particle system.

Ideally, one would desire the ligands to be resistant to nonspecific adsorption and dissociation from the particle surface while also being bio-conjugatable.

Methods for the attachment of biological materials to noble metal nanoparticles are mature enough to allow the synthesis and purification of discrete 1-1 bio-conjugated nanoparticles [42]. The more general topic of interfacing biological molecules with nanoparticle surfaces has been reviewed previously [41; 43]. The majority of these bio-conjugation methods, however, have been applied to smaller-diameter gold nanoparticles (diameter < 40 nm). The increased surface area per particle for larger diameter noble metal nanoparticles presents challenges for nonspecific absorption and colloid stability [19; 20].

Colloidal silver nanoparticles have an exceptionally strong and sharp plasmon resonance useful for different applications. Due to their utility, there have been numerous synthetic procedures using different reducing agents, such as citric acid [44], ascorbic acid [45], sodium borohydride [46], and hydrogen gas [47]. On the other hand, silver nanoparticles are even more susceptible to surface oxidation than gold nanoparticles, thereby impacting on the proper preparation for biological applications. This can be overcome by synthesizing the desired silver nanoparticles in house (rather than purchasing them from commercial sources) and passivating the freshly synthesized particles immediately. Once the silver nanoparticles are properly passivated, they can be stably stored for longer period of time and still retain the reactivity for further bio-conjugation.

This chapter details the practical procedures that have been routinely used in our laboratory for the preparation of nanoparticles for biological applications. Subheading 3 is conceptually divided into three parts. The first part describes the synthesis of three ligands for particle passivation. The first is a compact zwitterionic ligand containing a sulfobetaine moiety derived from lipoic acid (LA). The second is a dithiol ligand with a biotin terminal group, enabling the bio-conjugation to streptavidin-modified species. The third procedure describes the modification of single-stranded DNA with LA, for direct bio-conjugation of DNA onto larger gold nanoparticles using EDC chemistry. The relatively low cost of aminated primers allows for the increased preparative scale necessary to passivate large gold nanoparticles with DNA. Here, amine 5′-modified ssDNA is coupled to LA and then a complementary strand, 5′-modified with biotin, is annealed to the LA-modified strand. The second part describes a silver nanoparticle synthesis using a modified ascorbic acid reduction method and nanoparticle passivation with dithiol ligands. The last part describes methods to characterize the effectiveness of the passivation. Particles are characterized by absorption spectroscopy, gel electrophoresis, and column chromatography to ensure the robustness of the passivation. It should be pointed out that the latter two characterization methods are routinely used in preparing proteins and nucleic acids; ensuring that the passivated gold and silver nanoparticles are compatible with these standard biochemical methods is a necessary step to enable subsequent manipulation of the desired hybrid nano-materials.

2 Materials

2.1 Chemicals

1. 18.2 MΩ-cm water.
2. Room temperature is 20 °C.
3. (±)α-Lipoic acid, ≥99 %.
4. Thionyl chloride ($SOCl_2$), 99+ %.
5. 3-Dimethylamino-1-propanol, 99 %.
6. 1,3-Propanesultone, 99 %.
7. Sodium borohydride ($NaBH_4$).
8. Chloroform ($CHCl_3$), ACS grade.
9. Acetone, HPLC grade.
10. Sodium bicarbonate.
11. Ethyl acetate (EtOAc), ACS grade.
12. Agarose, bioreagent grade.
13. New England Biolabs Tridye 1 kb DNA ladder.
14. Ethanol, 200 proof.
15. 1-Ethyl-3-(3-dimethylaminopropyl)carbodiimide (EDC), stored in a desiccator at −20 °C.
16. Sulfo-NHS, stored in a dessicator at 4 °C.
17. Ethylenediaminetetraacetic acid (EDTA), bioreagent grade.
18. 0.1 M MES (2-(N-morpholino)ethanesulfonic acid sodium salt) pH 6, bioreagent grade.
19. Sephacryl S 300, size exclusion resin (GE Lifesciences).
20. Chromatography media sampler pack (Bio-Rad).
21. 80 nm gold nanoparticles (British BioCell International).
22. Silver nitrate, 99.9999 %.
23. Sodium acetate, bioreagent grade.
24. Sodium citrate dihydrate, granular/certified.
25. L-Ascorbic acid, ACS grade.
26. Sodium hydroxide, ACS grade.
27. DithiolalkanearomaticPEG6-NHS (CAS# 936115-55-8, SensoPath Technologies) (*see* **Note** 1).

28. EZ-Link Amine-PEG2-Biotin (Pierce).
29. 10 mM Hepes (4-(2-hydroxyethyl)piperazine-1-ethanesulfonic acid) pH 8, bioreagent grade.

2.2 Equipment

1. Aluminum foil.
2. Pasteur pipette.
3. Cotton balls.
4. Mini-tube rotator.
5. Rotary evaporator.
6. Speed-Vac, i.e., Savant SVC-100 or equivalent.
7. Schlenk line.
8. Temperature-controlled water bath.
9. Three-neck jacketed round bottom flask.
10. Syringe pumps.
11. Agarose gel electrophoresis apparatus.
12. UV–Vis absorption spectrophotometer.

3 Methods

3.1 Protocol for Dihydrolipoic Acid-Sulfobetaine (DHLA-SBE) Synthesis

1. All reactions performed under N_2 with a Schlenk line, unless otherwise stated.
2. Dilute lipoic acid (1,000 mg, 4.85 mmol) in 10 mL $CHCl_3$ and cool in an ice bath.
3. Dissolve $SOCl_2$ (5.34 mmol) in 2–3 mL $CHCl_3$ and add drop-wise.
4. After addition of $SOCl_2$, stir 30 min at room temperature.
5. Remove the solvent by rotary evaporation.
6. Re-dissolve the product in 10 mL $CHCl_3$, cool in an ice bath and purge with N_2.
7. Dilute 4.85 mmol 3-dimethylamino-1-propanol with 2–3 mL $CHCl_3$ and add drop-wise.
8. Stir the reaction vigorously overnight at room temperature.
9. Quench the reaction with $NaHCO_3$ (9.7 mmol) in 20 mL water.
10. Extract the quenched reaction with EtOAc (2 ×) and wash the combined organic extract with saturated aqueous $NaHCO_3$ (2 ×).
11. Combine organic layers, dry over $MgSO_4$, and filter.
12. Remove the solvent by rotary evaporation.
13. Dissolve the product in 10 mL acetone.

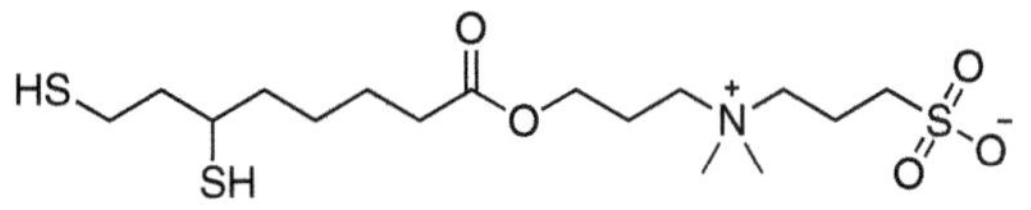

Fig. 1 The chemical structure of dihydrolipoic acid-sulfobetaine (DHLA-SBE)

14. Dissolve 4.85 mmol 1,3-propanesultone in 2–3 mL acetone and add dropwise at room temperature.
15. Reflux the reaction at 75 °C overnight under N_2. The final product will precipitate out of solution.
16. Filter the precipitate and wash with $CHCl_3$. Dissolve the product in water and extract the aqueous layer with $CHCl_3$. The desired product will be in the aqueous layer.
17. Filter through a pasteur pipette with a cotton plug and evaporate under reduced pressure.
18. Characterize the product. ^{1}H NMR (500 MHz, DMSO-d6): δ [ppm]: 3.46–3.50 (t, 2H), 3.38–3.43 (m, 2H), 3.28–3.35 (m, 2H), 2.98–3.06 (s, 6H), 2.77–2.85 (m, 2H), 3.46–3.50 (m, 4H), 2.19–2.23 (t, 1H), 1.87–2.08 (m, 4H), 1.79–1.86 (m, 2H), 1.27–1.63 (m, 6H). ESI MS *m/z*: 414.14451 (M+H)+.
19. Reduce dithiol bond with an equimolar amount of $NaBH_4$ just before use.
20. Store unreduced product as a solid in − 20 °C freezer until needed. The structure shown in Fig. 1 is of the product after reduction of the disulfide bond with $NaBH_4$.

3.2 Protocol for Dithiol-Biotin Synthesis

1. Store 10 mg (12.6 μmol) dithiolalkanearomatic PEG6-NHS ligand at − 20 °C, unopened in a dessicator until required.
2. Dissolve 6 mg (16 μmol) EZ-Link Amine-PEG2-Biotin in 500 μL 10 mM Hepes buffer at pH 8.
3. Add solution *immediately* to freshly opened bottle of dithiolalkanearomatic PEG6-NHS (*see* **Note 1**)
4. Let mixture react for 2 h at room temperature.
5. Store the final solution (dithiol-biotin) at − 20 °C.

3.3 Preparation of Lipoic Acid Functionalized Single-Stranded DNA

1. Prepare 10 mg/mL solution of lipoic acid in DMSO.
2. Equilibrate EDC and sulfo-NHS to room temperature.
3. Add 360 μL of 10 mg/mL lipoic acid (0.0174 mmol) solution to 640 μL 0.1 M MES at pH 6.

4. Add aqueous lipoic acid solution to an Eppendorf tube containing 14 mg (0.073 mmol) EDC and 22 mg (0.1 mmol) sulfo-NHS.
5. Let mixture react 5 min at room temperature to activate carboxylic acid.
6. Prepare 500 μL 0.5 mM aminated DNA in water and add to activated lipoic acid mixture.
7. Rotate on a mini-tube rotator at room temperature for 2 h.
8. Ethanol precipitate DNA using 1 mL 0.3 M sodium acetate in ethanol.
9. Incubate at 4 °C for 1 h.
10. Spin at 15,870 × *g* for 1 h at 4 °C in a microcentrifuge.
11. Carefully discard the supernatant.
12. Add 500 μL 70 % ethanol, cooled to 4 °C.
13. Centrifuge the solution at 15,870 × *g* for 20 min at 4 °C.
14. Carefully discard the supernatant.
15. Dry the DNA pellet with a Speed-Vac.
16. Dissolve the pellet in 50 μL 10 mM Hepes buffer at pH 8.
17. Quantitate the DNA by absorbance at 260 nm.
18. Dilute complementary ssDNA, 5′-modified with a biotin, in 10 mM HEPES buffer at pH 8.
19. Add an equimolar amount of complementary ssDNA to lipoic acid-functionalized single-stranded DNA.
20. Add sodium chloride to a concentration of 100 mM.
21. Heat DNA to 95 °C for 2 min.
22. Cool DNA on bench top at room temperature.
23. Store the lipoic acid-DNA at −20 °C until needed.

3.4 Large Gold Nanoparticle Passivation

1. Thaw DHLA-SBE from Subheading 3.1, dithiol-biotin from Subheading 3.2, and lipoic acid-DNA from Subheading 3.3 to room temperature.
2. Add 10 μL of 50 mM DHLA-SBE to either 1 μL of 10 mM dithiol-biotin or 10 μL of 500 μM lipoic acid-DNA (*see* **Notes 2** and **3**).
3. Add an equimolar amount of $NaBH_4$ to the mixture and incubate 5 min.
4. Add reduced thiols to 500 μL of 80-nm gold nanoparticle (10^{10} particles/mL).
5. Rotate the mixture on mini-tube rotator overnight at room temperature.
6. Save an aliquot of mixture for characterization by gel electrophoresis and UV–Vis absorption spectroscopy.

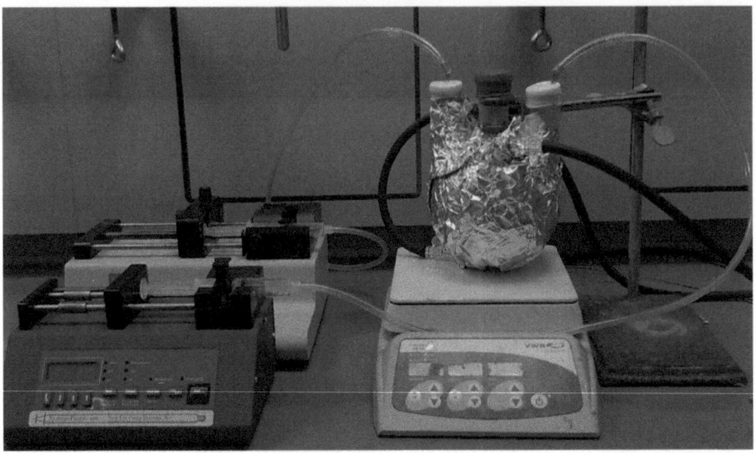

Fig. 2 Setup of the apparatus used for silver nanoparticle synthesis. Two syringe pumps are connected to a 500-mL thermally jacketed three-neck round bottom flask. The flask is placed on top of a stir plate. The temperature of the water flowing through the water-jacket is 75 °C

7. Dialyze gold nanoparticles in a 300-kDa membrane for 2 h at room temperature in 4 L water (*see* **Note 4**).
8. Exchange the 4 L water every 2 h for a total of three exchanges.

3.5 Silver Nanoparticle Synthesis and Passivation

1. Prepare 20 mg/mL silver nitrate stock solution fresh for each synthesis.
2. Prepare 70 mg/mL trisodium citrate, 2 mg/mL ascorbic acid, and 0.1 N NaOH stock solutions.
3. Connect a 500 mL three-neck jacketed round bottom flask to a circulating water bath set to 75 °C.
4. Wrap flask with aluminum foil. An example synthesis apparatus is depicted in Fig. 2.
5. Add 100 mL of ultra-pure 18.2 MΩ-cm water along with a stir bar.
6. Center the flask on a stir plate set to 750 rpm.
7. Cap the open necks of the flask with rubber stoppers.
8. Load two syringe pumps with NaOH and ascorbic acid stock solutions. Set syringe pumps to 250 μL/min for 2 min of NaOH addition, and 2 mL/min for 5 min of ascorbic acid addition.
9. Once the water reaches 75 °C, equilibrate for 5 min before adding reactants.
10. Quickly add 1 mL of silver nitrate. Wait 5–10 s, then quickly add 1 mL of trisodium citrate stock solution by syringe.

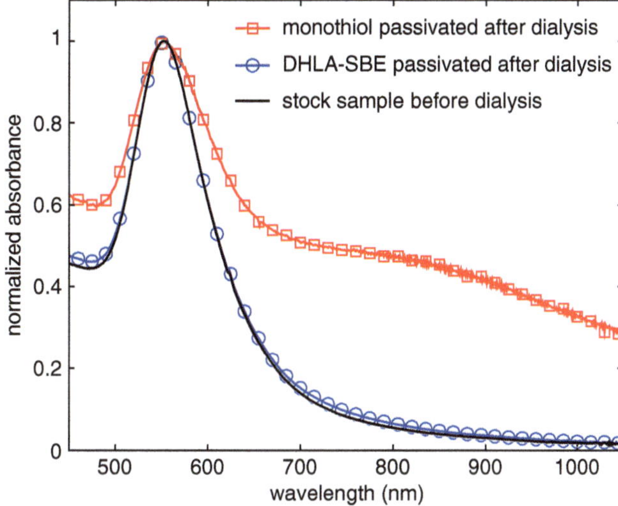

Fig. 3 The absorption spectrum of 80-nm gold nanoparticles passivated with a monothiol ligand (*red squares*) taken after dialysis shows increased near-infrared absorption with respect to the unpassivated particles before dialysis (*black curve*). The near-infrared absorption band is indicative of aggregation. The spectrum taken after dialysis of particles passivated with DHLA-SBE (*blue circles*) do not show the same increase. All spectra are normalized to the absorption maximum at the plasmon resonance wavelength

11. Add 500 μL of NaOH at 250 μL/min for 2 min and 10 mL of ascorbic acid at a rate of 2 mL/min for a total of 5 min.
12. Remove any injection needles and let react at the set temperature for 30 min.
13. Store product at 4 °C.
14. Spin 10 mL of 0.5 nM silver nanoparticles at 13,800 ×*g* for 30 min.
15. Remove the supernatant and re-suspend the pallet in methanol.
16. Add DHLA-SBE to a final concentration of 500 mM.
17. Rotate on a mini-tube rotator for 24 h at room temperature.

3.6 Absorption Spectra of Large Gold Nanoparticles

1. Record UV–Vis absorption spectrum for each particle sample before and after dialysis. If the particles are stable and monodisperse, they should still retain their characteristic plasmon band. If the particles are incompletely passivated, then the absorption spectra may display a large shift of the plasmon resonance and characteristic increase in near infra-red absorption [48]. A representative figure comparing particle aggregation after dialysis is shown in Fig. 3.

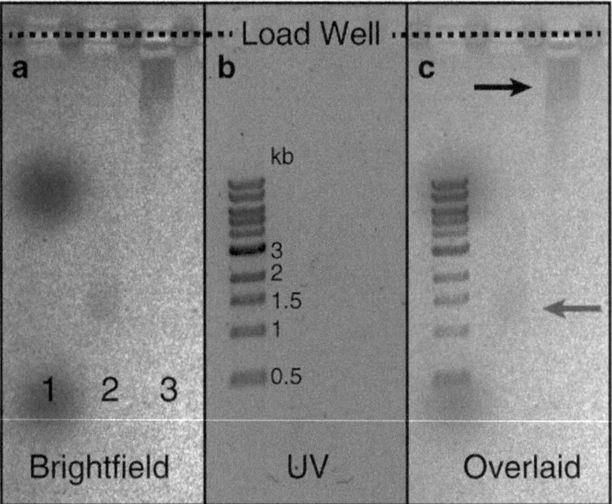

Fig. 4 The brightfield illuminated (**a**), UV illuminated (**b**), and overlaid images (**c**) of a 0.5 % 0.5 × TBE agarose ethidium bromide stained gel. Lane 1 is New England Biolab's Tridye 1 kb DNA ladder while lanes 2 and 3 are the passivated and unpassivated particles, respectively. The passivated particles (*red arrow*) move quickly into the gel while the unpassivated particles (*black arrow*) remain near the load well after 35 min at 6.5 V/cm

3.7 Gel Electrophoresis of Large Gold Nanoparticles

1. Cast a 0.5 % 0.5 ×TBE agarose gel. Use a comb with smallest teeth available. The teeth of the comb used in this protocol are 3 mm × 1.5 mm × 10 mm (*see* **Note** 5).
2. Mix 15 μL gold nanoparticle sample with 5 μL 15 % Ficoll solution as the loading buffer.
3. Load samples into the gel with 0.5 ×TBE as the running buffer. Reserve a lane in the gel for a DNA marker such as New England Biolab's Tridye 1 kb DNA marker. The Tridye marker contains three organic dyes which are a good visualization tool to monitor the progress of the gel (*see* **Note** 6).
4. Run samples at a field strength of 6.5 V/cm (*see* **Note** 7).
5. Record a brightfield image of the particle on a lightbox. The large gold nanoparticles, if properly passivated, should run as a tight pink band. A representative gel is seen in Fig. 4.

3.8 Column Test for Large Gold Nanoparticles

1. Prepare 10 mL of a 50 % slurry of Sephacryl S-300 size exclusion resin with 18.2 MΩ-cm water (*see* **Note** 8).
2. Pack a glass drip column (1.5 cm inner diameter) with 5 mL of slurry (*see* **Notes** 9 and 10).
3. Equilibrate the column with two column volumes of water.
4. Let the water drip through column until the meniscus of the mobile phase reaches the top of the packed resin bed.

Fig. 5 Incompletely passivated large gold nanoparticles will become immobilized at the top of various liquid chromatography resins. Here, unpassivated 80 nm diameter gold nanoparticles have become immobile on a Sephacryl S-300 column

5. Once the water has reached the top of the column, carefully load 500 μL of passivated particle sample from Subheading 3.4 to the top of the column.
6. Let gold sample move into the column.
7. After the gold has entered the top of the resin bed, carefully wash the column with enough water to wash the particles through the column. Incompletely passivated gold nanoparticles will remain at the top of the column, such as in Fig. 5.
8. Repeat column test with other anion exchange, cation exchange, DEAE, and hydroxyapatite resins from Bio-Rad's chromatography media sampler pack to determine scope of usable resins with the particle system. The passivated particles should not aggregate at the tops of the columns.

4 Notes

1. The NHS ligand is very reactive and will hydrolyze quickly if not used immediately. Store at −20 °C and use *immediately* after opening. Do not store unreacted stock solutions.
2. Nanoparticles passivated by lipoic acid-derived ligands appear to be most stable when a mixture of at least two ligands are used. In the presented protocol, DHLA-SBE is mixed with a dithiol biotin ligand. Other ligands, such as lipoic acid modified peptides, can also be used.

3. Optimizing the ratio of the relative ligands can help with particle stability. For large gold nanoparticles, millimolar concentrations of ligands are necessary to ensure adequate passivation.
4. Excess DNA can be difficult to remove by dialysis. Sephacryl S-300 size exclusion column described in Subheading 3.8 is used to separate large gold nanoparticles from unbound DNA. The large gold nanoparticles will be eluted in the excluded volume limit, whereas the DNA will enter the resin.
5. When casting a gel, combs with narrow teeth are useful for visualizing dilute samples. The path length of light passing through the sample increases with the total sample height in a transillumination configuration.
6. DNA can be visualized by ethidium bromide staining. Unlike small gold nanoparticles, large gold nanoparticles show some increased scattering by UV transillumination. Visualize the gel by UV transillumination before staining with ethidium bromide to account for the particle scattering signal.
7. Setting the voltage too high will cause particles to streak in the gel.
8. The pore size of the Sephacryl S-300 resin is such that nanoparticles greater than approximately 60 nm will run in the excluded volume limit. Particles smaller than this limit such as quantum dots and DNA molecules ≤ 90 bp can be resolved by the resin.
9. Use the minimal amount of resin necessary to achieve desired separation. Excessively large columns can dilute the recovered sample and make reconcentrating the sample difficult.
10. Acetone has a strong UV absorbance and can be used to determine the retention limit of a freshly packed column. One can use passivated large gold nanoparticles to determine the excluded volume limit.

Acknowledgements

This work was supported by the Department of Energy and the NIH-NIGMS.

References

1. Kreibig U, Vollmer M (1995) Optical properties of metal clusters. Springer, Berlin
2. Kelly KL, Coronado E, Zhao LL, Schatz GC (2003) The optical properties of metal nanoparticles: the influence of size, shape, and dielectric environment. J Phys Chem B 107:668–677
3. Cobley CM, Skrabalak SE, Campbell DJ, Xia Y (2009) Shape-controlled synthesis of silver nanoparticles for plasmonic and sensing applications. Plasmonics 4:171–179
4. Turkevich J, Stevenson PC, Hillier J (1951) A study of the nucleation and growth processes in

the synthesis of colloidal gold. Discuss Faraday Soc 11:55–75
5. Kinkhabwala A, Yu Z, Fan S, Avlasevich Y, Müllen K, Moerner WE (2009) Large single-molecule fluorescence enhancements produced by a bowtie nanoantenna. Nat Photon 3:654–657
6. Roth J, Bendayan M, Orci L (1978) Ultrastructural localization of intracellular antigens by the use of protein A-gold complex. J Histochem Cytochem 26:1074–1081
7. Link S, El-Sayed MA (2000) Shape and size dependence of radiative, non-radiative and photothermal properties of gold nanocrystals. Int Rev Phys Chem 19:409–453
8. Huang X, El-Sayed IH, Qian W, El-Sayed MA (2006) Cancer cell imaging and photothermal therapy in the near-infrared region by using gold nanorods. J Am Chem Soc 128: 2115–2120
9. Paciotti GF, Myer L, Weinreich D, Goia D, Pavel N, McLaughlin RE, Tamarkin L (2004) Colloidal gold: a novel nanoparticle vector for tumor directed drug delivery. Drug Deliv 11:169–183
10. Nam J-M, Thaxton CS, Mirkin CA (2003) Nanoparticle-based bio-bar codes for the ultrasensitive detection of proteins. Science 301:1884–1886
11. Nam J-M, Stoeva SI, Mirkin CA (2004) Bio-bar-code-based DNA detection with PCR-like sensitivity. J Am Chem Soc 126:5932–5933
12. Sperling R, Gil P, Zhang F, Zanella M, Parak WJ (2008) Biological applications of gold nanoparticles. Chem Soc Rev 37:1896–1908
13. Guo S, Dong S (2009) Biomolecule–nanoparticle hybrids for electrochemical biosensors. Trends Anal Chem 28:96–109
14. Hill HD, Mirkin CA (2006) The bio-barcode assay for the detection of protein and nucleic acid targets using DTT-induced ligand exchange. Nat Prot 1:324–336
15. Katz E, Willner I (2004) Integrated nanoparticle-biomolecule hybrid systems: synthesis, properties, and applications. Angew Chem Int Ed 43:6042–6108
16. Khlebtsov NG, Dykman LA (2010) Optical properties and biomedical applications of plasmonic nanoparticles. J Quant Spectrosc Radiat Transfer 111:1–35
17. Henglein A (1993) Physicochemical properties of small metal particles in solution: "microelectrode" reactions, chemisorption, composite metal particles, and the atom-to-metal transition. J Phys Chem 97:5457–5471
18. Ivanova OS, Zamborini FP (2009) Size-dependent electrochemical oxidation of silver nanoparticles. J Am Chem Soc 132:70–72
19. Wang D, Nap RJ, Lagzi I, Kowalczyk B, Han S, Grzybowski BA, Szleifer I (2011) How and why nanoparticle's curvature regulates the apparent pKa of the coating ligands. J Am Chem Soc 113:2192–2197
20. Hill HD, Millstone JE, Banholzer MJ, Mirkin CA (2009) The role radius of curvature plays in thiolated oligonucleotide loading on gold nanoparticles. ACS Nano 2:418–424
21. Wang Z, Tan B, Hussain I, Schaeffer N, Wyatt MF, Brust M, Cooper AI (2007) Design and polymeric stabilizers for size-controlled synthesis of monodisperse gold nanoparticles in water. Langmuir 23:885–895
22. Hurst SJ, Lytton-Jean AKR, Mirkin CA (2006) Maximizing DNA loading on a range of gold nanoparticle sizes. Anal Chem 78:8313–8318
23. Demers LM, Mirkin CA, Mucic RC, Reynolds RA III, Letsinger RL, Elghanian R, Viswanadham G (2000) A fluorescence-based method for determining the surface coverage and hybridization efficiency of thiol-capped oligonucleotides bound to thin films and nanoparticles. Anal Chem 72:5535–5541
24. Prasad BLV, Stoeva SI, Sorensen CM, Klabunde KJ (2002) Digestive ripening of thiolated gold nanoparticles: the effect of alkyl chain length. Langmuir 18:7515–7520
25. Cosgrove T (ed) (2005) Colloid science principles, methods and applications. Blackwell, Oxford
26. Pakiari AH, Jamshidi Z (2010) Nature and strength of M–S bonds (M = Au, Ag, and Cu) in binary alloy gold clusters. J Phys Chem A 114:9212–9221
27. Li Z, Jin R, Mirkin CA, Letsinger RL (2002) Multiple thiol-anchor capped DNA-gold nanoparticle conjugates. Nucl Acid Res 30:1558–1562
28. Mei BC, Oh E, Susumu K, Farrell D, Mountziaris TJ, Mattoussi H (2009) Effectors of ligand coordination number and surface curvature on the stability of gold nanoparticles in aqueous solutions. Langmuir 25: 10604–10611
29. Muro E, Pons T, Lequeux N, Fragola A, Sanson N, Lenkei Z, Dubertret B (2010) Small and stable sulfobetaine zwitterionic quantum dots for functional live cell imaging. J Am Chem Soc 132:4556–4557
30. Susumu K, Oh E, Delehanty JB, Blanco-Canosa JB, Johnson BJ, Jain V, Hervey WJ, Algar WR, Boeneman K, Dawson PE, Medintz IL (2011) Multifunctional compact zwitterionic ligands for preparing robust biocompatible semiconductor quantum dots and gold nanoparticles. J Am Chem Soc 133:9480–9496
31. Park J, Nam J, Won N, Jin H, Jung S, Cho S-H, Kim S (2011) Compact and stable quantum dots with positive, negative, or zwitterionic surface: specific cell interactions and non-specific adsorptions by the surface charges. Adv Funct Mater 21:1558–1566

32. Estephan ZG, Jaber JA, Schlenoff JB (2010) Zwitterion-stabilized silica nanoparticles: toward nonstick nano. Langmuir 26:16884–16889
33. Rouhana LL, Jaber JA, Schlenoff JB (2007) Aggregation-resistant water-soluble gold nanoparticles. Langmuir 23:12799–12801
34. Weber PC, Ohlendorf DH, Wendoloski JJ, Salemme FR (1989) Structural origins of high-affinity biotin binding to streptavidin. Science 243:85–88
35. Kawde A-N, Wang J (2003) Amplified electrical transduction of DNA hybridization based on polymeric beads loaded with multiple gold nanoparticle tags. Electroanalysis 16:101–107
36. Kerman K, Chikae M, Yamamura S, Tamiya E (2007) Gold nanoparticle-based electrochemical detection of protein phosphorylation. Anal Chim Acta 588:26–33
37. Zhu D, Tang Y, Xing D, Chem WR (2008) PCR-free quantitative detection of genetically modified organism from raw materials. An electrochemiluminescence-based bio bar code method. Anal Chem 80:3566–3571
38. Prime KL, Whitesides GM (1991) Self-assembled organic monolayers: model systems for studying adsorption of proteins at surfaces. Science 252:1164–1167
39. Marx KA (2003) Quartz crystal microbalance: a useful tool for studying thin polymer films and complex biomolecular systems at the solution–surface interface. Biomacromolecules 4:1099
40. Kim E-Y, Stanton J, Vega RA, Kunstman KJ, Mirkin CA, Wolinsky SM (2006) A real-time PCR-based method for determining the surface coverage of thiol-capped oligonucleotides bound onto gold nanoparticles. Anal Chem 34:e54
41. Park S, Hamad-Schifferli K (2010) Nanoscale interfaces to biology. Curr Opin Chem Biol 14:616–622
42. Zanchet D, Micheel CM, Parak WJ, Gerion D, Williams SC, Alivisatos AP (2002) Electrophoretic and structural studies of DNA-directed Au nanoparticle groups. J Phys Chem B 106:11758–11763
43. Surugau N, Urban PL (2009) Electrophoretic methods for separation of nanoparticles. J Sep Sci 32:1889–1906
44. Pillai ZS, Kamat PV (2004) What factors control the size and shape of silver nanoparticles in the citrate ion reduction method. J Phys Chem B 108:945–951
45. Qin Y, Ji X, Jing J, Liu H, Wu H, Yang W (2010) Size control over spherical silver nanoparticles by ascorbic acid reduction. Colloids Surf A Physicochem Eng Asp 372:172–176
46. Lee C, Meisel D (1982) Adsorption and surface-enhanced raman of dyes on silver and gold sols. J Phys Chem 86:3391–3395
47. Evanoff DD Jr, Chumanov G (2004) Size-controlled synthesis of nanoparticles. 1.'Silver-only' aqueous suspensions via hydrogen reduction. J Phys Chem B 108:13948–13956
48. Kreibig U, Genzel L (1985) Optical absorption of small metallic particles. Surf Sci 156:678–700

Chapter 19

Noncovalent Intracellular Drug Delivery of Hydrophobic Drugs on Au NPs

Tennyson Doane and Clemens Burda

Abstract

The successful delivery of hydrophobic drugs to cellular targets continues to present challenges to the pharmaceutical industry. The advances made by nanotechnology have generated new avenues for selectively loading, delivering, and targeting these drugs to their biological targets without compromising efficacy. Here, we describe how gold nanoparticles (Au NPs) functionalized with polyethylene glycol (PEG) can be evaluated for the delivery of hydrophobic drugs in aqueous systems. Specifically, we describe Au NP synthesis, ligand exchange, and delivery evaluation at-the-bench for screening of potential drug candidates.

Key words Gold nanoparticles, Noncovalent drug delivery, Ligand exchange, Hydrophobic

1 Introduction

Many promising drug candidates for the treatment of human diseases are hydrophobic and have limited effectiveness due to poor transport [1]. To overcome this problem, research is driven either to modify the drug to improve solubility or to design an effective transport vector for in vivo applications. Modifications to drug structure, however, can have the unwanted effect of limiting potency and subcellular localization in targeting a given disease. For this reason, the use of transporting vectors has become very popular for the delivery of hydrophobic drugs, such as the polyethylene glycol (PEG) liposomal formulation of doxorubicin known as Doxil® which currently has clinical applications [2].

One promising approach to improving drug delivery is through the use of nanoparticles (NPs) which have the potential for both simultaneous delivery and diagnostic capabilities (so-called theranostics) [3]. Gold nanoparticles (Au NPs), in particular, have been highly researched due to their highly tunable size [4], unique optical properties [4], facile platform based modification [5] and reported low toxicity [6]. Moreover, the ability to rationally design

Paolo Bergese and Kimberly Hamad-Schifferli (eds.), *Nanomaterial Interfaces in Biology: Methods and Protocols*, Methods in Molecular Biology, vol. 1025, DOI 10.1007/978-1-62703-462-3_19, © Springer Science+Business Media New York 2013

NPs to optimize the enhanced permeability and retention (EPR) effect [7] can provide further controls based on a solid material. Hydrophobic drugs have been successfully loaded via noncovalent techniques which improve solubility but still retain favorable release kinetics both in vitro and in vivo [8, 9]

The physical basis for hydrophobic drug loading relies on the use of amphiphilic ligand coatings, which are commonly based on oligomers of polyethylene glycol (PEG). PEG is commonly used in biomedical applications due to favorable protein resistance and "stealth" like properties [10]. Work by Lucarini et al. has investigated the partitioning of hydrophobic molecules in PEG oligomer termiminated alkyl monolayers via electron spin resonance [11, 12], and found that drug portioning is dependent on the packing of the surface ligands. The incorporation of a hydrophobic drug into the amphiphilic coating surrounding the nanoparticle provides a unique opportunity to study how diffusion related drug transport across membranes works in living systems, and due to the reported fluorescence quenching effect due to energy transfer, can provide important information pertaining to release behavior when a fluorescent compound is utilized.

The following protocol describes the synthesis, loading, and evaluation of Au NPs for the delivery of hydrophobic drugs through noncovalent drug loading. In general the two main methodologies for synthesizing Au NPs rely on chemical reduction of Au^{+3} ions with either citrate in water (the Turkevich method [13]) or sodium borohydride in organic solutions (the Brust method [14]). The following protocol is an adaptation of the later, which produces Au NPs with a diameter of ~6 nm. Further modification of the particles is undertaken by utilizing the strongly coordinating thiol modified PEG. This general approach has been investigated in our laboratory for a range of potential drug candidates, and analyzed in detail for the photodynamic therapy drug silicon phthalocyanine 4 [8, 15].

2 Materials

All chemicals reported here are commercially available (see below). Chemical storage must be in keeping with MSDS recommendations. Special caution should be employed for sodium borohydride ($NaBH_4$) which is a strong reducing agent and reacts violently with water. Store in a dry and flame-resistant place. Properly dispose of all waste according to regulations. The following are a list of chemicals needed for each step of preparation.

2.1 Preparations for NP Synthesis

2.1.1 Equipment

1. Small 20 mL vial with cap.
2. Clean small magnetic stir bar and stir plate.
3. 200–1,000 μL mechanical pipette.

4. 2–20 μL mechanical pipette.
5. Three sheets of weighing paper.
6. Small 2-dram vial with cap.
7. Four clean glass pipettes.
8. One rubber bulb.
9. Two plastic 50 mL centrifuge tubes.
10. One glass cuvette.

2.1.2 Solvents

1. ACS grade Toluene. Toluene is both toxic and flammable. Handle and dispose of accordingly.
2. ACS grade Chloroform. Chloroform is both toxic and flammable. Handle and dispose of accordingly.
3. 200 Proof Ethanol. Ethanol is flammable and should be stored accordingly.

2.1.3 Chemicals

1. Tetraoctylammonium bromide (TOAB) will be used as the phase transfer agent. TOAB is hazardous to the environment and should be disposed of accordingly.
2. 30 % Tetrachloroauric acid ($HAuCl_4$) 99 % purity (Sigma Aldrich Chemical Company, St. Louis, MO, USA.) will be used as the Au precursor. Solution should be stored at 4 °C.
3. Dodecylamine (DDA) will be used as the capping ligand. DDA is hazardous to the environment and should be disposed of accordingly. In addition, DDA can be caustic, and should be stored with similar agents.
4. ACS grade Sodium Borohydride ($NaBH_4$) will be used as the reduction agent. $NaBH_4$ reacts violently with water and releases H_2 gas upon reaction which is highly explosive. Observe caution when handling and store in a cool, dry, and flammable resistant place.

2.2 Preparations for Au NP Ligand Exchange

2.2.1 Equipment

1. Small 20 mL vial with cap.
2. Clean medium magnetic stir bar.
3. Stir plate.
4. 200–1,000 μL mechanical pipette.
5. 2–20 μL mechanical pipette.
6. One 2-dram vial with cap.
7. One glass pipette.
8. One glass cuvette (*see* **Note 1**).

2.2.2 Solvents

1. ACS grade chloroform (see above).

2.2.3 Chemicals

1. Thiolated-Methoxypolyethylene glycol 5000 (HS-mPEG 5k) (Laysan Bio, Arab, GA, USA). This compound is air, temperature, and light sensitive (*see* **Note 2**), and should be stored in an inert atmosphere at −18 °C during long term storage. Transfer to the reaction vessel should be done quickly and with minimal air contact.
2. Dry ice is recommended for transporting and storing the HS-mPEG 5k to the reaction vessel in the fume hood.

2.3 Preparations for Au NP Purification

2.3.1 Equipment

1. One 20 mL vial with cap.
2. One 250 mL separatory funnel.
3. Two VivaSpin 30,000 or 50,000 Da cutoff membrane centrifuge tubes (Sartorious AG, Goettingen, GER) (*see* **Note 1**).

2.3.2 Solvents

1. ACS grade Ethyl Acetate will be used for solvent extraction and purification, respectively. Caution: Ethyl Acetate has a very low flash point and should be handled with extreme caution. Proper storage and disposal is required according to regulations.
2. 18 MΩ deionized water will be used for solvent extraction.

2.4 Preparations for Drug Delivery Evaluation

2.4.1 Equipment

1. 2-Dram vial with cap.
2. One clean medium stir bar.
3. One clean flea-bar stir bar.
4. Magnetic stir-plate
5. One glass *or* quartz cuvette (*see* **Notes 1** and **3**).

2.4.2 Solvents

1. ACS grade Chloroform (see above) will be used during solvent extraction and drug loading.
2. 18 MΩ deionized water will be used during solvent extraction and drug loading.
3. ACS grade Toluene (*see* above) will be used to monitor drug release.

2.4.3 Chemicals

1. Sodium chloride (~0.5 g) will be used for solvent extraction (aqueous to organic layers).
2. Drug compound (*see* **Note 4**).

3 Methods

3.1 Au NP Synthesis

1. Combine 5 mL of toluene with 0.25 mmol (0.137 g) of tetraoctyl ammonium bromide (TOAB) and a stir bar in a 20 mL vial. Stir this solution slowly. Simultaneously, place 1 mL of water in an ice bath for later use.

2. Add 0.53 mmol $HAuCl_4$ (367 μL of 30 % stock solution) to the vial. After addition, the solution color should immediately turn red.

3. Add 0.6 mmol (0.111 g) of dodecylamine (DDA) to the mixture and increase to mild stirring. The solution should now look turbid and a pale orange color.

4. Measure out 2 mmol (0.076 g) of sodium borohydride ($NaBH_4$) and add to the ice water prepared in **step 1**. Shake (or vortex) the solution with frequent venting to prepare $NaBH_4$ solution. Caution: Addition of $NaBH_4$ will cause the evolution of H_2 gas and is highly exothermic which can cause an explosion. This can be extremely dangerous if the water is not properly chilled before addition.

5. Increase the stirring of the solution prepared in **step 3** and slowly add the reducing solution prepared in **step 4** to the reaction mixture at a rate of 3–4 drops per minute with a glass pipette. Instant color change at the entry of the drop should be visible to the eye. Stirring speed is intended to generate a highly uniform solution; if not mixed properly, the reduction of Au will take place in only one location and the resulting particles will aggregate.

6. As the amount of bubbling decreases with increasing $NaBH_4$ addition, the rate at which it is added can be increased. Caution: Allow the bubbling to subside before further addition to ensure that none of the reaction mixture is overflows from the container.

7. Continue stirring the solution for 20 min. Further stirring time can increase the size of the particles.

8. Discontinue stirring and add the contents of the vial to a 40 mL centrifuge vial (plastic is acceptable). Wash out the reaction vial with a 10 mL aliquot of ethanol two times. Add these washes to the centrifuge vial and bring up the total volume to ~45 mL (*see* **Note 5**).

9. Prepare counter vial (i.e., 45 mL of water) and centrifuge the mixtures for 10 min at 6,000 rpm (6,000 × *g*). Particles should be compressed into solid form at the bottom and sides of the centrifuge tube. Decant the ethanol from the centrifuge tube containing the Au NPs and add an additional 45 mL of ethanol to the centrifuge tube. Repeat purification with centrifugation for 10 min at 6,000 rpm (6,000 × *g*).

10. Decant final ethanol wash and dry the particles via an Ar stream. Redissolve the particles in 5 mL of chloroform, and obtain a UV–Vis spectra of the prepared particles by adding 10 μL of prepared solution to 2 mL of chloroform in a cuvette (dilution factor = 201). Obtain absorbance and determine

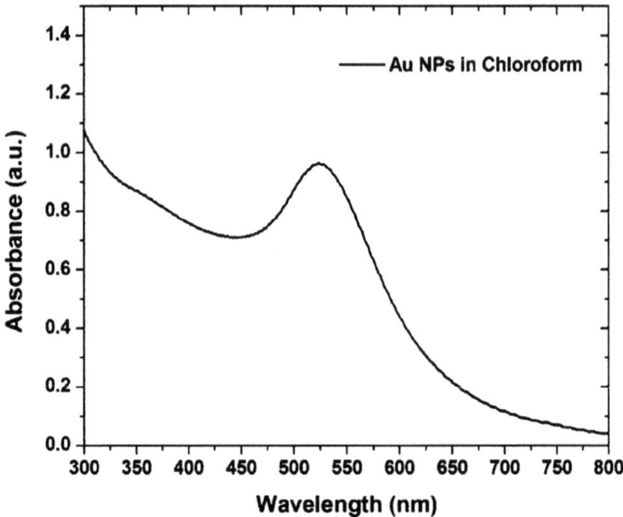

Fig. 1 Representative UV–Vis spectra of PEGylated Au Nps in chloroform

concentration of dilute sample using the extinction coefficient at 520 nm = $1.5 \times 10^7 cm^{-1}M^{-1}$ [8]. Typical stock concentrations are 13 μM.

3.2 Ligand Exchange for HS-mPEG

1. Add 2.5 mL of the synthesized DDA coated Au NPs in chloroform to a 10 or 20 mL vial. Dilute this volume with 2.5 mL of chloroform to bring the total volume up to 5 mL. The resulting concentration should now be half of the stock solution.
2. Using a HS-mPEG: Au NP ratio of 1,000:1, calculate the mass of PEG required for ligand exchange. Obtain an insulated container and fill with dry ice for the transport of PEG from the glove box to the reaction vessel.
3. Weigh out the required amount of HS-mPEG from the glove box and seal tightly in a gas container. Place this container in the dry ice and transport to the reaction vial. Add a stir bar to the reaction vial and place on a stir plate with gentle stirring.
4. Quickly add the PEG (oily flakes for > 1kDa PEG) to the reaction vial, using ~1 mL of chloroform to rinse out the vial and transfer to the Au NP solution.
5. Allow the mixture to stir at room temperature for 2 days in the capped vial. Secure the vial to prevent spilling.
6. After 48 h, take a 101:1 dilution of the PEGylated Au NPs and measure the UV–Vis spectra to ensure no changes in Au NP absorption (Fig. 1).

3.3 Purification of PEGylated Au NPs

1. Transfer the PEGylated Au NP solution to a 250 mL separatory funnel. Rinse the reaction vial with chloroform 2–3 times to ensure complete transfer. Add 5 mL of water to the separa-

tory funnel. A bilayer should be clearly visible with little transfer for Au NPs from the organic phase to the aqueous phase (*see* **Note 6**).

2. Add ~50–60 mL of ethyl acetate to the separatory funnel. Shake the funnel (with frequent venting) to ensure mixture of the ethyl acetate with the lower chloroform layer. Allow the layers to separate: the organic layer is now above the aqueous layer ($d_{\text{Ethyl Acetate}} = 0.8$ g/mL, $d_{\text{Water}} = 1$ g/mL) (*see* **Note 7**).
3. PEGylated Au NPs should now be in the lower aqueous phase. Extract lower aqueous layer and add fresh 5 mL water to the separatory funnel. Repeat 2–3 times until >90 % of particles have been removed.
4. Transfer 5 mL of particles to a cutoff membrane centrifuge tube (>30 kDa MW exclusion). Centrifuge at 6,000 rpm (6,000 × *g*) for 10 min. Once completed, drain off water from the lower part of the tube and add another 5 mL of solvent extracted Au NPs. Repeat as necessary until all of the particles are in the centrifuge tube.
5. Add fresh water to the top of the centrifuge tube (~5 mL) and centrifuge for 10 min at 7,000 rpm (8,200 × *g*). When completed, drain off the lower water and add another 5 mL of water. Repeat ten times to ensure that all of the loose PEG is removed from the Au NPs.
6. Add 3 mL of water and extract the PEGylated Au NPs in the top of the centrifuge tube. Further additions of 1 mL of water can be used to ensure as many particles are collected as possible. Store in a small dram vial.

4 Hydrophobic Drug Loading on PEGylated Au NPs and Confirming Delivery

1. Transfer the aqueous PEGylated Au NPs (~5–7 mL) to a 40 mL vial. Add 10 mL of chloroform to the vial.
2. Weigh out 0.25 g of sodium chloride and add to the 40 mL vial. Shake vigorously and allow the layers to separate. If a bilayer is not directly apparent after 10 min, add 10 mL water and 0.25 g of sodium chloride and shake again (*see* **Note 8**).
3. The lower organic layer should now be red in color and can be extracted using a Pasteur pipette. Add the extract to a 20 mL vial, taking care not to get water into the reaction mixture.
4. Place a stir bar in the reaction vessel and begin a gentle stirring. Add the hydrophobic drug at a 30:1 M ratio to the reaction mixture. Keep stirring the solution for ~2 h to promote incorporation (*see* **Note 9**).
5. After the allotted time, blow a compressed gas (argon, nitrogen, or air will suffice) over the mixture to evaporate off the

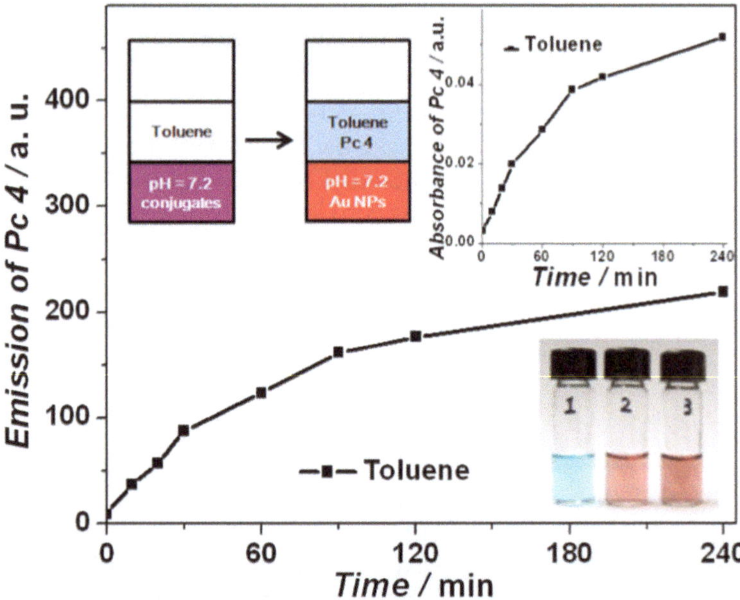

Fig. 2 Experimental data illustrating the release of a hydrophobic drug (Pc 4) from PEGylated Au NPs with increasing time in a water–toluene biphase experiment (*inset center*). The drug was monitored both by fluorescence and by changes in an absorbance (*upper right*) panel during the experiment, and as PEG is insoluble in toluene, the hydrophobic drug partitioned into the organic phase (*lower inset*, 1), while the Au NP remained in the aqueous phase (2). Reproduced with permission from ref. 8. Copyright 2008 American Chemical Society

chloroform, leaving the PEGylated Au NP-drug conjugates as a solid. When the mixture is completely dry, add 5 mL of water and vortex to solvate the particles.

6. Take a UV–Vis spectra of the Au NP-drug conjugates to determine the amount of drug which has now been incorporated on the Au NPs.
7. Add 1 mL of the Au NP-drug conjugates to a cuvette. When completed, add 1 mL of toluene to the cuvette and allow the solution to mix gently under magnetic stirring.
8. Monitor the absorbance of the upper organic layer at different time points (10, 20, 30, 60, 90, 120, and 240 min for example (Fig. 2)) to determine drug release kinetics (*see* **Note 10**).

5 Notes

1. The use of new glassware is to prevent contamination of the solution with other trace metals which may interfere with synthesis. Vials should be used in accordance with lab requirements

(i.e., smaller Eppendorf® centrifuge tubes could be used rather than two dram vials). Cleaning of stir bars and reusable glassware can be undertaken using small amounts of Aqua-Regia (four parts HCl: one part HNO_3); however, caution is advised as this mixture gives off dangerous fumes and could potentially cause chemical burns if accidently splashed. Work should be conducted in a fume hood with proper personal safety equipment (acid resistant gloves, goggles with side shields, and lab coat are minimal requirements).

2. These precautions are intended to minimize the risk of dimerization of HS-mPEG to mPEG-SS-mPEG which may change its reactivity. The molecular weight of the PEG can be varied from 550 to 10,000 Da successfully.
3. The choice of cuvette material is dependent on the photophysical properties of the drug compound of interest. If interested in a drug with primary absorption bands in the UV region, a quartz cuvette is required.
4. Compounds of interest should be both (1) hydrophobic and (2) be soluble in any chloroform and one other immiscible organic solvent. In addition, identification of major absorption and fluorescence bands will aid in determining delivery efficiency.
5. Simply adding ethanol to the solution of organic capped Au NPs can also cause precipitation [16], and if allowed to proceed overnight, is much gentler than centrifugation. This process is then repeated to ensure cleaning of excess TOAB and DDA. For the sake of time, however, we have adopted the centrifuge technique.
6. The use of solvent extraction is intended to remove the excess DDA which originally capped the Au NP. While the solubility of DDA in water is minimal, the presence of PEGylated particles may encapsulate excess toxic ligands.
7. Other solvents including toluene and hexanes were initially studied but were found to be inadequate in forcing PEGylated Au NP partitioning into the water phase. The use of ethyl acetate was chosen due to the very poor solubility of PEG with small ether substituent [17]. A quick roto-evaporation of the solution resulted in some white powder, indicating at least some effectiveness of removing leftover organic constituents.
8. The addition salt will cause the PEG coated Au NPs to preferentially partition into the lower chloroform layer and is used during manufacturing for PEG purification [18]. Pervious methods utilized vacuum based drying methods, but these were given to splattering and loss of material. When there is no chloroform available for partitioning, however, preliminary tests demonstrated particle stability.

9. Specific loading times are expected to vary based on the identity of the target compound.
10. As with loading, the release kinetics are expected to be drug property specific. Note: drugs which contain a free thiol group have been shown to bind to the particle surface and have minimal drug release efficiency [15].

References

1. Ferris DP, Lu J, Gothard C, Yanes R, Thomas CR, Olsen J-C, Stoddart JF, Tamanoi F, Zink JI (2011) Synthesis of biomolecule-modified mesoporous silica nanoparticles for targeted hydrophobic drug delivery to cancer cells. Small 7:1816–1826
2. Davis ME, Chen Z, Shin DM (2008) Nanoparticle therapeutics: an emerging treatment modality for cancer. Nat Rev Drug Discov 7:771–782
3. Kelkar SS, Reineke TM (2011) Theranostics: combining imaging and therapy. Bioconjugate Chem 22:1879–1903
4. Burda C, Chen X, Narayanan R, El-Sayed MA (2005) Chemistry and properties of nanocrystals of different shapes. Chem Rev 105:1025–1102
5. Boal AK, Rotello VM (2000) Fabrication and self-optimization of multivalent receptors on nanoparticle scaffolds. J Am Chem Soc 122: 734–735
6. Connor EE, Mwamuka J, Gole A, Murphy CJ, Wyatt MD (2005) Gold nanoparticles are taken up by human cells but do not cause acute cytotoxicity. Small 1:325–327
7. Maeda H, Wu J, Sawa T, Matsumura Y, Hori K (2000) Tumor vascular permeability and the EPR effect in macromolecular therapeutics: a review. J Control Release 65:271–284
8. Cheng Y, Samia AC, Meyers JD, Panagopoulos I, Fei B, Burda C (2008) Highly efficient drug delivery with gold nanoparticle vectors for in vivo photodynamic therapy of cancer. J Am Chem Soc 130:10643–10647
9. Kim CK, Ghosh P, Pagliuca C, Zhu Z-J, Menichetti S, Rotello VM (2009) Entrapment of hydrophobic drugs in nanoparticle monolayers with efficient release into cancer cells. J Am Chem Soc 131:1360–1361
10. Gref R, Luck M, Quellec P, Marchand M, Dellacherie E, Harnisch S, Blunk T, Muller RH (2000) 'Stealth' corona-core nanoparticles surface modified by polyethylene glycol (PEG): influences of the corona (PEG chain length and surface density) and of the core composition on phagocytic uptake and plasma protein adsorption. Colloids Surf B 18: 301–313
11. Lucarini M, Franchi P, Pedulli GF, Pengo P, Scrimin P, Pasquato L (2004) EPR study of dialkyl nitroxides as probes to investigate the exchange of solutes between the ligand shell of monolayers of protected gold nanoparticles and aqueous solutions. J Am Chem Soc 126: 9326–9329
12. Lucarini M, Franchi P, Pedulli GF, Gentilini C, Polizzi S, Pengo P, Scrimin P, Pasquato L (2005) Effect of core size on the partition of organic solutes in the monolayer of water-soluble nanoparticles: an ESR investigation. J Am Chem Soc 127:16384–16385
13. Turkevich J, Steveson PC, Hillier J (1951) A study of the nucleation and growth processes in the synthesis of colloidal gold. Discuss Faraday Soc 11:55–75
14. Brust M, Walker M, Bethell D, Schiffrin DJ, Whyman R (1994) Synthesis of thiol-derivatised gold nanoparticles in a two-phase liquid–liquid system. J Chem Soc Chem 7: 801–802
15. Cheng Y, Samia AC, Li J, Kenney ME, Resnick A, Burda C (2010) Delivery and efficacy of a cancer drug as a function of the bond to the gold nanoparticle surface. Langmuir 26:2248–2255
16. Prasad BLV, Stoeva SI, Sorensen CM, Klabundae KJ (2002) Digestive ripening of thiolated gold nanoparticles: the effect of alkyl chain length. Langmuir 18:7515–7520
17. Janda KP, Han H (1996) Combinatorial chemistry: a liquid phase approah. In: Abelson JN (ed) Methods in enzymology, vol 267: combinatorial chemistry. Academic Press, San Diego, CA, p 236 (seen on Googlebooks)
18. Personal Correspondence with Laysan Bio Technical Support (2011)

Chapter 20

Modification of Carbon Nanotubes for Gene Delivery Vectors

Victor Ramos-Perez, Anna Cifuentes, Núria Coronas, Ana de Pablo, and Salvador Borrós

Abstract

The surface modification of carbon nanotubes (CNTs) can be tailored to allow the formation of a complex between these potential carriers with DNA. In this chapter, protocols developed in our lab to prepare transfection vectors through the modification of MWCNTs are described. The protocol includes sections focused on the reduction of CNTs length, protocols to increase the dispersability of CNTs and finally, protocols for surface modification to attach through electrostatic interactions DNA to the CNTs.

Key words Carbon nanotubes, Plasma polymerization, DNA delivery

1 Introduction

Despite promising advantages, the use of carbon nanotubes (CNTs) for biomedical applications is limited due to their low biocompatibility. Naïve CNTs present highly hydrophobic surfaces, which result in poor water solubility and dispersability, making them unsuitable for biological applications. To improve the biocompatibility of CNTs, surface modification or functionalization is required to afford solubility in aqueous solutions [1, 2]. Modification of CNTs may be performed by covalent and non-covalent methods/techniques [3]. Covalent functionalization exploits the chemical reactivity of CNTs to form bonds between the CNT surface and a hydrophilic moiety. Covalent bonding of modifying moieties having hydrophilic groups in their structure results in increased water solubility of modified CNTs. In contrast, non-covalent modification techniques are based on hydrophobic interactions between the CNT surface and the hydrophobic region of an amphiphilic moiety. CNT wrapping with multiple amphiphilic moieties having surfactant-like activity results in solubilization of CNTs in aqueous media [4, 5] amphiphilic. Both techniques provide efficient CNT

Paolo Bergese and Kimberly Hamad-Schifferli (eds.), *Nanomaterial Interfaces in Biology: Methods and Protocols*, Methods in Molecular Biology, vol. 1025, DOI 10.1007/978-1-62703-462-3_20, © Springer Science+Business Media New York 2013

modification strategies for biological applications, resulting in increased biocompatibility and reduced toxicity.

In this sense, it should be noted that pristine CNTs lack appropriate functional groups to enable covalent binding of biocompatible groups, in order to make CNTs compatible with the biological milieu. Therefore, the surface of CNTs requires modification by different chemical synthetic methods to generate appropriate functional groups, enabling covalent attachment of suitable solubilizing, bioactive, or biocompatible moieties. Classically, researchers have differentiated CNTs into two zones in terms of their reactivity, i.e. sidewalls and tips. It has been shown that tips are more prone for modification than sidewalls.

It is clear that by adjusting the surface modification of CNTs it is possible to tailor their capability to bind genetic material and use their ability to cross the cellular membrane to transfect the genetic material to the cell [6, 7].

In this chapter, we describe the protocols developed in our lab to prepare transfection vectors through the modification of MWCNTs. The protocol includes sections describing how to reduce CNT length, increase the dispersability of CNTs, and modify the CNT surface modification to attach DNA by electrostatic interactions.

2 Materials

Prepare all solutions using ultrapure water (prepared by purifying deionized water to attain a sensitivity of 18 MΩ cm at 25 °C) and analytical grade reagents. Prepare and store all reagents at room temperature (unless indicated otherwise). Diligently follow all waste disposal regulations when disposing waste materials. We do not add sodium azide to the reagents.

2.1 Carbon Nanotubes and Shortening

1. Carbon nanotubes: The MWNT (multi-walled carbon nanotubes) used was obtained from Sigma-Aldrich. They have a length between 5 and 9 mm, and a diameter between 110 and 170 nm.
2. Nanotube shortening solutions: A 0.2 M solution of potassium permanganate (BioUltra with greater purity to 99.0 %, from Fluka). Sodium sulfite (Na_2SO_3) (purity 99 %, Panreac, Barcelona). Concentrated sulfuric acid (98 %, Panreac, Barcelona). Concentrated chlorhydric acid (Panreac Barcelona). 0.2 M NaOH solution.

2.2 Dispersion of Nanotubes

1. Cetyltrimethylammonium bromide CTAB (CTAB BioUltra, with a purity greater than 99 %, obtained from Sigma-Aldrich).
2. The amphiphilic polymer pHPMA, synthesized with a content of 2.5 and 5 % cholesterol (VP25 and VP50, respectively).

The polymers were synthesized in our laboratory (Materials Engineering Group (GEMAT) Institut Químic de Sarrià). For a detailed method to obtain the polymers Subheading 3.2.

3. Dispersability of nanotubes in ethanol: Ethanol purity 99.8 %.

2.3 Transfection Vectors

1. The MWNT were surface coated with allylamine by plasma polymerization: Allylamine (98 % purity).
2. Solution for transfection vector dispersion: Solution of 0.25 M sodium acetate buffer pH 5.
3. Gel agarose: Agarose standard agarose low molecular weight.
4. Ethidium bromide 1 % in water.
5. Green fluorescence protein plasmid (pmaxGFP, Amax).

3 Methods

Carry out all procedures at room temperature unless otherwise specified.

3.1 Shortening of Carbon Nanotubes by Oxidative Treatment with Potassium Permanganate (Adapted from Ref. 8)

1. CNTs are subjected to 590 °C for 3 h in a flask, amorphous carbon is oxidized and removed.
2. Take 25 mg of heat-treated carbon nanotubes and disperse in 30 mL of 0.2 M solution of potassium permanganate (KMnO4).
3. Add a few drops of sodium hydroxide (NaOH) 0.2 M until the pH of the dispersion is 10.
4. Placed the mixture in an ultrasound bath for 20 min and then reflux for 40 min at 100 °C.
5. Add 0.7 g of sodium sulfite (Na_2SO_3) and 7 mL of sulfuric acid (H_2SO4). Leave the mixture agitated for 20 h.
6. Add to the mixture Milli-Q water until the volume reaches approximately 100 mL.
7. Centrifuge the mixture for 20 min at 2,012 × *g*.
8. Remove the supernatant and filter the through a polytetrafluoroethylene membrane and filters part of the liquid is precipitated with a polytetrafluoroethylene membrane (Millipore FGLP 00 of 047) of 0.22 μM of pore size.
9. Wash thoroughly with a solution of 0.4 g/L NaOH.
10. Wash thoroughly with Milli-Q water and finally with concentrated HCl to remove the manganese oxide formed due to the reduction of potassium permanganate.
11. The washing procedure continues until there is no typical brown color of manganese oxides remains on the filter.
12. After filtration, the membrane with the filtered nanotubes is vacuum dried for 2 or 3 days.

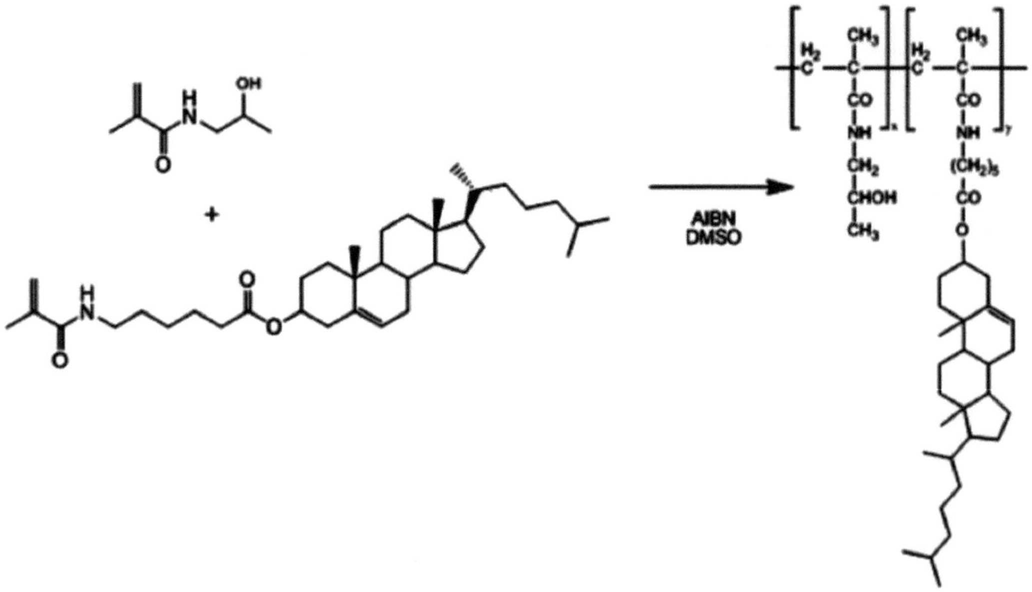

Fig. 1 Scheme of the amphiphilic polymers able to disperse carbon nanotubes in water

3.2 Dispersion of Carbon Nanotubes

3.2.1 Synthesis of Amphiphilic Polymers

The synthesis of the amphiphilic polymer poly-*N*-hydroxypropyl methacrylamide (pHPMA), is described. This polymer as well as CTAB, are used as dispersing agents for carbon nanotubes. The polymers obtained by radical polymerization of a monomer *N*-(2-hydroxypropyl) methacrylamide (HPMA) with propenamide monomer 2-methyl-2-*N*-[6-oxo-6-(cholesterol) hexyl] (Ma-ACAP-chol), as shown in Fig. 1. The total concentration of monomers is fixed at 12.5 % (w/w), and the concentration of the initiator AIBN in 2.0 % (w/w). This gives the pHPMA copolymer having a molecular weight average of 25.000 g/mol.

1. Dissolve HPMA, Ma-ACAP-chol, and AIBN in anhydrous DMSO. (The amounts to be used are shown in Table 1 for different desired concentrations of the cholesteryl group desired.)
2. The reaction mixture was purged with argon, sealed and inserted into a flask placed in a water bath at 60 °C for 6 h.
3. Cool the reaction mixture to room temperature.
4. Isolate the copolymer by precipitation in a mixture of acetone and diethyl ether (2:1).
5. Dry the precipitate in vacuum until a constant weight is obtained.

The polymer is obtained as a white solid product that does not require further purification.

Table 1
Concentrations of the different reagents to synthesize the amphiphilic polymers

Polymer	HPMA	Ma-ACAP-chol	AIBN	Product
VP25	1 g (6.98 mmol)	0.102 g (0.18 mmol)	0.176 g (0.11 mmol)	0.83 g
VP50	1 g (6.98 mmol)	0.209 g (0.37 mmol)	0.193 g (0.12 mmol)	0.67 g

3.2.2 Dispersion with Cationic Surfactant

The dispersions methodologies have been carried out with both shortened CNTs and untreated CNTs.

1. Weigh out ~0.1 mg of CNTs.
2. Prepare a solution of 0.3 mg/mL of CTAB in water (*see* **Note 1**).
3. Put the solid CNTs into the CTAB solution.
4. Treat the mixture in an ultrasound bath for 10 min.
5. The resulting dispersion is stable, with no sedimentation for more than 1 h when the CNTS have been shortened. For untreated CNTs, the dispersion is stable for more than 15 min.

3.2.3 Dispersion with Amphiphilic Polymers

The dispersions methodologies have been carried out with both shortened CNTs and untreated CNTs

1. Weigh out ~0.1 mg of CNTs.
2. Prepare a solution of 20 mg/mL of amphiphilic polymers in DMSO (Stock Solution) (*see* **Note 2**).
3. Prepare a solution of 40 μg/mL of the amphiphilic polymers in water from the stock solution.
4. Put the solid CNTs into the diluted amphiphilic polymer solution.
5. Treat the mixture in an ultrasound bath for 10 min.
6. When the CNTs have been shortened, the resulting dispersion is stable, with no sedimentation for more than 1 h. For untreated CNTs, the dispersion is stable for more than 15 min.

3.3 Preparation of the Transfection Vectors

3.3.1 Plasma Polymerization

The surface modification of CNTs is performed using the technique of plasma assisted CVD (plasma enhanced CVD or PECVD). The plasma reactor used has been developed in our research group and previously described in literature [9]. A scheme of the reactor is presented in Fig. 2.

1. Put 15 mg of CNTs on a 35 mm polystyrene capsule, maximizing the surface of CNTs exposed.
2. Collocate the polystyrene capsule inside the reactor underneath the copper coil (where the plasma is generated).
3. Close the reactor and slowly open the vacuum pump and expected and wait until the pressure reaches 0.02–0.03 mbar.

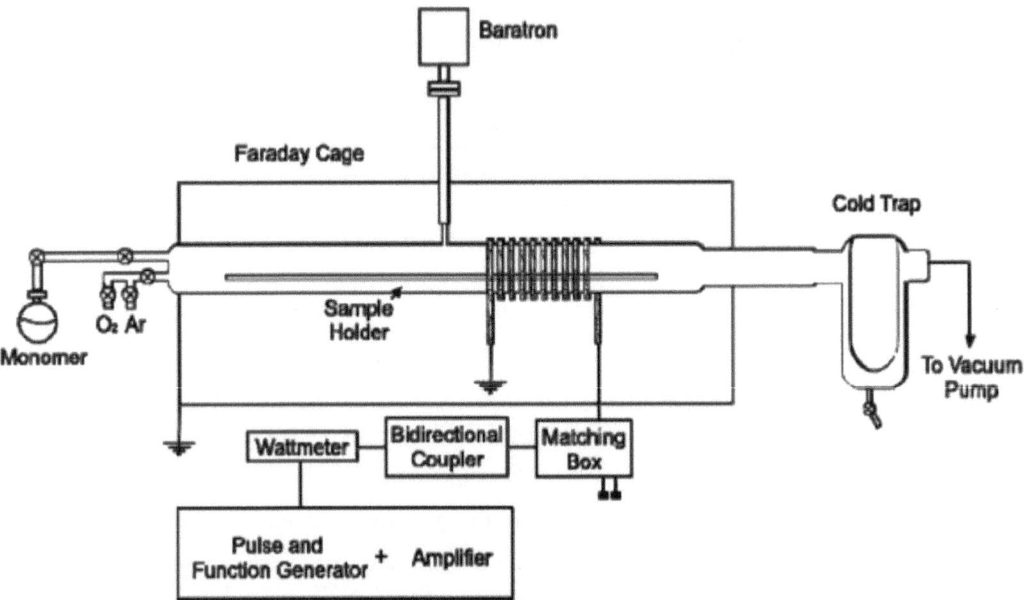

Fig. 2 Scheme of the plasma reactor, designed to surface modification of carbon nanotubes

4. Connect a round flask containing allylamine to one of the entrance connections of the reactor. Cover the flask with an aluminum foil to protect the monomer from light.
5. Activate the cold trap by adding a mixture of dry ice with acetone (temperature −78 °C).
6. Open the needle valve which connects the flask with the monomer with the reactor. Wait until the pressure stabilizes around 0.13 mbar.
7. Wait for 15 min to ensure that there is no air inside the balloon, and that only monomer allylamine vapor enters the reactor.
8. Connect the RF generator at a power of 15 W and with a duty cycle ($DC = (t_{on}/t_{on} + t_{off})$) of 2/4 (*see* **Note 3**).
9. Observe the plasma generation (pink color around the copper coil).
10. Keep the plasma connected for 5 min. After that, turn off the RF generator and wait for another 5 min to assure a complete reaction of the allylamine monomer with the active sites generated on CNTs surface.
11. Turn off the vacuum pump and carefully open the plasma reactor.
12. Shake the CNTs inside the polystyrene capsule and reintroduce it in the plasma reactor.
13. Repeat **steps 7–10** but using a pressure of 0.3 mbar and a RF maximum power of 10 W.

Table 2
Concentrations studied

Ratio MWNT/DNA	[MWNT] (mg/mL)
200:1	6
150:1	4.5
100:1	3
75:1	2.25
50:1	1.5
25:1	0.75
10:1	0.3
1:1	0.03

3.3.2 Complex Formation Between Genetic Material and Modified Carbon Nanotubes

Optimization of the Ratio Between CNTs and DNA ***(See Note 4)***

Sample preparation

1. Disperse CNTs using the amphiphilic polymer method described in Subheading 3.2 using a 25 mM acetate buffer.
2. Prepare the sample for electrophoresis by mixing 5.5 mL of pGFP at a concentration of 0.03 g/mL (25 mM acetate buffer at pH 5) and 5.5 mL CNTs dispersion (*see* Table 2 for the range of concentrations tested).

Gel preparation

1. Prepare an agarose gel at a concentration of 0.8 %. Dissolve 0.6 g of agarose in 75 mL of TAE 1× buffer solution (prepared from 2.42 g Tris-base and Trizma® base, 0.57 mL of glacial acetic acid, 0.185 g of EDTA, and 500 mL of distilled water), and then add 6 μL of ethidium bromide.
2. Once the support is gelled, put 10 μL of each sample in the wells in increasing order.
3. Connect the electrodes, applying a potential of 65 V for about 5 min.
4. Cover the TAE 1× gel and apply a potential of 65 V for 1 h.
5. Observe how each plasmid run in the gel using an UV light.

Preparation of the transfection vector

1. Disperse CNTs using the amphiphilic polymer method described in Subheading 3.2 using also a 25 mM acetate buffer.
2. Mix equal volumes of the plasmid at concentration of 0.03 μg/μL in acetate buffer solution (pH 5) and the CNTS dispersion, prepared according to Subheading 3.2 at a concentration optimized according to Subheading 3.3.

4 Notes

1. CTAB is a cationic surfactant and stabilizes the CNTs dispersion by electrostatic repulsion.
2. The amphiphilic polymer is poorly soluble in water. Thus, the stock solution is prepared in DMSO and diluting the stock solution in water makes the working solution. The final amount of DMSO in the working solution is always less than 0.5 % (w/w).
3. The term "duty cycle" used means the ratio between the time the power is connected and the total time, and is a characteristic of pulsed plasma mode. In these conditions the plasma is on only half the time, preventing the destruction of the monomer under the active gas.
4. Once the CNTS are successfully modified with polyallylamine (Subheading 3.3) the complex with the genetic material is formed through ionic interaction between negatively charged phosphate groups on DNA, and the positively charged amines on the surface of the modified CNTs at pH 5. However, before the formation of the plasmid vector, you need to determine the load capacity of DNA presented by the CNTs. The aim is to determine the amount of genetic material that can be charged per unit mass of carbon nanotubes. Therefore, agarose gel electrophoresis is used to probe the CNTs when the same amount of DNA is exposed to increasing amounts of CNTs. When the DNA does not migrate towards the positive electrode indicates that it is complexed with the CNTs, as they are unable to pass through the pores of the gel due to the large size of the complex.

References

1. Chen J, Chen Q, Ma Q (2012) Influence of surface functionalization via chemical oxidation on the properties of carbon nanotubes. J Colloid Interface Sci 370(1):32–38
2. Chen RJ, Zhang Y, Wang D, Dai H (2001) Noncovalent sidewall functionalization of single-walled carbon nanotubes for protein immobilization. J Am Chem Soc 123(16): 3838–3839
3. Balavoine F, Schultz P, Richard C, Mallouh Â, Ebbesen TW, Mioskowski C (1999) Helical crystallization of proteins on carbon nanotubes: a first step towards the development of new biosensors. Angew Chem Int Ed 38(13–14):1912–1915
4. Arnold MS, Guler MO, Hersam MC, Stupp SI (2005) Encapsulation of carbon nanotubes by self-assembling peptide amphiphiles. Langmuir ACS J Surfaces Colloids 21(10):4705–4709
5. Clark MD, Subramanian S, Krishnamoorti R (2011) Understanding surfactant aided aqueous dispersion of multi-walled carbon nanotubes. J Colloid Interface Sci 354(1):144–151
6. Dovbeshko GI, Repnytska OP, Obraztsova ED, Shtogun YV, Andreev EO (2003) Study of DNA interaction with carbon nanotubes. Semicond Phys Quantum Electron Optoelectron 6(1):105–108
7. Guo Z, Sadler PJ, Tsang SC (1998) Immobilization and visualization of DNA and proteins on carbon nanotubes. Adv Mater 10(9):701–703
8. Aitchison TJ, Ginic-Markovic M, Matisons JG, Simon GP, Fredericks PM (2007) Purification, cutting, and sidewall functionalization of multiwalled carbon nanotubes using potassium permanganate solutions. J Phys Chem C 111:2440–2446
9. Yagüe JL, Agulló N, Borrós S (2009) Plasma polymerization of polypyrrole-like films on nanostructured surfaces. Plasma Process Polym 5:433–443

Chapter 21

Lipid-Based Nanoparticles as Nonviral Gene Delivery Vectors

Daniele Pezzoli, Anna Kajaste-Rudnitski, Roberto Chiesa, and Gabriele Candiani

Abstract

Efficient delivery of nucleic acids into cells is a promising technique to modulate cellular gene expression for therapeutic and research applications. Cationic lipid-based liposomes represent one of the most intensively studied and employed nonviral vectors. They are positively charged at physiological pH and spontaneously self-assemble with polyanionic nucleic acids forming nanoscaled complexes named lipoplexes. Here, we draft a simple protocol for the development, characterization, optimization, and screening of liposomal formulations for in vitro gene delivery. In particular, we report as a practical example a quick method to formulate and extrude nanometer-sized unilamellar cationic vesicles composed of DOTAP as cationic lipid and DOPE as zwitterionic helper lipid at 1:1 molar ratio. The physico-chemical characterization of liposomes and lipoplexes involves the measurement of mean diameter and overall surface charge using Dynamic Light Scattering (DLS) and Laser Doppler Microelectrophoresis. The outlined transfection procedure takes into account several experimental parameters affecting the in vitro performance of gene delivery systems, paying special attention to the charge ratio (CR). Gene delivery effectiveness is evaluated both in terms of transfection efficiency and cytotoxicity of the vector to find the optimal transfection conditions. Importantly, the proposed protocol can be easily shifted to different types of nonviral vectors.

Key words Gene delivery, Cationic liposomes, Extrusion, Charge ratio, Transfection efficiency, Cytotoxicity, DLS, Firefly luciferase

1 Introduction

Gene delivery consists in the introduction into host cells of exogenous nucleic acids and represents a promising technique for the modulation of gene expression. The therapeutic application of gene delivery, named gene therapy, uses nucleic acids as pharmaceutical agents and aims to eradicate causes rather than symptoms of diseases, offering new approaches for the treatment of diseases that conventional clinical therapeutic strategies cannot cure [1]. Nonviral synthetic gene carriers (transfectants) are considered as promising vectors for future gene therapy and have been subject of

Paolo Bergese and Kimberly Hamad-Schifferli (eds.), *Nanomaterial Interfaces in Biology: Methods and Protocols*, Methods in Molecular Biology, vol. 1025, DOI 10.1007/978-1-62703-462-3_21,

increasing attention because of their relative safety and simplicity of use; however, their clinical application is still hindered by their relatively low efficiency [2, 3]. In addition to gene therapy applications, nonviral gene delivery vectors are fundamental tools in the study of gene expression, gene function, protein function, protein–protein interactions, and cell signaling, and most of the researchers who employ cell-based techniques exploit gene delivery techniques in their experiments.

Nonviral gene delivery vectors can be divided in two main classes: cationic lipids and cationic polymers. They rely on the basics of supramolecular chemistry named "self-assembling": at physiological pH, their positive charges allow the interaction and spontaneous self-organization with polyanionic nucleic acids, leading to the formation of nanoscaled complexes named lipoplexes and polyplexes for lipids and polymers, respectively [4].

Since their introduction as gene delivery systems in 1987 [5], cationic liposomes have become one of the most intensively studied nonviral vectors, owing to their tunable chemical structure, formulation, and dimensions. Along this line, over the last two decades liposomes containing cationic lipids such as 1,2-dioleoyl-3-trimethylammonium-propane (DOTAP) [6] in combination with zwitterionic helper lipids such as 1,2-dioleoyl-*sn*-glycero-3-phosphatidylethanolamine (DOPE) [7] have been extensively exploited for gene delivery purposes.

There are several experimental parameters affecting the in vitro transfection efficiency of cationic liposomes: the charge ratio (CR) of lipoplexes, the complexation buffer, the cell type, the cell seeding density, the dose of lipoplexes administered to cells, the presence of serum during transfection. Moreover, the optimization of transfection effectiveness relies on tuning of physicochemical properties of vector–nucleic acids complexes, such as size and surface charge [8]. Unfortunately, well-defined universal correlations between these parameters and the transfection effectiveness of liposomes have not been found yet, making it difficult for liposome developers and users to select a priori optimal transfection conditions [9].

With this in mind, the aim of this book chapter is to draft a simple protocol for the development, characterization, optimization and screening of liposomal formulations for gene delivery (Fig. 1). In particular, herein we describe an easy and quick method to formulate and extrude nanometer-sized two-component unilamellar cationic vesicles. Liposomes are characterized in terms of mean diameter and overall surface charge (ζ-potential) using Dynamic Light Scattering (DLS) and Laser Doppler Microelectrophoresis (LDM) techniques. The ability of cationic liposomes to condense DNA is evaluated by a fluorescence-exclusion assay, using SYBR® Green I as fluorescent DNA-binding dye. In addition to preparing and characterizing lipoplexes we present a procedure for the rapid screening of the best CR in transfection experiments using the

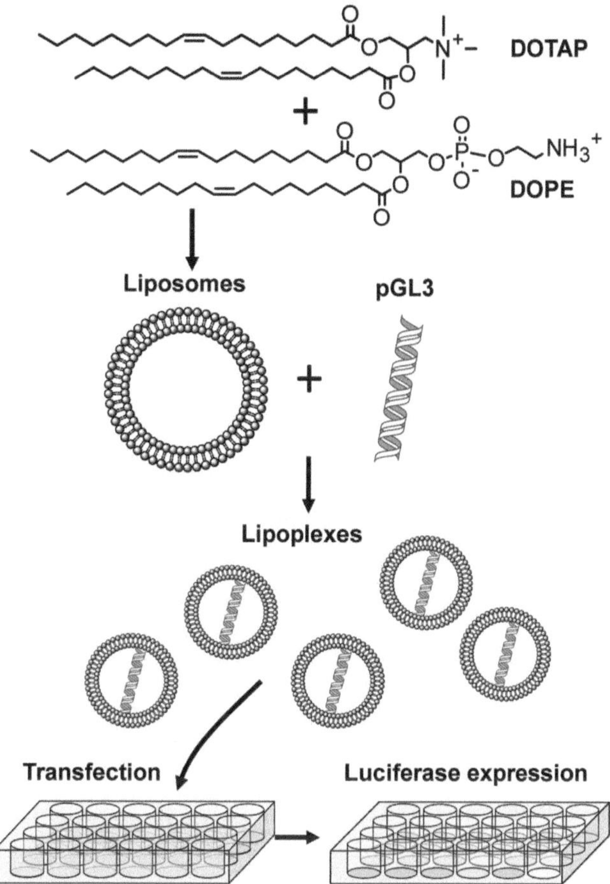

Fig. 1 Schematic representation of the protocol described herein for the development of liposomal formulations for gene delivery. A cationic lipid and a zwitterionic helper lipid (here DOTAP and DOPE, respectively) are formulated in unilamellar cationic liposomes. Liposomes are mixed with plasmid DNA encoding for the modified firefly luciferase (pGL3), thus forming liposome–DNA complexes (lipoplexes) that are delivered to cells. Transgene expression is finally evaluated by luciferase expression assay

modified firefly luciferase as easily detectable reporter gene (pGL3-Control Vector) and alamarBlue® assay to test lipoplex cytotoxicity. Importantly, the vector/DNA complexes characterization and transfection protocol outlined here can also be used for the screening of polymeric gene delivery systems [10].

As a practical example, the development of DOTAP:DOPE = 1:1 (molar ratio) liposomes prepared in deionized water and extruded at 100 nm is reported here (Fig. 1). The transfection procedure suggests specific cell seeding density, lipoplex dose and transfection conditions for HeLa model cell line, often used in gene delivery experiments [9]. However, the proposed protocol can be easily shifted to other formulations following the advices reported in Subheading 4.

2 Materials

2.1 Cationic Liposome and Lipoplex Preparation and Characterization

1. DOTAP chloride (Sigma-Aldrich, St. Louis, MO, USA). Store at −20 °C.
2. DOPE (Sigma-Aldrich). Store at −20 °C.
3. Chloroform.
4. Ultrapure water (dH_2O) with resistivity values greater than 5 MΩ × cm at 25 °C.
5. Plasmid DNA encoding for the modified firefly luciferase pGL3-Control Vector (Promega, Madison, WI, USA).
6. SYBR® Green I (Sigma-Aldrich).
7. LiposoFast™ apparatus equipped with two 1.0 ml gas tight syringes and polycarbonate membranes with pore size of 100 nm (Avestin, Ottawa, Canada; *see* **Note 1**).
8. DLS and LDM apparatus: Malvern Zetasizer Nano ZS apparatus (Malvern Instruments Ltd, Worcestershire, UK).
9. Disposable capillary cell for Z-potential measurements (Malvern Instruments Ltd).
10. Absorbance and fluorescence microplate reader: Tecan GENios Plus (Tecan, Mannedorf, Switzerland).

2.2 Cell Culture

1. HeLa, human cervix adenocarcinoma cell line (American Type Culture Collection (ATCC), Manassas, VA, USA).
2. High-glucose (4.5 g/l) Dulbecco's Modified Essential Medium stored at 4 °C.
3. Fetal bovine serum (FBS). Aliquot under sterile conditions and store at −20 °C.
4. 100× Penicillin–streptomycin solution, stock solution. Aliquot under sterile conditions and store at −20 °C.
5. 200 mM L-Glutamine, stock solution. Aliquot and store at −20 °C.
6. 1 M HEPES buffer pH 7.0–7.6, sterile solution. Store at 4 °C.
7. Sodium pyruvate 100 mM, sterile solution. Store at 4 °C.
8. Cell Culture Medium: high-glucose DMEM supplemented with 10 % (v/v) FBS, 2 mM L-glutamine, 100 IU penicillin, and 100 μg/ml streptomycin, 10 mM HEPES, 1 mM sodium pyruvate. Store at 4 °C and warm to 37 °C prior to use.
9. Dulbecco's Phosphate Buffered Saline (PBS). Store at 4 °C and warm to 37 °C prior to use.

2.3 Transfection Experiments and Post-transfection Assays

1. Transfection Medium: high-glucose DMEM supplemented with 10 % FBS, 2 mM L-glutamine, 10 mM HEPES, 1 mM sodium pyruvate (*see* **Note 2**). Store at 4 °C and warm to 37 °C prior to use.

2. 10× AlamarBlue® solution (Life Technologies, Paisley, UK).
3. BCA Protein Assay Kit (Pierce Chemical, Rockford, IL, USA).
4. Luciferase Assay System (Promega).
5. 5× Cell Culture Lysis Reagent (Promega).
6. Microplate luminometer: Centro LB 960 (Berthold Technologies, Bad Wildbad, Germany).

3 Methods

3.1 Cationic Liposomes Preparation

1. Dissolve the cationic lipid DOTAP and the zwitterionic helper lipid DOPE in chloroform to a final concentration of 20 mM in two separate glass vials (*see* **Note 3**).
2. Add equal volumes (e.g., 500 μl) of lipid solutions in a round-bottomed flask using a glass syringe (*see* **Note 4**) and mix well.
3. Remove chloroform on a rotary evaporator at 40 °C until a dry lipid film can be observed on the flask inner surface.
4. Dry under vacuum overnight to remove all the organic solvent.
5. Add dH_2O to a final lipid concentration of 20 mM (*see* **Note 5**).
6. Vortex thoroughly to hydrate the lipid film until a clear solution of large multilamellar vesicles is obtained (*see* **Note 6**).
7. Freeze/thaw at least five times and bring to room temperature.
8. Extrude for an odd number of times (at least 21 passages, *see* **Note 7**) the lipid dispersion through two 100 nm-pore polycarbonate membranes (*see* **Note 8**) using a LiposoFast™ apparatus. Nanometer-sized unilamellar liposomes are obtained.
9. Harvest the liposome dispersion in a sterile plastic vial and store at 4 °C (*see* **Note 9**).

3.2 Liposome Characterization

3.2.1 Dynamic Light Scattering (DLS)

1. Dilute 100 μl of liposome suspension 1:10 in dH_2O (*see* **Note 10**) inside a disposable 12 mm square polystyrene cuvette.
2. Measure particle size using a DLS apparatus verifying that the polydispersity index (PI) of the Cumulant analysis is below 0.2 (*see* **Note 11**).

3.2.2 Laser Doppler Microelectrophoresis (LDM)

1. Move 750 μl of the liposome suspension used for DLS measurement to a disposable capillary cell for ζ-potential measurements.
2. Measure ζ-potential using a LDM apparatus, verifying that liposomes have a positive ζ-potential (*see* **Note 12**).

3.3 Lipoplex Preparation and Characterization

1. Dilute plasmid DNA (in this case, modified firefly luciferase pGL3-Control Vector) in dH_2O (*see* **Note 10**) to a final concentration of 0.04 μg/μl, corresponding to a phosphate (P) concentration of 121.2 μM (DNA solution) (*see* **Note 13**).

2. Dilute liposome suspension in dH_2O so that the final concentration of net positive charges is 1, 2, 4, 8, 12 times higher than P concentration in the DNA solution (*see* **Note 14**).
3. Mix equal volumes of DNA solution with liposome suspensions at different concentrations and incubate at room temperature for 30 min to obtain lipoplex solutions at CR 1, 2, 4, 8, 12.
4. Measure particle size and ζ-potential of the lipoplex solutions as described above (*see* **Note 15**).

3.4 Evaluation of DNA Complexation/Condensation Ability

1. Prepare 20 μl of lipoplexes at different CR as described above and 20 μl of DNA at 0.02 μg/μl in dH_2O (CR = 0).
2. Add 100 μl of dH_2O (*see* **Note 10**), containing 2× SYBR® Green I and incubate for 10 min at room temperature (*see* **Note 16**).
3. Prepare a blank sample by adding 100 μl of dH_2O, containing 2× SYBR® Green I to 20 μl of dH_2O.
4. Place 3 × 35 μl aliquots of the resulting solutions in a black polystyrene 384 well plate (*see* **Note 17**).
5. Read fluorescence with an excitation wavelength $\lambda_{ex} = 495$ nm and an emission wavelength $\lambda_{em} = 520$ nm using a fluorescence microplate reader.
6. Plot the samples average fluorescence signal (subtracted of blank average signal) against CR to find the minimal fluorescence plateau, corresponding to maximum DNA condensation (*see* **Note 18**).

3.5 Transfection Experiments

1. Seed HeLa cells at a density of 2.0×10^4 cells/cm^2 (*see* **Note 19**) in a 96 well-cell culture plate in 100 μl per well of Cell Culture Medium (*see* **Note 20**).
2. 24 h after seeding, remove old medium.
3. Rinse cells with 100 μl of PBS (*see* **Note 21**).
4. Add 100 μl/well of Transfection Medium.
5. Add 6.4 μl/well of lipoplex solutions at different CR, at least four replicates per CR are suggested (*see* **Note 19**).
6. Place cells in a cell culture incubator at 37 °C, 5 % (v/v) CO_2 in a humidified atmosphere.
7. Facultative: after 4 h of incubation with cells, remove lipoplex-containing medium and add FBS-supplemented medium (*see* **Note 22**).

3.6 Cytotoxicity Assay

1. 24–48 h after transfection, remove old medium from cells (*see* **Note 23**).
2. Add 100 μl/well of 1× alamarBlue® in complete medium (*see* **Note 24**).

3. After 1, 2 and 3 h (*see* **Note 25**), read fluorescence at $\lambda_{em} = 520$ nm, exciting at $\lambda_{ex} = 540$ nm, (*see* **Note 26**) using a fluorescence microplate reader.
4. Subtract the average fluorescence of blank sample (100 μl of 1× alamarBlue® incubated in empty wells) from the fluorescence signal of samples.
5. Calculate cell viability as the percentage of samples net fluorescence signal with respect to the net signal of controls (non-transfected cells) and cytotoxicity as:

$$\text{Cytotoxicity}\,[\%] = 100\% - \text{Viability}\,[\%].$$

3.7 Cell Lysis and Protein Quantitation

1. Remove alamarBlue® solution from cells.
2. Rinse cells with 100 μl of PBS (*see* **Note 27**).
3. Add 110 μl of 0.25× solution of Cell Culture Lysis Reagent (5× solution of Cell Culture Lysis Reagent diluted 1:19 in dH_2O; *see* **Note 28**) and mix by pipetting (*see* **Note 29**).
4. Facultative: freeze-thaw samples to enhance cell lysis (*see* **Note 30**).
5. Prepare a set of bovine serum albumin (BSA) standards by diluting BSA in 0.25× Cell Culture Lysis Reagent; BSA concentration range should be between 0 (blank standard) and 2 mg/ml (*see* **Note 31**).
6. Prepare BCA Working Reagent (WR) by adding 1 part of reagent B to 50 parts of reagent A (*see* **Note 32**).
7. Pipette 20 μl of each standard or unknown sample replicate into a clear polystyrene 96 well microplate.
8. Add 200 μl of WR to each well and place the plate on an orbital shaker to thoroughly mix the contents of the wells.
9. Incubate the microplate at 37 °C for 2 h (*see* **Note 33**).
10. Cool the plate to room temperature.
11. Measure the Optical Density at 560 nm (OD_{560}) using an absorbance microplate reader.
12. Subtract the average OD_{560} of the blank standard replicates from the OD_{560} of all other standard and unknown sample replicates.
13. Plot the average Blank-corrected OD_{560} for each BSA standard against its concentration and fit a standard curve.
14. Calculate the protein concentration of each unknown sample using the standard curve.

3.8 Luciferase Expression Assay

1. Place 20 μl of cell lysate in an opaque white polystyrene 96 well plate (*see* **Note 34**).
2. Read luminescence signal by adding 100 μl/well of Luciferase Assay Substrate using a microplate luminometer.
3. Normalize the luminescence value of each sample with the corresponding protein concentration to obtain luminescence expression values expressed as Relative Luminescence Units (RLU)/mg of protein.

3.9 Choice of the Best Transfection Conditions

1. The choice of the best CR for transfection experiments should be done taking into account both cytotoxicity and luciferase expression results (*see* **Note 35**).
2. Among low cytotoxic conditions, choose the one leading to the highest luciferase expression (*see* **Note 36**).

4 Notes

1. LiposoFast™ is a low cost, hand driven extrusion apparatus with low capacity suggested for researchers who use only small amounts of liposomes. Vesicles prepared with LiposoFast™ are comparable to those obtained with large volume commercial apparatus that may need gas in pressure to work [11]. The apparatus can be autoclaved, thus allowing to obtain sterile liposomes.
2. It is recommended to carry out transfection experiments in antibiotics-free medium as increase in cell membrane permeability by cationic liposomes during transfection could lead to higher antibiotic uptake causing toxicity.
3. Use glass or stainless steel in presence of organic solutions; do not use vials made of polymeric materials such as polystyrene or polyethylene as impurities could leach out of the container.
4. The preparation of an equimolar formulation of cationic lipid–helper lipid is reported here as example but desired molar ratios can be easily obtained by simply changing volume ratio of the starting lipid solutions.
5. The aqueous solution in which liposomes are prepared can influence their final physico-chemical and transfection properties; in addition to dH_2O, other buffers such as 10 mM Hepes buffer pH 7 and PBS can be used. Do not use buffers containing ethylenediaminetetraacetic acid (EDTA) as it could cause liposome aggregation.
6. If detaching the lipid film from the glass surface by vortexing results difficult, bath sonicate the lipid dispersion for 2–5 min, until clarity is obtained.
7. An odd number of passages is needed to find extruded liposomes in the originally empty syringe avoiding contamination with any unextruded large vesicles which might not have passed through the filter.

8. Using two stacked membranes helps yielding monodisperse, nanometric-sized liposomes. Liposome dimension can be tuned by simply using polycarbonate membranes with different pore sizes (e.g., 50, 100, 200, 400 nm; available from Avestin).
9. Usually liposomes are stable at 4 °C up to 6–12 months; however, liposomes stability can be checked periodically by mean size (hydrodynamic diameter) and ζ-potential measurements (*see* Subheading 3.2).
10. Liposome formulations should be diluted in the same aqueous solution they have been prepared for both characterization experiments and lipoplex formation.
11. A high PI indicates a high width of the size distribution. If PI is higher than 0.2 or mean size is much higher than membrane pore size, it may be necessary to extrude again liposomes. In these cases, reducing lipid concentration during extrusion could be useful.
12. Cationic liposomes should have a positive ζ-potential in dH_2O.
13. DNA phosphate density is assumed to be 3.03 nmol P/μg DNA.
14. DOTAP is assumed to carry one cationic charge/molecule at pH 7.
15. When using cationic liposomes as transfection reagents, positively charged liposome–DNA complexes are necessary to assure DNA complexation and lipoplex interaction with negatively charged cell surfaces.
16. SYBR® Green I stain has been shown to be much less mutagenic and much more sensitive than ethidium bromide; therefore, it is considered as the DNA stain of choice for these experiments.
17. The use of black polystyrene well microplates is suggested for fluorescence analysis as they have low background fluorescence and minimize light scattering and well-to-well crosstalk.
18. Maximum DNA condensation is usually necessary to obtain efficient lipoplexes, if a plateau is not observed, try with higher CR ratios.
19. Cell type can strongly influence the transfection effectiveness. If high cytotoxicity or poor transfection results are obtained, please find out optimal cell seeding density and/or lipoplex dose.
20. When transfecting other cell lines or primary cells use appropriate culture medium, according to existing literature (*see* **Note 2**). Since serum may affect the transfection efficiency of gene delivery vectors, it is sometimes preferable to carry out transfection experiments in serum-free transfection medium.
21. A PBS washing step is often used in transfection experiments; however, it can be avoided to reduce cell detachment if culture medium is the same before and during transfection.

22. This step can be used to reduce cytotoxicity and should be done if lipoplexes are incubated with cells in serum-free medium.
23. Luciferase expression usually peaks 24–48 h after transfection but longer incubation times could be necessary for some liposomal reagents to optimize protein expression.
24. AlamarBlue® is a nontoxic nondestructive fluorimetric cell growth indicator based on the detection of cellular metabolic activity. Using alamarBlue® as cell viability assay allows to quantify total protein content in the same sample.
25. Reading at different time points allows avoiding problems connected to low sensitivity (at early time points of incubation) or to signal saturation (at late time points of incubation).
26. If cells will not be cultured anymore, it is not necessary to move conditioned alamarBlue® solution in another microplate but fluorescence can be read directly in the cell-containing culture plate, with negligible errors.
27. A PBS washing step is necessary to completely remove serum proteins.
28. Cell Culture Lysis Reagent is supplied as a 5× solution. Reducing the working concentration of Cell Culture Lysis Reagent from the suggested 1× to 0.25× allows to lower lysis buffer interference in the BCA assay maintaining high cell lysis ability.
29. After lysis reagent addition, samples can be stored at −80 °C.
30. Freeze-thaw step can be adopted to increase cell lysis when, after lysis buffer addition, adherent cells are still observed.
31. BSA dilution scheme for standard microplate procedure suggested in the datasheet of BCA assay kit can be followed.
32. The volume of WR to be prepared can be calculated using the following formula:

$$V_{\mathrm{WR}} = (N_{\mathrm{standards}} + N_{\mathrm{samples}}) \times (N_{\mathrm{replicates}}) \times 200\,\mu\mathrm{l},$$

where 200 μl is the volume of reagent per sample if the test is performed in a 96 well microplate.
33. Given the interference of 0.25× Cell Culture Lysis Reagent with the BCA assay, a 2 h incubation time is needed to enhance signal to noise ratio.
34. Opaque white microplates should be used for luciferase assays as they maximize luminescence signal.
35. Conditions leading to high cytotoxicity (cytotoxicity > 40 %) are considered unuseful.
36. In general, conditions leading to a good compromise between high transfection efficiency and low cytotoxicity should be chosen.

Acknowledgments

This work was partly supported by the Politecnico di Milano, 5xmille Junior, "SURGES" Project and by the Italian Ministry for Education, University and Research (MIUR)—FIRB Futuro in Ricerca 2008, Grant RBFR08XH0H (to Dr. Candiani, both), as well as by the Italian Ministry of Health Grant Giovani Ricercatori 2009, GR-2009-1471693 (to Dr. Kajaste-Rudnitski).

References

1. Candiani G, Pezzoli D, Cabras M, Ristori S, Pellegrini C, Kajaste-Rudnitski A, Vicenzi E, Sala C, Zanda M (2008) A dimerizable cationic lipid with potential for gene delivery. J Gene Med 10:637–645
2. Candiani G, Frigerio M, Viani F, Verpelli C, Sala C, Chiamenti L, Zaffaroni N, Folini M, Sani M, Panzeri W, Zanda M (2007) Dimerizable redox-sensitive triazine-based cationic lipids for in vitro gene delivery. ChemMedChem 2:292–296
3. Pezzoli D, Candiani G, Cabras S, Giordano C, Daniele F, Cigada A (2008) Liposomes versus micelles as gene delivery vectors. Biomed Pharmacother 62:493
4. Candiani G, Pezzoli D, Ciani L, Chiesa R, Ristori S (2010) Bioreducible liposomes for gene delivery: from the formulation to the mechanism of action. PLoS One 5:e13430
5. Felgner PL, Gadek TR, Holm M, Roman R, Chan HW, Wenz M, Northrop JP, Ringold GM, Danielsen M (1987) Lipofection -- a highly efficient, lipid-mediated DNA-transfection procedure. Proc Natl Acad Sci U S A 84:7413–7417
6. Ciani L, Ristori S, Calamai L, Martini G (2004) DOTAP/DOPE and DC-Chol/DOPE lipoplexes for gene delivery: zeta potential measurements and electron spin resonance spectra. Bba-Biomembranes 1664: 70–79
7. Hui SW, Langner M, Zhao YL, Ross P, Hurley E, Chan K (1996) The role of helper lipids in cationic liposome-mediated gene transfer. Biophys J 71:590–599
8. Ristori S, Ciani L, Candiani G, Battistini C, Frati A, Grillo I, In M (2012) Complexing a small interfering RNA with divalent cationic surfactants. Soft Matter 8:749–756
9. van Gaal EVB, van Eijk R, Oosting RS, Kok RJ, Hennink WE, Crommelin DJA, Mastrobattista E (2011) How to screen non-viral gene delivery systems in vitro? J Control Release 154:218–232
10. Pezzoli D, Olimpieri F, Malloggi C, Bertini S, Volonterio A, Candiani G (2012) Chitosan-graft-branched polyethylenimine copolymers: influence of degree of grafting on transfection behavior. PLoS One 7:e34711
11. Macdonald RC, Macdonald RI, Menco BPM, Takeshita K, Subbarao NK, Hu LR (1991) Small-volume extrusion apparatus for preparation of large unilamellar vesicles. Biochim Biophys Acta 1061:297–303

Chapter 22

Stabilizing Gold Nanoparticle Bioconjugates in Physiological Conditions by PEGylation

Joan Comenge and Víctor F. Puntes

Abstract

Stability of gold nanoparticles (AuNPs) is often compromised in physiological conditions. The loss of colloidal stability might lead to undesired biological responses for drug delivery nanosystems. Here a methodology to confer additional stability to the AuNPs by the addition of PEG is presented. Also, protocols to prepare and characterize the composition and conformation of mixed layers with PEG and alkanethiols are described here. Finally, methods to assay the stability of the link between the NP conjugates and a model drug are shown.

Key words Gold nanoparticles, Conjugates, Biological stability, PEG, Mixed layers, Protein corona, Link stability, Cisplatin

1 Introduction

Gold nanoparticles (AuNPs) present unique physicochemical properties to be used as drug delivery scaffolds such as biocompatibility, easy surface functionalization, excellent size and morphology control, and a surface plasmon resonance (SPR) in the range of visible and infrared light [1, 2]. Drugs, proteins, nucleotides and organic molecules can be attached to AuNPs which act then as nanocarriers. However, the stability of these systems might be compromised when they are assayed in biological media [3]. Biological fluids are a complex mixture of salts and proteins which may lead to different phenomena such as the aggregation of NPs, normally due to the compacting of the stem layer of the conjugates induced by the high ionic strength of this media [4]. Avoiding aggregation of the nanocarriers is a key point for not only in vitro applications but also in vivo applications, because size strongly influences the way the organisms deal with nanosystems. For example, aggregation leads to a modification of the biodistribution and a rapid clearance of the nanoparticles by the immune system [5]. In addition, the uncontrolled release of molecules (e.g., drugs)

Paolo Bergese and Kimberly Hamad-Schifferli (eds.), *Nanomaterial Interfaces in Biology: Methods and Protocols*, Methods in Molecular Biology, vol. 1025, DOI 10.1007/978-1-62703-462-3_22, © Springer Science+Business Media New York 2013

from the nanocarrier before reaching the target might be favored by nucleophilic attack from species present in biological media [3]. This induced unstability of the nanoparticles and/or link in biological media would explain the in vivo lack of benefits of some nanocarriers which worked properly in preliminary in vitro assays.

For that reason, it is important to know methodologies to stabilize the nanoparticles and to evaluate the stability of the nanoparticles in biological media before the final assays. We present here a stabilization process of AuNP conjugates carrying an antitumoral drug by means of pegylation. Although we focused the work in the stabilization of 11-mercaptoundecanoic acid (MUA)-capped AuNPs, these strategies can be used with other conjugates (i.e., other capping agents and other NPs) since they are not element specific. Moreover, protocols to test the stability of the nanosystem in biological environment are also provided. Finally the stability of the link between drug and nanoparticle is assayed with two model bonds: a strong coordination bond and electrostatic adsorption of cisplatin.

2 Materials

1. AuNPs. 14 nm AuNPs (4.7×10^{12} NP/mL) are synthesized by a seeding growth approach: Au seeds (8 nm, 4.7×10^{12} NP/mL) were synthesized by adding an aqueous solution of $HAuCl_4$ (1 mL, 25 mM) to a boiling sodium citrate solution (150 mL, 2.2 mM). The reaction is completed after 10 min. The temperature was then decreased to 90 °C, $HAuCl_4$ (1 mL, 25 mM) was added to the previously synthesized AuNP seeds, and the reaction allowed to run for 20 min. This step is repeated two more times in order to get the final AuNP size.
2. SH-PEG and SH-PEG-FITC (3.4 kDa). Prepare 0.3 and 1 mM solutions of SH-PEG by dissolving 3.40 mg and 1.02 mg, respectively, in 1 mL of milliQ water.
3. MUA. Prepare 0.3 and 1 mM solutions of MUA by dissolving 0.69 and 2.3 mg of MUA, respectively, in 9.925 mL of milliQ water. Add 75 μL of 2 M NaOH and sonicate until total dissolution.
4. CAPS buffer. To prepare a 0.2 M solution, 4.43 g are dissolved in 90 mL of milliQ water and the pH is brought to 10.6 by adding 75 μL of 2 M NaOH. Finally, milliQ water is added up to 100 mL.
5. NaCN. To prepare a 0.1 M solution, 49 mg of NaCN are dissolved in 10 mL of milliQ water.
6. Bovine serum albumin (BSA). 86 mg are dissolved in 1 mL of milliQ water.

7. AuNP-cisplatin. AuNPs were firstly functionalized with MUA and later with commercial cisplatin ($[Pt(Cl)_2(NH_3)_2]$) or aquated cisplatin ($[Pt(H_2O)_2(NH_3)_2]^{2+}$) as described in the literature [6].

3 Methods

3.1 Direct PEGylation of AuNPs (See Note 1)

1. Add 125 μL of an aqueous solution of 3.4 kDa SH-PEG (0.3 mM) to 1 mL of citrate-stabilized AuNPs (*see* **Note 2**) in a 1.5 mL plastic tube.
2. The conjugation was allowed to run for 6 h at room temperature.
3. 1 mL of the solution is transferred to a 12 mm square polystyrene cuvette and the hydrodynamic diameter of the NPs measured by dynamic light scattering (DLS). Adjust the settings with refractive index = 0.200 and absorbance = 1.32. Perform the measure in triplicate and at least 12 runs per measurement.
4. Repeat **step 3** with citrate-capped AuNPs.
5. The hydrodynamic diameter after conjugating 3.4 kDa SH-PEG increases about 12 nm with respect to citrate-capped AuNPs (Fig. 1) (*see* **Note 3**).

3.2 Mixed Layers of SH-PEG and Alkanethiols (See Note 4): Co-conjugation of SH-PEG and MUA (See Note 5)

1. Choose the appropriate mixture/s of SH-PEG and MUA depending on the coverage density of your interest. Always the total concentration must be kept constant (i.e., [SH-PEG] + [MUA] = 1 mM).
2. Add 125 μL of this mixture to 1 mL of citrate-stabilized AuNPs in a 1.5 mL plastic tube.
3. The conjugation was allowed to run for 6 h at room temperature.

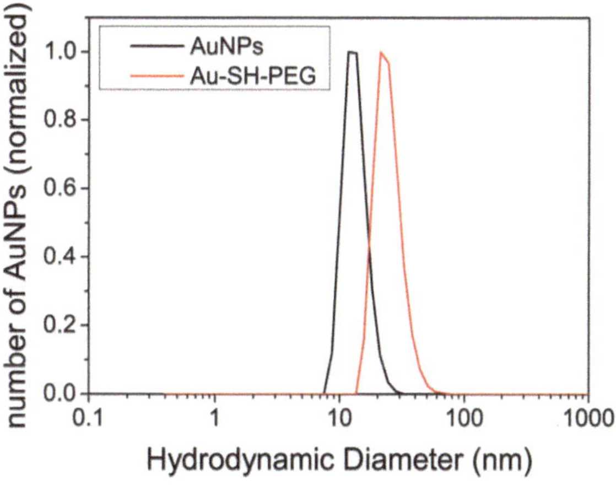

Fig. 1 Hydrodynamic diameter as measured by DLS. The hydrodynamic diameter increased from 13.1 ± 0.3 to 24.5 ± 0.4 nm after SH-PEG conjugation

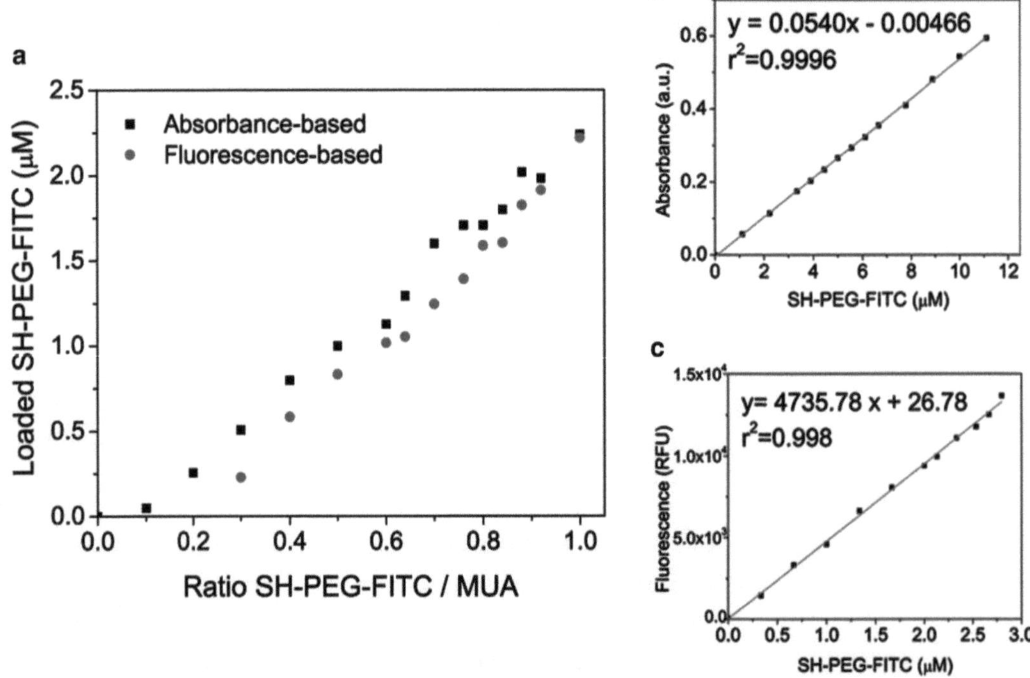

Fig. 2 Loading of SH-PEG in function of the ratio SH-PEG/MUA added. The loading on the AuNPs was quantified with two different methodologies (absorbance and fluorescence after digestion of AuNPs) giving similar results. Note that in the case of the fluorescence-based measures, the lower ratios could not be measured due to the impossibility of digesting completely the AuNPs, likely due to the compactness of the MUA shell

3.3 Quantification of SH-PEG on the Surface of the AuNPs by UV–VIS Spectroscopy

1. Repeat **steps 1–3** in Subheading 3.2 but using SH-PEG-FITC instead of SH-PEG.
2. Remove the AuNPs by two centrifugation steps (35,000 × *g*, 30 min).
3. The absence of AuNPs is confirmed by the disappearance of the typical SPR band of AuNPs at around 520 nm.
4. 125 μL of 200 mM CAPS buffer (pH 10.6) were added to 875 μL of the resulting supernatant.
5. To prepare the standards, known concentrations of SH-PEG-FITC are dissolved in 2.2 mM sodium citrate solution. 125 μL of 200 mM CAPS buffer (pH 10.6) are added to 875 μL of every standard solution.
6. The standards and samples are analyzed by UV–VIS spectroscopy. The absorbance at 495 nm is recorded.
7. The concentration of SH-PEG-FITC found in the supernatant is subtracted from the known concentration of SH-PEG-FITC added to know the amount loaded on the AuNPs (Fig. 2) (*see* **Note 6**).

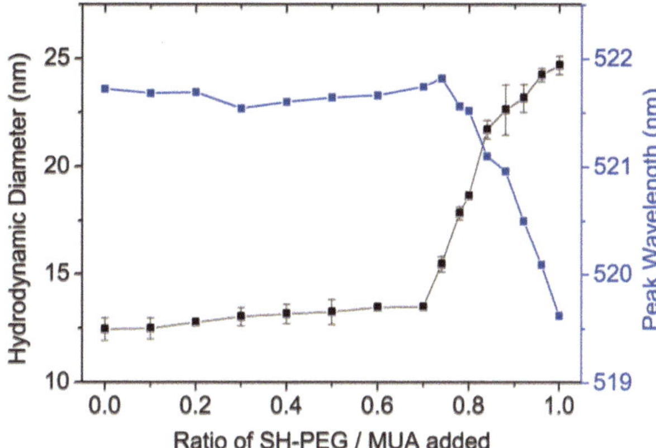

Fig. 3 Conformational change of SH-PEG. The stretching of the SH-PEG molecules represents an increase on the size of the AuNPs which can be followed by measuring the hydrodynamic diameter by DLS. A similar trend was observed by analyzing the shift of the SPR band

3.4 Quantification of SH-PEG on the Surface of the AuNPs by Fluorescence

1. Repeat **steps 1–3** in Subheading 3.3 but using SH-PEG-FITC instead of SH-PEG.
2. The corresponding pellets are digested using 1 mL of a 100 mM NaCN aqueous solution and stirred at 37 °C for 6 h (500 rpm, thermomixer compact®).
3. The total digestion of AuNPs is again proved by the disappearance of the SPR band.
4. 125 μL of 200 mM CAPS buffer (pH 10.6) were added to 875 μL of the corresponding samples.
5. Standards are prepared using same conditions than samples.
6. 100 μL of samples and standards were transferred to a 96-well opaque plate. The fluorescence spectra (λ_{ex}=494, λ_{em} = 521) is taken. The concentration of SH-PEG-FITC in the samples is calculated from the calibration curve (Fig. 2) (*see* **Note 6**).

3.5 Assay the Change in SH-PEG Conformation

1. Repeat **steps 1–3** in Subheading 3.2.
2. Measure the hydrodynamic diameter of the different conjugates as explained in **step 3**, Subheading 3.1.
3. Measure the SPR band by UV–VIS spectroscopy. Measure absorption at wavelengths from 300 to 800 nm.
4. The change from mushroom configuration to brush configuration (*see* **Note 7**) is identified by a great change in the hydrodynamic diameter as well as a larger shift of the SPR band (Fig. 3). Note that the changes in composition of the layer follow a linear trend (Fig. 2) and therefore the great changes of hydrodynamic diameter cannot be due to great changes in the layer composition.

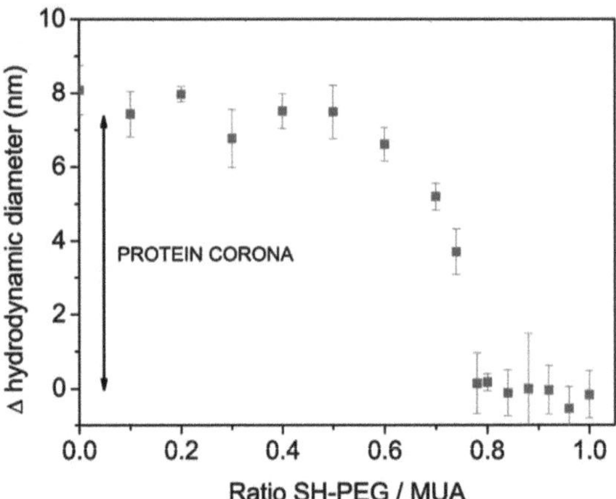

Fig. 4 Inhibition of protein adsorption. The formation of the protein corona results in an increase of the hydrodynamic diameter. The change on the SH-PEG conformation leads to minimize and ultimately to inhibit the adsorption of protein which is seen by no differences of the hydrodynamic diameter after addition of BSA

3.6 Assay the Protein Adsorption on the Mixed Layers by DLS

1. Repeat **steps 1–3** in Subheading 3.2.
2. Remove the excess of ligands by centrifugation (18,000 × *g*, 14 min) and resuspend the pellet in freshwater (*see* **Note 8**).
3. Add 8 μL of BSA 1.3 mM to 1 mL of the different AuNP solutions. Stir gently overnight (200 rpm, thermomixer®).
4. Measure the hydrodynamic diameter of AuNPs with and without BSA as explained in **step 3**, Subheading 3.1. The differences in the hydrodynamic diameter are indicative of protein adsorption [7]. An example is seen in Fig. 4. At lower ratios the difference in size is around 8 nm due to the absorption of a monolayer of BSA. This difference becomes smaller for higher SH-PEG/MUA ratios (less BSA is adsorbed) and finally at ratios >0.78 there is no difference indicating the total inhibition of BSA adsorption. Note that the decrease of protein adsorption is strongly related with the conformational change of SH-PEG.

3.7 Stability of the Link in Physiological Media

1. 1 mL AuNPs loaded with commercial cisplatin (AuNP-cisplatin) are dialyzed against 1 L of water. Independently, the same procedure is followed with 1 mL of AuNPs loaded with aquated cisplatin (AuNP-aq.cisplatin) (*see* **Note 9**).
2. Mix 1 mL of both AuNPs (4.7×10^{12} NP/mL) with 9 mL of DMEM supplemented with 10 % FBS (v/v) (*see* **Note 10**). Stir gently at room temperature.
3. Take 1 mL of sample at the desired times (e.g., 0.5, 6, 24, and 48 h). Immediately, remove the AuNPs by two centrifugation

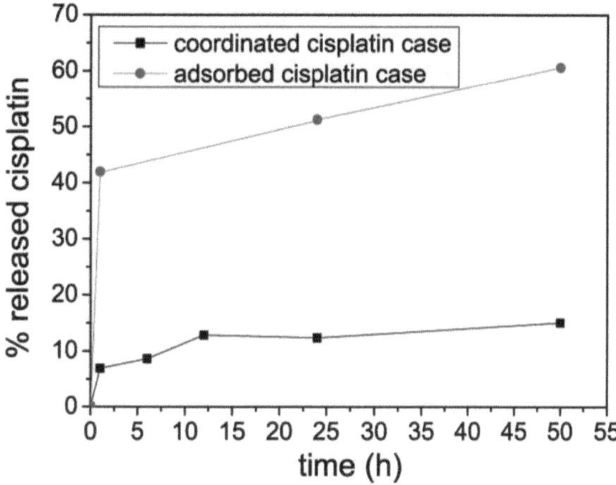

Fig. 5 Release of cisplatin depending on the strength of the bond. In the case of coordination bond, the unspecific release is minimal (ca. 15 % after 50 h), whilst a burst release up to 42 % in the first hour followed by a continuous unspecific release up to 62 % is produced when cisplatin was weakly linked to the AuNPs

steps ($35{,}000 \times g$, 30 min). The amount of Pt in the resultant supernatant is analyzed by inductively coupled plasma mass spectroscopy (ICP-MS). The amount of Pt loaded in the AuNPs is also analyzed by ICP-MS (*see* **Note 11**).

4. The release is calculated dividing the amount of Pt in the supernatant by the amount of Pt loaded on the corresponding AuNPs (Fig. 5). Unspecific release is identified by a high release at very short times. This is normally undesired. On the other hand, a strong bond will lead to minimal release unless a triggering stimulus is produced. In this example the trigger is a pH decrease.

4 Notes

1. Direct PEGylation is used when the initial material is bare nanoparticles (e.g., citrate-stabilized) and there is no need to have additional ligands. This is commonly used to assay general properties of AuNPs such as biodistribution and cell uptake [8]. However, it is not recommended for further loading of drugs or other molecules of interest on the AuNPs because PEG tends to nonspecifically and weakly adsorb small molecules. Moreover, the shell is not as compact as using other smaller ligands (i.e., alkanethiols). Therefore, the number of ligands per nanoparticle will be lower and consequently the number of available chemical groups to bind a drug is lower than in the compact layers achieved when using alkanethiolates.
2. The amounts of reagent added are calculated for 14 nm AuNPs at 4.7×10^{12} NP/mL. The same protocol can be followed for

other sizes and concentrations maintaining the proportions and adjusting the amounts to the total Au surface.

3. Obviously, the increase of the diameter will be different if an SH-PEG with a different molecular weight is used.

4. Mixed layers are interesting whenever the functionality given by the alkanethiol monolayer will be needed in further steps (e.g., linking a drug) and the PEG molecules to confer stability to the system. There are two options to achieve these mixed layers: co-conjugation of both components or displacement of the SH-PEG by the alkanethiol (despite that both are linked by an SH group to the Au surface, the high packing in the self-assembled monolayers (SAM) confers an additional stability making the bond stronger and allowing SH-PEG displacement). This section focuses on the first option since it leads to a better control and more reproducibility according to our experience. SH-PEG (3.4 kDa) and MUA have been chosen as model molecules to explain these processes.

5. SH-PEG and MUA have different affinities for the AuNP surface. The affinity of the molecule for the NP surface takes into account not only the chemical bond but also the stability of the formed monolayer. Therefore the proportion added will not be the proportion of the different species on the surface. This statement is valid for every co-conjugation of two different species. The protocol and characterization techniques used here can be also applied to other alkanethiol/SH-PEG mixtures.

6. Knowing the amount of SH-PEG loaded on the NP, one can know the percentage of the surface occupied by this ligand simply by knowing the amount of SH-PEG when the surface is saturated. For example, for the ratio SH-PEG-FITC/MUA = 0.5, the amount of SH-PEG-FITC loaded is ca 0.9 μM which would correspond to the 39 % of the surface (i.e., [0.9/2.3] × 100). Thus, the proportions of SH-PEG/MUA added in solution are not maintained on the NP surface. This is explained by the different affinity of the molecules to the AuNPs: MUA is a smaller molecule than SH-PEG, with a hydrophobic chain and a hydrophilic head that favors the formation of dense SAM on AuNPs [9] more than SH-PEG does.

7. It is well known that SH-PEG can adopt two conformations. A folded conformation, known as mushroom conformation [10], is adopted when the density of SH-PEG is low. On the other hand, a more compact layer of SH-PEG leads to a stretching of the molecules in what is known as brush conformation [10]. In the low density case, the SH-PEG polymer attaches the NPs via multiple weak bonds while in the high density, the surface is saturated with the more stable SH–Au bonds and therefore, there is no room for the polymer to wrap on the surface.

8. Note that the centrifugation conditions here are softer than in the previous sections. This is because we need these AuNPs for further steps, whilst in the other cases a potential aggregation of the AuNPs in the centrifugation step would not represent a problem. Thus, in this section it is more important to preserve the colloidal stability of the AuNPs, whilst in the previous sections it is more important to ensure that there are no AuNPs in the supernatant.
9. The reason to use these cisplatin derivatives is the different bond that they form with MUA-capped AuNPs. Whilst aquated cisplatin forms a strong coordination bond with carboxylic groups of MUA, commercial cisplatin binds the AuNPs via an ion–dipole interaction [6].
10. Physiological media are complex and reactive media. They contain proteins, amino acids, carbohydrates, and high concentration of salts, among other components. This can affect the chemical stability of the nanoparticle with the molecule of interest and therefore has to be studied before any possible biological application. The stability of the NPs must be assessed in the same media than the final application.
11. An easier experiment of dialysis and analysis of the outer solution (cell media) could have been performed if the free drug had not interacted with protein of the media which would have retained the drug in the dialysis bag [11].

References

1. Daniel MC, Astruc D (2004) Gold nanoparticles: assembly, supramolecular chemistry, quantum-size-related properties, and applications toward biology, catalysis, and nanotechnology. Chem Rev 104:293–346
2. Rana S, Bajaj A, Mout R, Rotello VM (2012) Monolayer coated gold nanoparticles for delivery applications. Adv Drug Deliv Rev 64:200–216
3. Ruenraroengsak P, Cook JM, Florence AT (2010) Nanosystem drug targeting: facing up to complex realities. J Control Release 141: 265–276
4. Lim IS, Goroleski F, Mott D, Kariuki N, Ip W et al (2006) Adsorption of cyanine dyes on gold nanoparticles and formation of J-aggregates in the nanoparticle assembly. J Phys Chem B 110:6673–6682
5. Dobrovolskaia MA, McNeil SE (2007) Immunological properties of engineered nanomaterials. Nat Nanotechnol 2:469–478
6. Comenge J, Romero FM, Sotelo C, Dominguez F, Puntes V (2010) Exploring the binding of Pt drugs to gold nanoparticles for controlled passive release of cisplatin. J Control Release 148:e31–32
7. Casals E, Pfaller T, Duschl A, Oostingh GJ, Puntes V (2010) Time evolution of the nanoparticle protein corona. ACS Nano 4:3623–3632
8. Lipka J, Semmler-Behnke M, Sperling RA, Wenk A, Takenaka S et al (2010) Biodistribution of PEG-modified gold nanoparticles following intratracheal instillation and intravenous injection. Biomaterials 31:6574–6581
9. Love JC, Estroff LA, Kriebel JK, Nuzzo RG, Whitesides GM (2005) Self-assembled monolayers of thiolates on metals as a form of nanotechnology. Chem Rev 105: 1103–1170
10. Simpson CA, Agrawal AC, Balinski A, Harkness KM, Cliffel DE (2011) Short-chain PEG mixed monolayer protected gold clusters increase clearance and red blood cell counts. ACS Nano 5:3577–3584
11. Dhar S, Gu FX, Langer R, Farokhzad OC, Lippard SJ (2008) Targeted delivery of cisplatin to prostate cancer cells by aptamer functionalized Pt(IV) prodrug-PLGA-PEG nanoparticles. Proc Natl Acad Sci 105: 17356–17361

Index

Paolo Bergese and Kimberly Hamad-Schifferli (eds.), *Nanomaterial Interfaces in Biology: Methods and Protocols*, Methods in Molecular Biology, vol. 1025, DOI 10.1007/978-1-62703-462-3, © Springer Science+Business Media New York 2013

Q

S

T

U

Z

MIX
Papier aus verantwortungsvollen Quellen
Paper from responsible sources
FSC® C105338

If you have any concerns about our products,
you can contact us on
ProductSafety@springernature.com

In case Publisher is established outside the EU,
the EU authorized representative is:
Springer Nature Customer Service Center GmbH
Europaplatz 3, 69115 Heidelberg, Germany

Printed by Libri Plureos GmbH
in Hamburg, Germany